ENVIRONMENTAL CHEMISTRY OF DYES AND PIGMENTS

ENVIRONMENTAL CHEMISTRY OF DYES AND PIGMENTS

Edited by

ABRAHAM REIFE

HAROLD S. FREEMAN

A Wiley-Interscience Publication

JOHN WILEY & SONS, INC.

New York / Chichester / Brisbane / Toronto / Singapore

CHEM

Library of Congress Cataloging in Publication Data:

Environmental chemistry of dyes and pigments / edited by Abraham Reife &
   Harold S. Freeman.
      p.   cm.
    ''A Wiley-Interscience publication.''
    Includes index.
    ISBN 0-471-58927-6 (acid-free paper)
    1. Sewage—Purification—Color removal. I. Reife, Abraham, 1931– .
  II. Freeman, Harold S. III. Title: Environmental chemistry of dyes
  and pigments.
  TD758.5.C65E54 1996
  628.3—dc20   95-6113

Printed in the United States of America

10 9 8 7 6 5 4 3 2 1

CONTRIBUTORS

L. DON BETOWSKI, U.S. Environmental Protection Agency, Las Vegas, Nevada

GREGORY D. BOARDMAN, Department of Civil Engineering, Virginia Polytechnic Institute and State University, Blacksburg, Virginia

MICHAEL BREWSTER, CSK Technical, Inc., Tonawanda, New York

E. A. CLARKE, ETAD, Basel, Switzerland

MICHAEL M. COOK, Morton International, Inc., Danvers, Massachusetts

SHARON F. DUBROW, Science Applications International Corp., McLean, Virginia

JOHN ELLIOTT, Ciba Corporation, Toms River, New Jersey

HAROLD S. FREEMAN, North Carolina State University, College of Textiles, Raleigh, North Carolina

DAVID G. HUTTON, D.G. Hutton, Inc., Newark, Delaware

MASAKI MATSUI, Gifu University, Department of Chemistry, Faculty of Engineering, Yanagido 1-1, Gifu, Japan

DANIEL C. MCINTYRE, Ciba Corporation, Toms River, New Jersey

JOHN A. MEIDL, Zimpro Environmental, Inc., Rothchild, Wisconsin

DONALD J. MICHELSEN, Department of Chemical Engineering, Virginia Polytechnic Institute and State University, Blacksburg, Virginia

v

G. J. O'BRIEN, E.I. du Pont de Nemours & Co., Deepwater, New Jersey

G. PECK, Andco Environmental Processes, Buffalo, New York

JOHN J. PORTER, Clemson University, School of Textiles, Clemson, South Carolina

ABRAHAM REIFE, 60 Oakside Drive, Toms River, New Jersey

HUGH M. SMITH, Sun Chemical Corp., Cincinnati, Ohio

JOLANTA SOKOLOWSKA-GAJDA, Institute of Dyes, Technical University (Politechnika), Lodz, Poland

ROBERT D. VOYKSNER, Research Triangle Institute, Analytical and Chemical Sciences, Research Triangle Park, North Carolina

ANNE E. WILCOX, University of Guelph, Department of Consumer Studies, Guelph, Ontario, Canada

JEHUDA YINON, Department of Environmental Sciences and Energy Research, Weizmann Institute of Science, Rehovot, Israel

CONTENTS

7 OPERATING EXPERIENCE WITH THE PACT® SYSTEM 165

Daniel C. McIntyre

IV REGULATORY ISSUES   293

12 U.S. SAFETY, HEALTH, AND ENVIRONMENTAL REGULATORY AFFAIRS FOR DYES AND PIGMENTS   295

Hugh M. Smith

PREFACE

Over the past two decades, manufacturers and users of synthetic colorants have faced increasingly stringent regulations promulgated by agencies established to safeguard human health and the environment. This has meant that much of the emerging technology in these two areas arose from the need to comply with the regulations (e.g., the 1976 Toxic Substances Control Act) enacted. A significant proportion of this technology has been designed, specifically, to analyze for and remove color and priority pollutants from wastewater effluents and to circumvent pollution problems by eliminating their source.

The inspiration for this book stems in part from the recognition that although an appreciable volume of information pertaining to the environmental chemistry of synthetic colorants can be generated from the open literature, except for a recent chapter on this subject in Volume 8 of the *Kirk-Othmer Encyclopedia of Chemical Technology*, there is no book available that provides a consolidated treatise covering (1) fundamental principles and applications of the various known wastewater treatment methods, (2) pollution prevention, (3) analytical methods, (4) basic terminology, and (5) regulatory issues associated with the environmental chemistry of organic dyes and pigments. Although not all-encompassing, this book is presented as a step toward the achievement of that worthwhile goal.

The present volume consists of 13 chapters, 7 of which pertain to the scope and limitations of the most commonly used wastewater treatment methods. The first 7 chapters address factors to be considered in maximizing the effectiveness of the treatment methods and provide data from studies (laboratory, pilot plant, and production level) involving a wide variety of dyes and dye intermediates. These chapters also cover the physicochemical principles of the processes and the mechanisms associated with color removal for a given process. This section of the book includes a description of the fundamental principles and applications of the PACT® system.

The next three chapters, while having a pollution prevention theme, are closely related to the preceding seven in that they double as wastewater treatment methods. The key virtue of two of these methods is the opportunity to generate an acceptable effluent while simultaneously recovering the cost of the waste treatment equipment by recycling/reusing water and chemicals utilized in various manufacturing processes.

The utility of mass spectrometry in the analysis of wastewater containing dyes and dye intermediates is described. It is clear from Chapter 11 that the sensitivity and wide variety of ionization techniques associated with mass spectrometry make it the method of choice for characterizing trace quantities of hydrophobic and hydrophilic organic priority pollutants in an effluent.

The final two chapters contain a description of regulations governing the manufacture, handling, use, and disposal of synthetic dyes and pigments, and the impact of those regulations on the associated industries. International and U.S. perspectives are presented by experts on the subject.

The number of different wastewater treatment methods that have emerged is indicative of the diversity of types of wastewater encountered and the ingenuity of the scientists and engineers working in this area. It is hoped that the contents of this book will stimulate further research and development needed to keep academic and industrial scientists and technologists abreast of new discoveries in the field. It is also anticipated that a key outcome of the present publication and additional research will bridge the gap between the views of environmentalists and industrialists in a manner that protects the integrity of the environment and the economic well-being of the industries involved.

ABRAHAM REIFE
HAROLD S. FREEMAN

Toms River, New Jersey
Raleigh, North Carolina

ENVIRONMENTAL CHEMISTRY OF DYES AND PIGMENTS

PART 1

WASTEWATER TREATMENT METHODS

CHAPTER 1

CARBON ADSORPTION OF DYES AND SELECTED INTERMEDIATES

ABRAHAM REIFE and HAROLD S. FREEMAN

North Carolina State University, Raleigh, North Carolina

INTRODUCTION

Adsorption is an effective method for lowering the concentration of dissolved organics in an effluent. In this regard, activated carbon has been evaluated extensively for the waste treatment of the different classes of dyes, that is, acid, direct, basic, reactive, disperse, and so forth (1–28), and is now the most widely used adsorbent for dyes (29).

Commercial activated carbon can be prepared from lignite and bituminous coal, wood, pulp mill residue, coconut shell, and blood, having a surface area ranging from 500 to 1400 m^2/g (30). The feasibility of adsorption on carbon for the removal of dissolved organic pollutants has been demonstrated by adsorption isotherms (31). Carbon adsorption isotherms are generated by contacting a fixed quantity of dye wastewater with different amounts of activated carbon for a fixed length of time in order to remove impurities. The Freundlich empirical equation can be used to express the mathematical relationship between the quantity of impurity remaining in solution versus the quantity adsorbed:

$$X/M = KC^{1/n}$$

where X = amount of impurity adsorbed
      M = mass of carbon
      C = equilibrium concentration of impurity in solution
   K, n = constants

From a logarithmic plot of the data, it is possible to determine the carbon's capacity to adsorb the particular impurities at a specified equilibrium concentration. The quantity of adsorbate in solution can be determined by total organic carbon

Environmental Chemistry of Dyes and Pigments, Edited by Abraham Reife and Harold S. Freeman.
ISBN 0-471-58927-6   © 1996 John Wiley & Sons, Inc.

(TOC), chemical oxygen demand (COD), biological oxygen demand (BOD), high-pressure liquid chromatography (HPLC), and so forth. Applying the Freundlich equation to the analyzed values on log-log paper, a carbon adsorption isotherm is obtained, making it possible to determine whether or not the desired degree of adsorbate is plausible with the particular activated carbon tested. Information such as the theoretical capacity of the carbon exhaustion can also be determined.

The theoretical quantity of liquid (liters) that can be treated per gram of carbon can be calculated using the following formula:

$$V_{C_o} = \frac{(X/M)C_o\ (V)}{C_o}$$

where C_o = influent concentration, i.e., analyses in ppm
   $(X/M)C_o$ = ultimate capacity per gram of carbon at influent concentration C_o in mg/g
   V = volume of liquid used in the isotherm test

If one liter of solution is divided by V_{C_o}, the theoretical carbon column dosage (or usage rate) required to treat each liter of wastewater may be obtained. This figure, when multiplied by the conversion factor 8.337, gives the carbon exhaustion rate in pounds per 1000 gallons. If a straight line is obtained upon plotting X/M vs. C_o, the slope of this line ($1/n$ of the Freundlich equation) provides adsorption intensity.

APPLICATIONS

Several pilot plant and commercial-scale systems using activated carbon adsorption columns have been developed (32–34).

Disperse and Cationic Dyes

In a study conducted by Hall (2), the adsorption characteristics of Disperse Yellow 54, Disperse Red 73, and Disperse Blue 7, using powdered activated carbon and the following four variables, were investigated as to their effect upon its removal from effluent: concentration, pH, surfactants, and alkali salts. Experimental data led to the conclusion that by controlling these variables a high degree of adsorption was possible.

Disperse Yellow 54

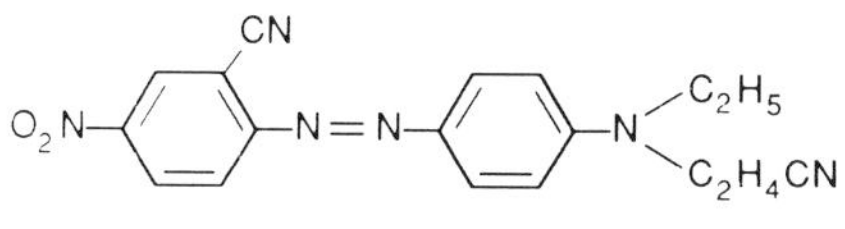

Disperse Red 73

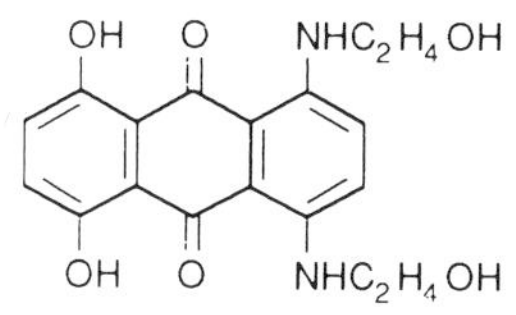

Disperse Blue 7

The rate processes for the adsorption of Disperse Blue 7 on activated carbon have been reported (5) and found to vary with agitation, initial dye concentration, carbon particle size, and temperature of the dye solution. The activation energy of adsorption was determined to be 34 kJ/mol.

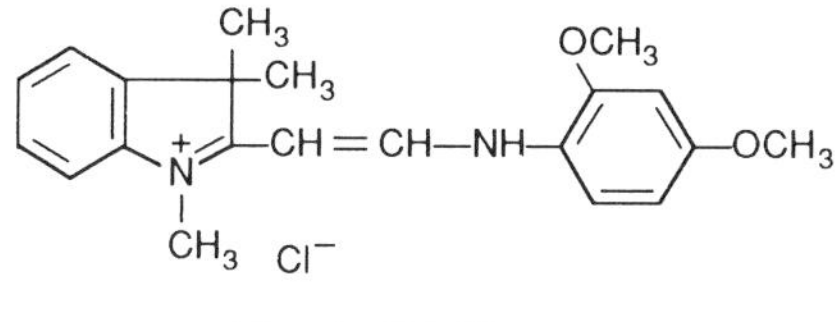

Basic Yellow 11

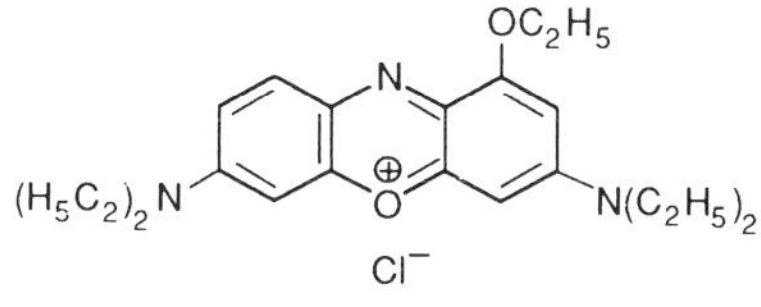

Basic Blue 3

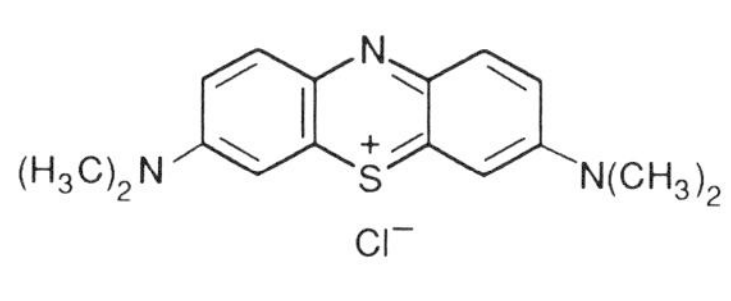

Basic Blue 9

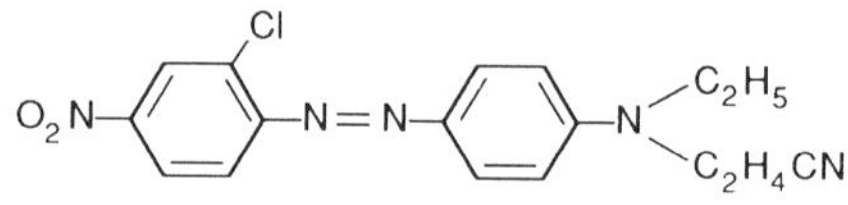

Disperse Red 78

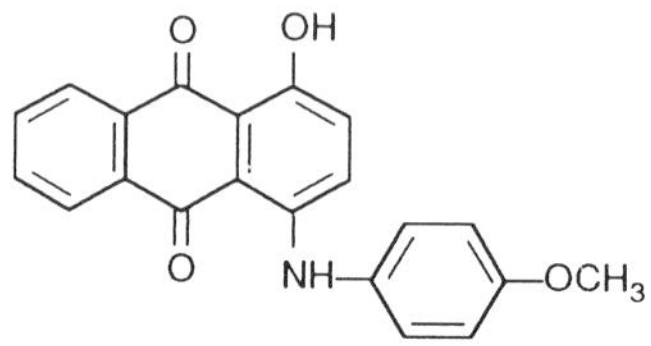

Disperse Blue 64

A group of basic and disperse dyes (cf. Basic Yellow 11, Basic Blue 3, Basic Blue 9, Disperse Red 78, and Disperse Blue 64) were studied to better define factors affecting the removal of textile dyes from wastewater by carbon adsorption (3). It was shown that dyes of the same application class, that is, disperse or basic, exhibit similar absorption characteristics. Within each particular application class, differences in chemical structure (e.g., oxazine, methine, thiazine, azo, or anthraquinone) influence the affinity of specific dyes for the absorbent. In the case of the basic dyes, the relative affinity can be correlated to the basicity of the dye molecules. Disperse dyes did not exhibit as wide a variation in adsorption properties between dyes of different chemical structures as the basic dyes. The poor carbon adsorption behavior of disperse dyes is a consequence of their colloidal nature. A study of disperse dye–carrier interactions on activated carbon (4) demonstrated that the affinity of Disperse Yellow 54 and Disperse Red 73 increased, in the presence of dye carrier, to a level comparable to that of the basic dyes. Higher carrier concentration probably enhances dye adsorption by increasing the solubility of disperse dyes in the bulk solution and in the immediate area surrounding the carbon granules. In turn, this reduces the repulsive forces between the colloidal dye and the carbon surface, allowing adsorption to take place. Disperse Blue 58 adsorbed to a greater extent than either Disperse Yellow 64 or Disperse Red 73, an observation that is probably due to the flat, planar shape of Disperse Blue 58.

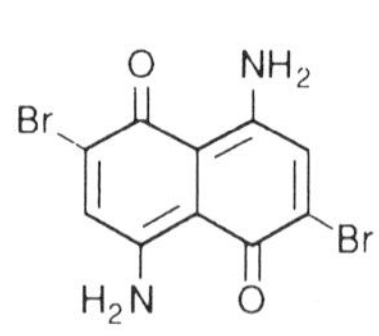

Disperse Blue 58

In general, carbon adsorption of dyes is neither very efficient nor economical. Disperse dyes, vat dyes, and pigments have such low solubility that their rate of adsorption onto carbon is prohibitively slow at room temperature (1). On the other hand, water-soluble dyes such as acid, basic, direct, metallized, mordant, and reactive dyes are also not readily adsorbed on carbon. One of the main reasons for the observed poor adsorption is the polar nature of these dyes versus the nonpolar nature of carbon (29). Therefore, while no single effluent treatment is effective for dye removal, a combination of physical and chemical or biological treatments is recommended (35), with carbon adsorption specifically recommended as a finishing, polishing step.

The fact that carbon adsorption is relatively ineffective in removing disperse dyes has been attributed to the low solubilities and colloidal dispersion properties of the dyes, characteristics that prevent the adsorption and migration of dye particles to the carbon surface. It is important to note that in dyeing hydrophobic fibers, a sparingly water-soluble disperse dye must be blended with anionic surfactants, to reduce agglomeration and "solubilize" the dye. It appears, from Table 1.1, that the presence of an anionic surfactant in the dispersions of disperse dyes Blue 79, Blue 281, Red 167, and Orange 30 has the same effect on the adsorption of dye onto carbon as it does on hydrophobic fibers; that is, carbon adsorption is

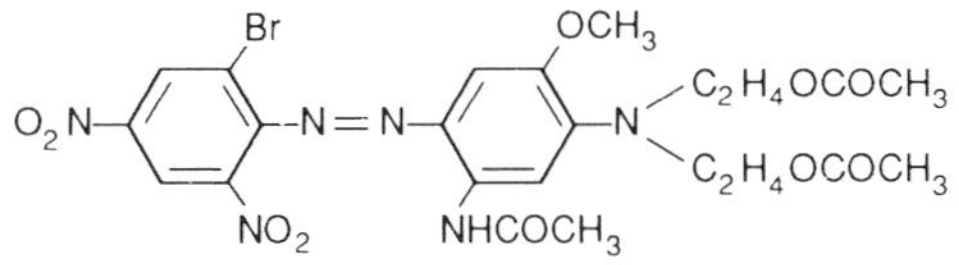

Disperse Blue 79

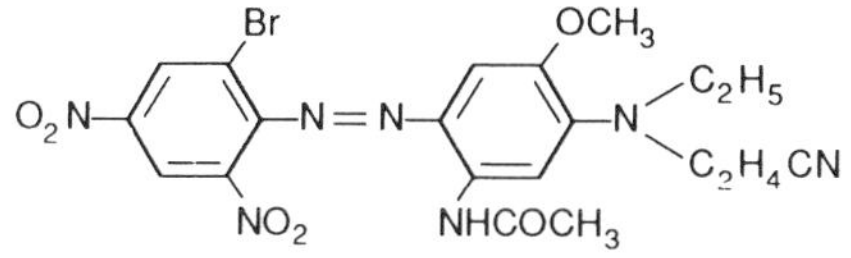

Disperse Blue 281

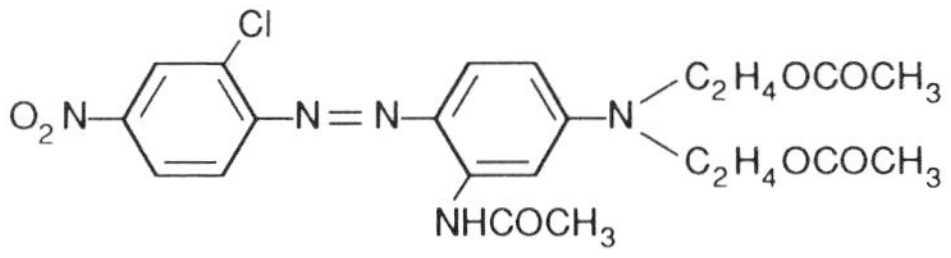

Disperse Red 167

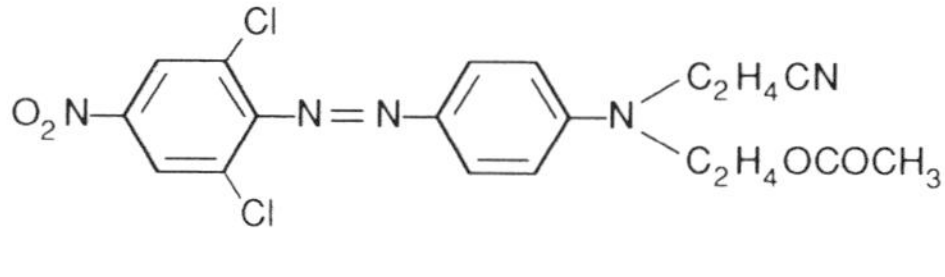

Disperse Orange 30

TABLE 1.1 Adsorbability of Disperse Dye Dispersions from Wastewaters

Product	Percentage of Dye in Effluent	Adsorption Capacity (mg/L)	Grams of Carbon Required to Treat 1 Liter Solution	Carbon Exhaustion Rate (lb/1000 gal)	Adsorbability by Carbon Percent											
					Percentage TOC Reduction						Percentage APHA Color Reduction					
					0.5%	1.0%	2.0%	5.0%	7.5%	10.0%	0.5%	1.0%	2.0%	5.0%	7.5%	10.0%
Disperse Blue 79	0.02	50.0	2.38	19.8	34.5	41.2	45.4	53.8	57.1	61.3	53.7	58.1	60.1	70.2	74.5	79.0
Disperse Blue 281	0.005	90.0	0.94	7.9	50.6	51.8	61.2	64.7	71.8	83.5	75.4	77.6	81.7	87.9	88.9	92.8
Disperse Orange 30	0.012	23.0	5.10	42.4	34.2	38.5	47.9	59.0	65.0	66.7	70.2	74.9	80.1	85.0	88.1	89.8
Disperse Red 167	0.007	60.0	1.70	14.2	52.9	61.8	67.6	76.5	79.4	77.5	88.4	89.4	91.1	93.2	94.9	95.7

enhanced by preventing agglomeration and keeping the disperse dye well dispersed. This is especially prevalent in disperse azo dyes such as Disperse Red 167 in which 88.4 and 95.7% APHA (American Public Health Association) color is effectively removed using 0.5 and 10.0% activated carbon, respectively. The lower TOC values are due to nondyeing, nonadsorbable auxiliaries such as surfactants, antifoaming agents, and so forth (27).

Carbon adsorption isotherms were generated from the effluents of the four afore-mentioned disperse dyes, and the Freundlich empirical equation was used for log-arithmic plotting of the data. This resulted in the steeply sloped lines in Figure 1.1, indicating low adsorbability. However, significant removal of these disperse dyes from wastewater effluents, as measured by TOC, was achieved by using a polymer flocculation pretreatment followed by carbon adsorption (see Table 1.2).

Anionic Dyes

In addition to polymer flocculation, one of the methods commonly used in combination with carbon adsorption for the removal of soluble azo dyes in wastewater is chemical reduction. Reduction by agents such as sodium hydrosulfite, thiourea dioxide, and sodium borohydrite cleaves the azo linkage of the dye, producing the corresponding aromatic amines and resulting in decolorization. This process was developed by Asahi Chemical Company and served as the basis for its Japanese patent (36). According to the patent, effluents containing azo reactive dyes are very difficult to treat in environ-mental systems, due to the sulfonic acid groups, which make the dyes very water soluble and polar. Therefore, such dyes are not readily adsorbed onto the nonpolar carbon surface. The dye must first be degraded (reduced), and its degradation products adsorbed. The example cited in the patent was Reactive Yellow 3.

Reactive Yellow 3

Subsequent lab work on reactive dyes confirmed Asahi's claim that, besides being "uncolored," the individual degradation products of reactive dyes have much better adsorbability on carbon than the parent dyes themselves. Therefore, sodium hydro-sulfite reduction and carbon adsorption isotherms were conducted on the following group of acid, direct, and reactive dyes (29):

Acid Yellow 219

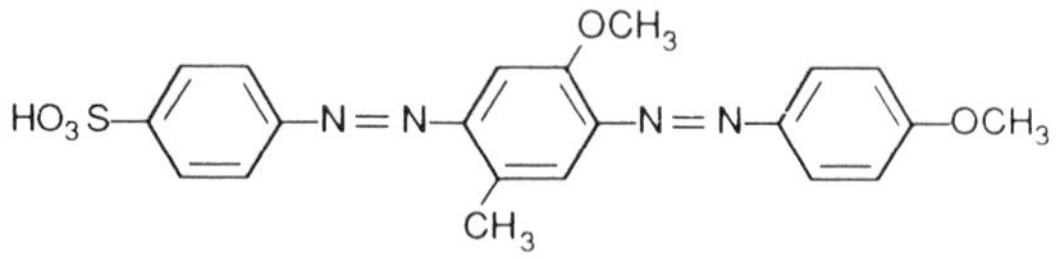

Acid Orange 156

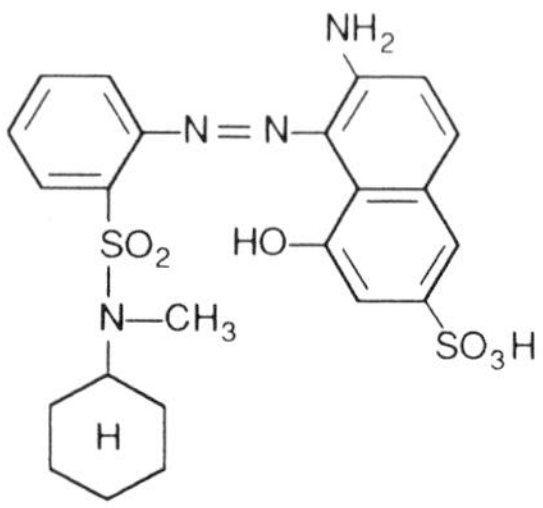

Acid Red 361

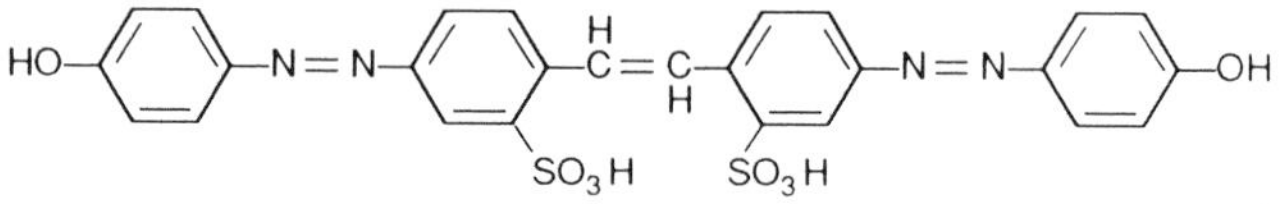

Direct Yellow 4

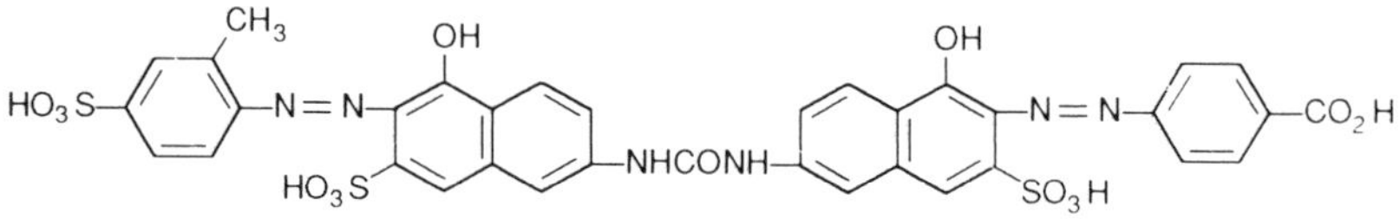

Direct Orange 102

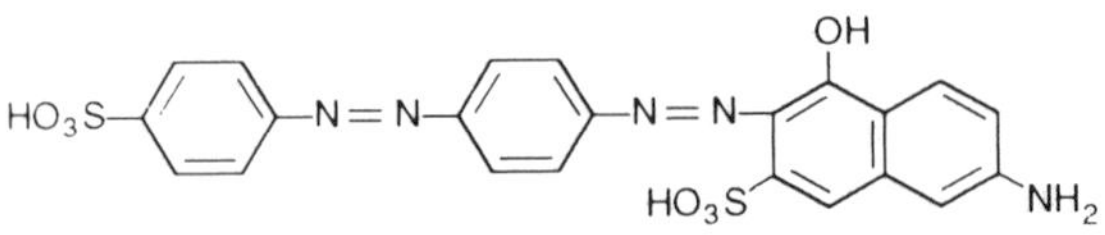

Direct Red 254

Direct Blue 281

Reactive Orange 95

Reactive Red 24

Reactive Red 43

Reactive Red 120

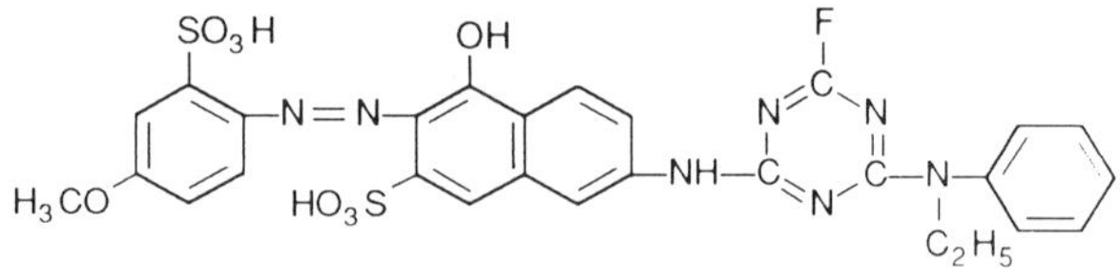

Reactive Red 183

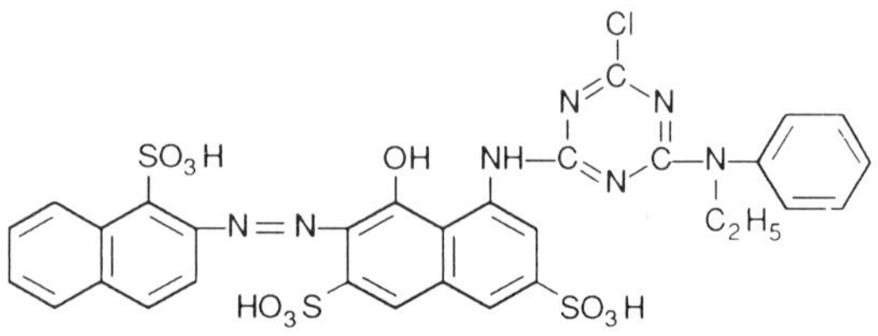

Reactive Red 218

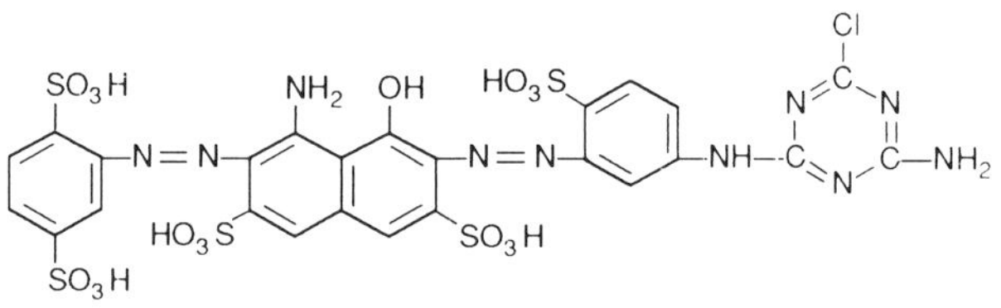

Reactive Black 39

The three acid dye solutions underwent an 82–100% reduction in APHA color before carbon adsorption and an 18–88% increase in adsorption capacity, in addition to a 15–69% decrease in carbon adsorption rate, after carbon adsorption. The amount of sodium hydrosulfite required to decolorize a 0.4% acid dye solution was 5.0–7.0 g. The four direct dye solutions underwent a 71–99% reduction in APHA color before carbon adsorption, a 26–188% increase in adsorption capacity, and a 20–63% decrease in carbon adsorption rate following carbon adsorption. The amount of sodium hydrosulfite required to decolorize a 0.4% direct dye solution was 5.0–7.0 g. The seven reactive dye solutions underwent a 69–99% reduction in APHA color prior to carbon adsorption, an 83–122% increase in adsorption capacity, and a 42–68% decrease in carbon adsorption rate following carbon adsorption. The amount of sodium hydrosulfite required to decolorize a 0.4% reactive dye solution was 3.0–5.0 g. The results are summarized in Tables 1.3–1.5.

It is clear that the molecular structure of a compound has a significant effect on the extent to which it will be adsorbed on carbon. In general, a compound's adsorbability is enhanced by increasing molecular size and aromaticity, and by de-

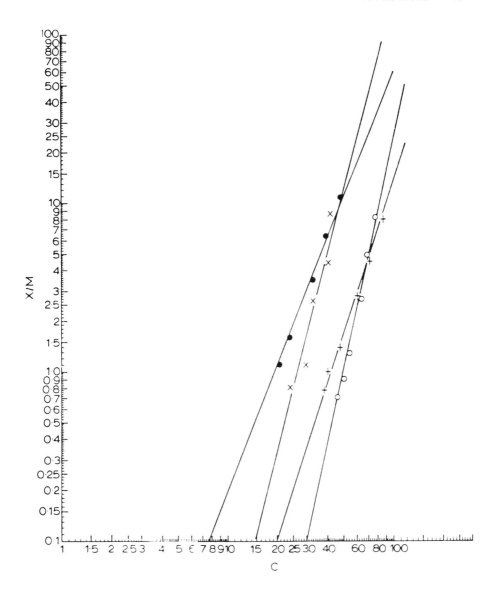

FIGURE 1.1   Carbon adsorption isotherms for selected disperse dyes: (○) Disperse Blue 79, (×) Disperse Blue 281, (+) Disperse Orange 30, and (●) Disperse Red 167.

creasing solubility, polarity, and carbon chain branching. Dyes of high molecular weight are adsorbed into the transitional pores of an activated carbon, whereas smaller dyes penetrate into the micropores. Surface area in these micropores is drastically reduced by thermal regeneration (37).

TABLE 1.2 Polymer Flocculation/Carbon Adsorption of Disperse Dye Dispersions

Dye	Filtrate TOC (ppm)	Filtrate TOC after Cationic Polymer Treatment (ppm)	TOC Reduction (%)	Filtrate TOC after Cationic Polymer +1% Carbon Treatments (ppm)	New TOC Reduction (%)	Percentage Dye in Effluent Prior to Treatments	Percentage Dye in Effluent after Cationic Polymer +1% Carbon Treatments
Disperse Blue 79	267	103	61.4	9.8	96.3	0.02	0
Disperse Blue 281	115	98	14.9	13.5	88.3	0.005	0
Disperse Orange 30	146	110	24.7	25.4	82.6	0.012	0
Disperse Red 167	132	113	14.4	26.4	80.0	0.007	0

TABLE 1.3 Comparison of APHA Color before and after Na$_2$S$_2$O$_4$ Reduction and Comparison of Adsorption Capacity and Carbon Adsorption Rate before and after Reduction and Carbon Adsorption

Dye (0.4% soln)	APHA Color before and after Na$_2$S$_2$O$_4$ Reduction			Following Na$_2$S$_2$O$_4$ Reduction					
				Adsorption Capacity (mg/g) before and after Carbon Adsorption			Carbon Adsorption Rate (lb/1000 gal) before and after Carbon Adsorption		
	Before	After	Percent Reduction	Before	After	Percent Increase	Before	After	Percent Decrease
Acid Yellow 219	575,602	1,033	100	170	200	18	100	85	15
Acid Orange 156	579,676	16,298	97	160	300	88	110	60	45
Acid Red 361	177,792	32,447	82	110	160	45	140	43	69

TABLE 1.4 Comparison of APHA Color before and after Na$_2$S$_2$O$_4$ Reduction and Comparison of Adsorption Capacity and Carbon Adsorption Rate before and after Reduction and Carbon Adsorption

| | APHA Color before and after Na$_2$S$_2$O$_4$ Reduction | | | Following Na$_2$S$_2$O$_4$ Reduction | | | | | |
| | | | | Adsorption Capacity (mg/g) before and after Carbon Adsorption | | | Carbon Adsorption Rate (lb/1000 gal) before and after Carbon Adsorption | | |
Dye (0.4% soln)	Before	After	Percent Reduction	Before	After	Percent Increase	Before	After	Percent Decrease
Direct Yellow 4	733,392	10,190	99	70	110	57	206	133	35
Direct Orange 102	406,699	5,778	99	66	190	188	201	74	63
Direct Red 254	585,232	27,410	95	95	120	26	147	117	20
Direct Blue 281	79,266	22,742	71	60	150	150	146	58	60

TABLE 1.5 Comparison of APHA Color before and after Na$_2$S$_2$O$_4$ Reduction and Comparison of Adsorption Capacity and Carbon Adsorption Rate before and after Reduction and Carbon Adsorption

				Following Na$_2$S$_2$O$_4$ Reduction					
	APHA Color before and after Na$_2$S$_2$O$_4$ Reduction			Adsorption Capacity (mg/g) before and after Carbon Adsorption			Carbon Adsorption Rate (lb/1000 gal) before and after Carbon Adsorption		
Dye (0.4% soln)	Before	After	Percent Reduction	Before	After	Percent Increase	Before	After	Percent Decrease
Reactive Orange 95	422,256	5,556	99	43	84	95	266	137	48
Reactive Red 24	110,194	34,077	69	45	100	122	213	98	54
Reactive Red 43	205,942	2,767	99	60	110	83	150	87	42
Reactive Red 120	108,898	9,482	91	19	38	100	434	139	68
Reactive Red 183	320,396	7,038	98	66	130	97	208	98	53
Reactive Red 218	101,860	16,112	84	52	114	119	221	114	49
Reactive Black 39	92,600	21,928	76	35	70	100	205	95	54

Dye Intermediates

A number of articles have been published on the amenability of aliphatic and aromatic organic compounds such as alcohols, aldehydes, amines, esters, ethers, ketones, and so forth for adsorption onto activated carbon (38–44). Although some work has also been conducted on sulfonated organics (45), which shows that the presence of a sulfonic acid group usually decreases adsorbability on activated carbon (44,46), there is no related literature pertaining to sulfonated azo dye intermediates. Of special interest are letter acids such as H acid (1-naphthol-8-amino-3,6-disulfonic acid), J acid (1-naphthol-6-amino-3-sulfonic acid), R acid (2-naphthol-3,6-disulfonic acid), G acid (2-naphthol-6,8-disulfonic acid), and so forth, due to the importance of these sulfonated compounds as intermediates for the manufacture of azo acid, direct, and reactive dyes. Such dyes represent the major classes of dyestuffs for dyeing natural and synthetic fibers, leather, and paper. Carbon adsorption isotherms were recorded on the most important of these dye intermediates to determine the feasibility of removing them from wastewater using virgin activated carbon (see Tables 1.6–1.9).

The introduction of a sulfonic group onto aniline to produce orthanilic acid (aniline-2-sulfonic acid), metanilic acid (aniline-3-sulfonic acid), and sulfanilic acid (aniline-4-sulfonic acid) lowers the absorbability of the amine, as judged by the reduction in TOC, adsorption capacity, and rate of carbon exhaustion. Also, the position of the sulfonic acid group in the aniline molecule has an effect on adsorbability in that adsorption of the isomers decreased in the following order:

orthanilic > sulfanilic > metanilic

Similarly, carbon adsorption properties of mono and disulfonated naphthols, naphthylamines, and aminonaphthols were compared. Whereas β-naphthol has excellent adsorption on carbon, the introduction of a sulfonic group to give, for example, Schaeffer's salt (2-naphthol-6-sulfonic acid), lowers carbon adsorption, as indicated by a reduction of TOC, adsorption capacity, and carbon exhaustion rate. The introduction of two sulfonic groups to give G acid (2-naphthol-6, 8-disulfonic acid) lowers adsorption further. Substituting an amino group for the hydroxy group of Schaeffer's acid, to give Broenner's acid (2-naphthylamine-6-sulfonic acid), has little effect on reduction of TOC; however, there is an increase in adsorption capacity and a decrease in carbon exhaustion rate. A comparison of Broenner's acid (2-naphthylamine-6-sulfonic acid) and Tobias acid (2-naphthylamine-1-sulfonic acid) reveals that shifting the sulfonic acid from the sixth to the first position lowers adsorbability on carbon.

The addition of a hydroxy group to Broenner's acid to form Gamma acid (2-naphthylamine-8-hydroxy-6-sulfonic acid) lowers carbon capacity slightly. A comparison of the isomers Gamma acid and J acid (3-naphthylamine-8-hydroxy-6-sulfonic acid) indicates that shifting the amino group from the second to the third position decreases adsorbability. Attaching a phenyl or benzoyl group to the *N* atom of J acid to give *N*-phenyl J acid (3-anilino-8-hydroxy-6-sulfonic acid) and

TABLE 1.6 Comparison of Adsorbability of 0.1% Solution of Aniline and Sulfonated Derivatives on Calgon Filtersorb F400–Virgin, at pH 7.0 and 30°C

Product	Adsorption Capacity (mg/g)	Carbon Exhaustion Rate (lb/1000 gal)	Adsorbability by Carbon Percent, Percent Reduction for 0.1% Product						Structure
			0.5%	1.0%	2.0%	5.0%	7.5%	10%	
Aniline	280.0	23.2	77.9	93.7	98.0	98.9	99.0	99.6	$-NH_2$ on benzene ring
Orthanilic acid	34.0	104.2	31.8	54.6	80.7	92.2	92.2	98.4	benzene with SO_3H and NH_2
Metanilic acid	21.0	169.5	21.8	40.3	68.1	90.9	91.8	98.6	benzene with HO_3S and NH_2
Sulfanilic acid	25.0	150.0	25.6	40.4	62.4	91.1	92.9	97.8	benzene with HO_3S and NH_2

TABLE 1.7 Comparison of Adsorbability of 0.1% Solution of β-Naphthol and Sulfonated Derivatives on Calgon Filtersorb F400–Virgin, at pH 7.0 and 30°C

Product	Adsorption Capacity (mg/g)	Carbon Exhaustion Rate (lb/1000 gal)	Adsorbability by Carbon Percent, Percent Reduction for 0.1% Product						Structure
			0.5%	1.0%	2.0%	5.0%	7.5%	10%	
β-Naphthol	200	35.6	97.1	98.8	98.8	98.9	98.8	99.5	
Schaeffer's acid	66	70.0	56.7	81.8	93.8	93.8	95.6	95.6	
Broenner's acid	72	62.3	60.4	87.2	94.8	95.0	95.4	98.3	
Tobias acid	55	83.1	45.0	76.1	93.8	94.5	94.7	94.7	
Gamma acid	52	85.9	46.5	75.0	89.6	95.1	95.7	95.9	
J acid	21	163.6	26.2	48.5	73.8	87.4	90.0	90.5	

TABLE 1.8 Comparison of Adsorbability of 0.1% Solution of J Acid, Phenyl J Acid, and Benzoyl J Acid on Calgon Filtersorb F400–Virgin, at pH 7.0 and 30°C

Product	Adsorption Capacity (mg/g)	Carbon Exhaustion Rate (lb/1000 gal)	Adsorbability by Carbon Percent, Percent Reduction for 0.1% Product						Structure
			0.5%	1.0%	2.0%	5.0%	7.5%	10%	
J acid	21.0	163.6	26.2	48.5	73.8	87.4	90.0	90.5	
N-phenyl J acid	52.0	55.2	73.5	91.0	94.0	95.6	96.2	96.2	
N-benzoyl J acid	62.0	54.2	74.9	93.5	94.0	94.3	94.3	94.7	

TABLE 1.9 Comparison of Adsorbability of 0.1% Solution of Some Naphthalene Disulfonic Acids on Calgon Filtersorb F400–Virgin, at pH 7.0 and 30°C

Product	Adsorption Capacity (mg/g)	Carbon Exhaustion Rate (lb/1000 gal)	Adsorbability by Carbon Percent, Percent Reduction for 0.1% Product						Structure
			0.5%	1.0%	2.0%	5.0%	7.5%	10%	
Naphthalene 2,6-disulfonic acid	40.0	137.4	29.0	46.1	70.6	93.3	93.9	94.1	naphthalene with SO_3H and HO_3S
Amino G acid	22.0	183.8	19.6	38.1	62.9	90.5	93.4	93.4	naphthalene with NH_2, SO_3H, HO_3S
G acid	29.0	115.9	28.3	43.2	58.1	82.9	90.3	92.1	naphthalene with OH, SO_3H, HO_3S
R acid	36.0	87.1	33.3	55.0	73.1	88.8	91.5	91.8	naphthalene with OH, SO_3H, HO_3S
H acid	21.0	142.9	30.3	50.3	73.9	89.2	91.9	92.5	naphthalene with NH_2, HO, SO_3H, HO_3S
Chromotropic acid	12.0	159.8	23.4	35.2	58.7	84.3	84.3	84.8	naphthalene with OH, HO, SO_3H, HO_3S

N-benzoyl J acid (6-benzoyl-8-hydroxy-6-sulfonic acid) greatly enhances adsorption on carbon.

The presence of two sulfonic groups on the naphthalene ring significantly lowers adsorbability, as seen in the adsorption behavior of G acid (2-naphthol-6-8-disulfonic acid), R salt (2-naphthol-3,6-disulfonic acid), amino G acid (2-naphthylamine-6,8-disulfonic acid), H acid (1-naphthol-8-amino-3,6-disulfonic acid), and chromotropic acid (1-naphthol-8-hydroxy-3,6-disulfonic acid).

PROPERTIES OF CARBON ADSORBENTS

Several factors must be taken into consideration when generating carbon adsorption isotherms on wastewater: temperature, pH, contact time, carbon dosage, and choice of activated carbon. The choice of carbon is especially important. To illustrate this point, carbon adsorption isotherms were generated using Acid Blue 277 and four different types of activated carbon, viz. Calgon virgin and regenerated Filtersorb F400 (a carbon manufactured from bituminous coal), Norit Darco (made from wood), and Westvaco Nuchar (made from pulp mill residue).

When Calcon Filtersorb F400–virgin carbon was used, the maximum adsorbate capacity was found to be 36.6 mg/g carbon. The amount of carbon required to treat 1 liter of solution was 13.0 g, and the carbon exhaustion rate was 108.2 lb/1000 gal. A sample of 2% carbon at pH 7.0 and 70°C removed 96.1% TOC and 96.8% APHA color (cf. Table 1.10).

When Calgon Filtersorb F400–regenerated carbon was used, the maximum carbon capacity was 58.0 mg/g carbon. The quantity of carbon required to treat 1 liter of solution was 8.9 g, and carbon exhaustion rate was 74.2 lb/1000 gal. A sample of 2% carbon at pH 7.0 and 70°C removed 98.4% TOC and 98.2% APHA color (cf. Table 1.11).

When Norit Darco was used, the maximum carbon capacity was 117.0 mg/g carbon. The quantity of carbon required to treat 1 liter of solution was 4.4 g, and carbon exhaustion rate was 37.1 lb/1000 gal. A sample of 1% carbon at pH 7.0 and 70°C removed 99.4% TOC and 98.2% APHA color (cf. Table 1.12).

When Westvaco Nuchar was used, the maximum carbon capacity was 93.0 mg/g carbon. The quantity of carbon required to treat 1 liter of solution was 5.6 g, and carbon exhaustion rate was 46.6 lb/1000 gal. A sample of 0.5% carbon at pH 7.0 and 70°C removed 97.8% TOC and 96.6% APHA color (cf. Table 1.13)

TABLE 1.10   Treatment of Acid Blue 277 (0.1%) with Calgon Filtersorb F400–Virgin Carbon at pH 7.0 and 70°C

% Carbon	TOC	% Reduction	APHA Color	% Reduction
0.0	519	—	4000	—
0.5	329	36.6	2556	36.1
1.0	199	59.5	1519	62.0
2.0	19	96.1	130	96.8
5.0	17	96.5	67	98.3
7.5	17	96.5	33	99.2
10.0	16	96.7	19	99.5

Isotherm Data Tabulation

% Carbon	M	C_o	X	X/M
0.0	0.0	519	—	—
0.5	5.0	329	190	38.0
1.0	10.0	199	292	29.2
2.0	20.0	19	472	23.6
5.0	50.0	17	474	9.5
7.5	75.0	17	474	6.3
10.0	100.0	16	475	4.8

TABLE 1.11   Treatment of Acid Blue 277 (0.1%) with Calgon Filtersorb F400–Regenerated Carbon at pH 7.0 and 70°C

% Carbon	TOC	% Reduction	APHA Color	% Reduction
0.0	519	—	4111	—
0.5	261	49.4	2148	47.7
1.0	98	80.6	778	83.3
2.0	8	98.4	70	98.2
5.0	9	98.2	56	98.6
7.5	9	98.2	37	99.1
10.0	9	98.2	19	99.5

Isotherm Data Tabulation

% Carbon	M	C_o	X	X/M
0.0	0.0	516	—	—
0.5	5.0	261	255	51.0
1.0	10.0	98	408	40.8
2.0	20.0	8	498	24.9
5.0	50.0	9	497	9.9
7.5	75.0	9	497	6.6
10.0	100.0	9	497	5.0

TABLE 1.12   Treatment of Acid Blue 277 (0.1%) with Norit Darco Carbon at pH 7.0 and 70°C

% Carbon	TOC	% Reduction	APHA Color	% Reduction
0.0	521	—	4111	—
0.5	39	92.5	367	91.1
1.0	3	99.4	74	98.2
2.0	0	100.0	19	99.5
5.0	0	100.0	19	99.5
7.5	0	100.0	19	99.5
10.0	0	100.0	19	99.5

Isotherm Data Tabulation

% Carbon	M	C_o	X	X/M
0.0	0.0	521	—	—
0.5	5.0	39	482	96.4
1.0	10.0	3	518	51.8
2.0	20.0	0	521	26.1
5.0	50.0	0	521	10.4
7.5	75.0	0	521	6.9
10.0	100.0	0	521	5.2

TABLE 1.13   Treatment of Acid Blue 277 (0.1%) with Westvaco Nuchar Carbon at pH 7.0 and 70°C

% Carbon	TOC	% Reduction	APHA Color	% Reduction
0.0	520	—	4167	—
0.5	11	97.8	141	96.6
1.0	0	100.0	41	99.0
2.0	0	100.0	19	99.5
5.0	0	100.0	19	99.5
7.5	0	100.0	19	99.5
10.0	0	100.0	19	99.5

Isotherm Data Tabulation

% Carbon	M	C_o	X	X/M
0.0	0.0	520	—	—
0.5	5.0	11	509	101.8
1.0	10.0	0	475	47.5
2.0	20.0	0	475	23.8
5.0	50.0	0	475	9.5
7.5	75.0	0	475	6.3
10.0	100.0	0	475	4.8

Based on Table 1.14, the best carbon, in terms of TOC reduction of 0.1% Acid Blue 277 (dried cake) using 0.5% carbon, was Nuchar (97.8%), followed by Darco (92.5%), regenerated Filtersorb F400 (49.4%), and virgin Filtersorb F400 (36.6%).

Examination of the absorption isotherms of the four different carbons shows Nuchar to be the most efficient for adsorption of the dye (cf. Figure 1.2). Since the slope of the line is an indication of the relative adsorbability of the product, the slightly sloped lines of Darco and both virgin and regenerated Filtersorb F400 carbon reflect high adsorbability, whereas the horizontal line of Nuchar carbon represents infinite adsorbability.

Finally, it is of interest to note that Acid Blue 277 dried cake is better adsorbed by Calgon regenerated carbon rather than virgin Filtersorb F400 carbon. The reason for this may be partially explained from the work of DeJohn and Hutchins (47). The authors found that when granular activated carbon is regenerated, the surface area, in the small pore or micropore range, is drastically reduced and transitional pore surface is increased slightly. This change in pore structure changes the adsorptive performance of a regenerated carbon, with the degree of change dependent on the substance to be adsorbed. If the compound is a relatively large molecule (Acid Blue 277 has a molecular weight of 546), the adsorptive performance of the regenerated carbon is equal to or better than that of the virgin carbon. In the case of 0.1% Acid Blue 277 on 0.5% carbon, a TOC reduction of 49.4 vs. 36.6% is observed (cf. Table 1.14). On the other hand, if the organic compound is a relatively small molecule, the adsorbability of the regenerated carbon decreases substantially. Whereas large molecules are adsorbed into transitional pores that are essentially unaffected by the thermal regeneration process, small molecules are adsorbed into the micropores, a good portion of which are lost during carbon reactivation.

ALTERNATIVE ADSORBENTS

Investigations have been undertaken to evaluate inexpensive alternative materials as potential adsorbents for dyes, using activated carbon as a reference. The alternatives considered included peat (2,48), corn stalks (49), chitin (50), carbonized wool (51,52), sawdust (53), cellulosic graft polymers (54–56), fly ash (pulverized fuel ash) (57,58), bagasse pith (59), bentonite (60–62), calcium metasilicate (Wallastonite) (63), organosilicon (64), clays and Fuller's earth (65,66), activated alumina (67), pig and human hair, meat, bone meal, wheat and rice bran, and turkey feathers (68), and activated sludge (69). The adsorption properties of activated sludge were found to be similar to activated carbon in studies involving acid, direct, reactive, disperse, and basic dyes (70). The adsorption capacity of activated sludge for the above-mentioned classes of dyes has been determined by the Freundlich equation and adsorption isotherms (69–73). Like other substrates, the adsorption of dyes by activated sludge is mainly dependent on dye properties (molecular structure and type, number, and position of the substituents in the dye molecule). Adsorption is increased by the presence of hydroxy, nitro, and azo groups but decreased by sulfonic acid groups.

TABLE 1.14 Comparison of Adsorbability of Acid Blue 277 (0.1%) Using Different Activated Carbons at pH 7.0 and 70°C

Carbon	Adsorption Capacity (mg/g)	Grams of Carbon Required to Treat 1-liter Solution	Carbon Exhaustion Rate (lb/1000 gal)	Adsorbability by Carbon Percent, % TOC Reduction for Acid Blue 277 (0.1%)					
				0.5%	1.0%	2.0%	5.0%	7.5%	10.0%
Calgon Filtersorb F400 Virgin	40.0	13.0	108.2	36.6	59.5	96.1	96.5	96.5	96.7
Calgon Filtersorb F400 Regenerated	58.0	8.9	74.2	49.4	80.6	98.4	98.2	98.2	98.2
Norit Darco	117.0	4.4	37.1	92.5	99.4	100.0	100.0	100.0	100.0
Westvaco Nuchar	93.0	5.6	46.6	97.8	100.0	100.0	100.0	100.0	100.0

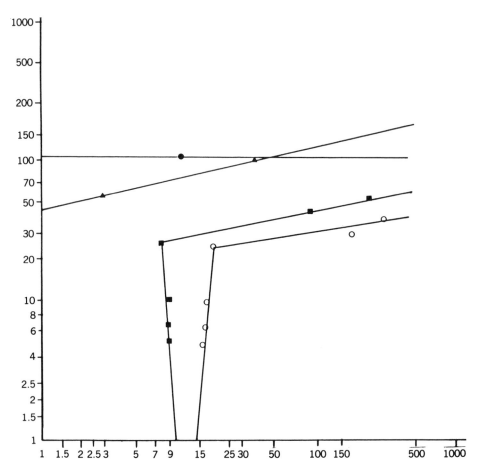

FIGURE 1.2 Comparative carbon adsorption isotherms for Acid Blue 277 using (○) Calgon Filtersorb F400–virgin, (■) Calgon Filtersorb F400–regenerated, (▲) Norit Darco, and (●) Westvaco Nuchar.

SUMMARY

Activated carbon adsorption has been extensively studied as a waste treatment method for the removal of different classes of dyes from wastewater. Although carbon adsorption of dyes is neither efficient nor economical when used alone, when used in tandem with polymer flocculation, chemical reduction, or biodegradation, it becomes a very useful polishing step for efficient dye removal.

Factors such as choice of activated carbon, temperature, pH, contact time, and dosage must be taken into consideration for optimum removal of dyes from wastewater. Also, the molecular structure of a dye has a significant effect on the extent to which it will be adsorbed, with decreasing solubility and polarity of dye favoring adsorbability on carbon.

REFERENCES

1. J. Porter, *Am. Dyest. Reptr.* **60**(8), 17 (1971); **61**(8), 24 (1972).

2. S. G. Hall, Ph.D. Thesis, The Adsorption of Disperse Dyes on Powdered Activated Carbon, University of North Carolina, Greensboro, North Carolina, 1975 Xerox, Univ. Microfilms, Ann Arbor, Michigan.

3. A. F. DiGiano, W. H. Frye, and A. S. Natter, *Am. Dyest. Reptr.* **64**(8), 15 (1975).

4. A. F. DiGiano and A. S. Natter, *J. Water Poll. Control Fed.* **49**(2), 235 (1975).

5. G. McKay, M. S. Otterburn, and A. G. Sweeney, *J. Soc. Dyers Colour* **96**, 576 (1980).

6. G. McKay and B. Al-Duri, *Colourage* **36**(13), 15 (1989); **35**(20), 24 (1988).

7. N. V. Berseneva, *Tekstil Naya Promyshlennost* **49**(7), 14 (1989).

8. Ya I. Tarasevich, V. E. Doroshenko, W. M. Rudenko, and Z. G. Ivanova, *Sov. J. Water Chem. Tech.* **10**(4), 34 (1988).

9. G. S. Gupta, G. Prasad, and V. N. Singh, *Environ. Technol. Lett.* **9**(12), 1413 (1988).

10. J. Perkowski, *L. Kos Przeglad Wlokienniczy* **43**(3), 113 (1989).

11. S. J. Allen, G. McKay, and K. Y. H. Khader, *J. Chem. Tech. Biotech.* **45**(4), 291 (1989).

12. P. Broglio and E. Scaglia, *Tinctoria* **85**(10), 71 (1988).

13. G. McKay and B. Al-Duri, *Chem. Eng. Proc.* **24**(1), 1 (1988).

14. G. S. Gupta, G. Prasad, and V. N. Singh, *Environ. Technol. Lett.* **9**(2), 153 (1988).

15. A. A. Mamontova, T. B. Kondratova, and N. A. Klimenko, *Sov. J. Water Chem. Tech.* **9**(2), 24 (1987).

16. R. J. Posey, *J. Water Poll. Control Fed.* **59**(1), 47 (1987).

17. S. Kuwabara, *Sen-I Gakkaishi* **41**(11), 1485 (1985).

18. G. McKay, *J. Chem. Tech. Biotech.* **32**, 759, 773 (1982); **33A**, 196, 205 (1983).

19. G. McKay, *Colourage* **29**(25), 11 (1982).

20. R. H. Horning, *Textile Dyeing Wastewaters—Characterization End Treatment*, U.S. Environmental Protection Agency, EPA-600/2-78-098, Washington, D.C., 1978.

21. O. G. Rhys, *J. Soc. Dyers Colour* **94**, 293 (1978).

22. M. Mitchell, W. R. Ernst, and G. R. Lithsey, *Bull. Environ. Contam. Toxicol.* **19**, 307 (1978).

23. C. A. Rodman, *Tex. Chem. Color* **3**, 239 (1971).

24. P. B. DeJohn and R. A. Hutchins, *Tex. Chem. Color* **8**, 69 (1976).

25. E. L. Shunney, A. E. Perrotti, and C. A. Rodman, *Am. Dyest. Reptr.* **50**(6), 32 (1971).

26. R. A. Davies, H. J. Kaempf, and M. M. Clemens, *Chem. Ind.* **17**, 827 (1973).

27. A. Reife, Reduction of Toxic Wastewaters in Disperse Azo Dye Manufacture, in *Colour Chemistry*, A. T. Peters and H. S. Freeman, eds., Elsevier Applied Science, London and New York, 1991.

28. A. Reife, Reduction of Toxic Components and Wastewaters of Disperse Blue 79, *41st Southeast Regional American Chemical Society Meeting*, October 9–11, 1989, Raleigh, NC.

29. A. Reife, Waste Treatment of Soluble Azo Acid, Direct and Reactive Dyes Using a Sodium Hydrosulfite Pretreatment Followed by Carbon Absorption, *AATCC Book of Papers*, 1990 International Conference and Exhibition, October 1–3, 1990, Philadelphia.

30. P. N. Cheremisinoff, F. Ellerbusch, eds., *Carbon Absorption Handbook*, Ann Arbor Science, Ann Arbor, MI, 1978.

31. W. G. Schuliger, Purification of Industrial Liquids with Granular Activated Carbon: Techniques for Obtaining and Interpreting Data and Selecting the Type of Commercial System, in *Carbon Absorption Handbook*, P. N. Cheremisinoff and F. Ellerbusch, eds., Ann Arbor Science, Ann Arbor, MI, 1978, pp. 55–83.

32. E. L. Shunney, A. E. Perrotti, and C. A. Rodman, *Am. Dyest. Reptr.* **80**, 32 (1971).

33. W. A. Johnston, *Chem. Eng.* **79**(26), 87 (1972).

34. G. M. Lukchis, *Chem. Eng.* **80**(16), 83 (1973).

35. M. L. Richardson, *J. Soc. Dyers Colour*, **99**, 198 (1983).

36. Asahi Chemical Company, Jpn. Pat. 52-166,644, Process for Purifying Reactive Dye Containing Industrial Wastewater, 1977.

37. P. B. DeJohn and R. A. Hutchins, Treatment of Dye Wastes with Granular Activated Carbon, *AATCC Book of Papers*, 1975 International Conference and Exhibition, 327, 1975.

38. D. M. Giusti, R. A. Conway, and C. T. Lawson, *J. Water Poll. Control Fed.* **46**(5), 947 (1974).

39. K. S. Al-Bahrani and R. J. Martin, *Water Res.* **10**, 731 (1976).

40. T. M. Ward and F. W. Getzen, *Environ. Sci. Technol.* **4**(1), 64 (1970).

41. G. Belfort, *Environ. Sci. Technol.* **13**(8), 939 (1979).

42. V. H. Cheldelin and R. J. Williams, *J. Am. Chem. Soc.* **64**, 1513 (1942).

43. R. J. Martian and K. S. Al-Bahrani, *Water Res.*, **11**, 991 (1977).

44. R. S. Reimers, A. J. Englande, D. D. Leftwich, C. P. Lo, and R. J. Kainz, Evaluation of Organics Adsorption by Activated Carbon, in A. J. Rubin, ed., *Chemistry of Wastewater Technology*, Ann Arbor Science, Ann Arbor, MI, 1980, pp. 123–142.

45. J. C. Morris and W. J. Weber, Adsorption of Biochemically Resistant Materials from Solution, U.S. Public Health Science, AWTR-9 Rept. 999-WP-11 (1964) and AWTR-16 Rept. 999-WP-33 (1966).

46. D. C. Ford, Paper presented at the Open Forum on Management of Petroleum Refinery Wastewaters, U.S. Environmental Protection Agency, American Petroleum Institute, National Petroleum Refiners Association and University of Tulsa, January 1976.

47. P. B. DeJohn and R. A. Hutchins, Treatment of Dye Wastes with Granular Activated Carbon, *Textile Chem. Color.* April 1976, pp. 69–73.

48. W. Ottemeyer, *Gesundh.-Ing.* **53**, 185 (1930).

49. F. Perineau, J. Molinier, and A. Gaset, *Tribune Du Cebedeau* **35** (468), 453 (1982).

50. G. McKay, H. S. Blair, and J. R. Gardner, *J. Appl. Poly. Sci.* **27**, 3043 (1982); **28**, 1767 (1983); **29**, 1499 (1984).

51. G. Malmary, F. Perineau, J. Molinier, and A. Gaset, *J. Chem. Tech. Biotech.* **35A**(8), 431 (1985).

52. F. Perineau, J. Molinier, and J. Gaset, *Water Res.* **17**(5), 559 (1983).

53. H. M. Asfour, O. A. Fadali, M. M. Nassar, and M. S. El-Geundi, *J. Chem. Technol. Biotech.* **35A**(1), 21, 28 (1985).

54. N. Miyata, *Cellulose Chem. Technol.* **21**(5), 551 (1987).

55. N Miyata, *Sen-I Gakkaishi* **42**(8), 1468 (1986).

56. K. Dimov, E. Terlemesyan, and D. Dimitrov, *Textil* **40**(2), 39 (1985).

57. M. R. Balasubramanian and I. Muralisankar, *Indian J. Tech.* **29**(10), 471 (1987).

58. S. K. Khare, K. K. Panday, and R. M. Srivastava, *J. Chem. Technol. Biotech.* **38**(2), 99 (1987).

59. G. McKay, M. El-Geundi, and M. M. Nassar, *Water Res.* **21**(12), 1513 (1987).

60. H. Hoppe and H. J. Taglich, *Acta Hydrochim. Hydrobiol.* **10**, 69 (1982); *Chem. Tech.* **34**, 636 (1982).

61. I. Arvanitoyannis, E. Eleftheriades, and E. Tsotsaroni, *Chemosphere* **18**(9), 1707 (1989).

62. G. McKay, G. Ramprasad, and P. O. Mowli, *Water Air Soil Poll.* **29**(3), 273 (1986).

63. S. K. Khare, R. M. Srivastova, K. K. Panday, and V. N. Singh, *Environ. Technol. Lett.* **9**(10), 1163 (1988).

64. T. I. Denisova, S. I. Meleshevich, and I. A. Sheka, *Zur. Prik. Khim.* **62**(5), 1182 (1989).

65. G. McKay, *Am. Dyest. Reptr.* **68**(4), 29 (1979).

66. G. McKay, M. S. Otterburn, and J. A. Aga, *Water Air Soil Poll.* **24**(3), 307 (1985); **33**(3/4), 419 (1987); **36**(3/4), 381 (1987).

67. C. A. Rodman, *Am. Dyest. Reptr.* **60**, 45 (1971).

68. R. M. Woodby and D. L. Michelsen, Emerging Technologies for Hazardous Waste Treatment Symposium, I&EC Division of ACS, June 4–7, 1990, Atlantic City, NJ.

69. M. Dohanyas, V. Madera, and M. Sedlacek, *Prog. Water Tech.* **10**(5/6), 559 (1978).

70. M. Nakaoka, S. Tamura, Y. Maeda, and T. Azumi, *Sen-I Gakkaishi* **39**(2), 69 (1983).

71. G. M. Shaul, R. J. Lieberman, C. R. Dempsey, and K. A. Dostol, Fate of Azo Dyes in the Activated Sludge Process, presented at the 41st Annual Purdue Industrial Waste Conference, Purdue University, West Lafayette, Indiana, May 13–15, 1986.

72. G. M. Shaul, R. J. Lieberman, C. R. Dempsey, and K. A. Dostol, Treatability of Water Soluble Azo Dyes by the Activated Sludge Process, Industrial Wastes Symposia Proceedings, 59th Water Pollution Control Federation Annual Conference, Los Angeles, California, October 5–9, 1986.

73. G. M. Shaul, C. R. Dempsey, and K. A. Dostol, Fate of Water Soluble Azo Dyes in the Activated Sludge Process, Report Water Engineering Research Laboratory, Office of Research and Development, USEPA, Cincinnati, Ohio.

CHAPTER 2

SODIUM BOROHYDRIDE DYE REDUCTION IN WASTEWATER

MICHAEL M. COOK

Morton International, Danvers, Massachusetts

Wastewater treatments utilizing sodium borohydride reduction have been commercialized for the removal of various metal cations such as copper (2+), nickel (2+), lead (4+), mercury (2+), and silver (1+) (1–10). These chemical reductions occur with both complexed and noncomplexed metal cations, even when the complexing agent(s) is present in excess relative to the cation. Employing bisulfite in conjunction with borohydride at reaction pHs in the 4–7 range has been shown to dramatically improve the efficiency of sodium borohydride in waste treatment applications (11).

Technologies based on bisulfite-catalyzed borohydride reduction are now being applied to water-soluble dyes containing azo or other reducible groups and to copper-based metallized dyes (11–14). This particular reductive treatment chemically cleaves the dye into smaller molecules. In comparison to the parent dye compounds, such smaller molecules are virtually colorless and can be more readily metabolized by activated sludges, chemically oxidized, adsorbed on carbon, or precipitated by polycationic agents.

CHEMICAL ASPECTS OF THIS TECHNOLOGY

Water-soluble dyes and/or closely related by-products containing azo groups can be chemically reduced by a variety of agents including dithionite (15–17), formamidine sulfinic (FAS) acid (15,18), or tin chloride (19) to the corresponding aromatic amines [1].

$$NaO_3S \quad \text{———} \quad N = N \quad \text{———} \quad SO_3Na + 6H^+ + 4e^-$$

$$\longrightarrow \quad 2 NaO_3S \quad \text{———} \quad N^+H_3 \tag{1}$$

Environmental Chemistry of Dyes and Pigments, Edited by Abraham Reife and Harold S. Freeman.
ISBN 0-471-58927-6 © 1996 John Wiley & Sons, Inc.

Sodium borohydride, one of the strongest water-soluble reducing agents commercially available [2] (20),

$$BH_4^- + 8OH^- \rightarrow B(OH)_4^- + 4H_2O + 8e^- \qquad E°_{298} = 1.24 \text{ V} \qquad [2]$$

has not been extensively employed in dye wastewater decolorization due to its rapid reaction with water at pHs $\leq$ 8 [3] (20) and its tendency to reduce via a hydride transfer mechanistic pathway.

$$BH_4^- + 3H_2O + H^+ \rightarrow B(OH)_3 + 4H_2\uparrow \qquad \Delta F°_{298} = -88.8 \text{ kcal/mol } BH_4^- \quad [3]$$

These deficiencies have been overcome by employing an effective level of bisulfite in conjunction with borohydride. In the absence of other reducible species, borohydride quantitatively and rapidly reduces bisulfite to dithionite in the pH range of 5–8 [4] (21,22).

$$BH_4^- + 8HSO_3^- + H^+ \xrightarrow[\text{pH } 5-8]{} 4S_2O_4^{2-} + B(OH)_3 + 5H_2O \qquad [4]$$

The optimized yield of $S_2O_4^{2-}$ (based on BH_4^-) is $\geq$ 90%.

Although the exact mechanism of this reaction has not been fully characterized, the $[BH_5]$ and $[SO_2^-]$ species are likely intermediates [5 and 6] (10).

$$BH_4^- + H^+ \rightarrow [BH_5] \qquad [5]$$

$$[BH_5] + 8HSO_3^- \rightarrow \rightarrow 8[SO_2^-] + B(OH)_3 + 5H_2O \qquad [6]$$

The SO_2 radical anion is also a very strong single electron reducing agent whose oxidation product is bisulfite [7].

$$[SO_2^-] + H_2O \rightarrow HSO_3^- + H^+ + e^- \qquad [7]$$

The overall borohydride chemistry [8] can be viewed as involving bisulfite as a catalyst (e.g., [8a] and [8b]).

$$BH_4^- + 8HSO_3^- + H^+ \rightarrow 8[SO_2^-] + B(OH)_3 + 5H_2O \qquad [8a]$$

$$8[SO_2^-] + 2R_1R_2Ar=NArR_3R_4 + 4H^+ + 8H_2O \rightarrow$$

$$8HSO_3^- + 2R_1R_2Ar\overset{+}{N}H_3 + 2R_3R_4Ar\overset{+}{N}H_3 \qquad [8b]$$

$$BH_4^- + 2R_1R_2ArN=NArR_3R_4 + 5H^+ + 3 H_2O \xrightarrow[\text{pH } 6]{HSO_3^-}$$

$$2R_1R_2Ar\overset{+}{N}H_3 + 2 R_3R_4Ar\overset{+}{N}H_3 + B(OH)_3 \qquad [8]$$

Thus, in the overall borohydride reduction chemistry, bisulfite is not consumed but acts as a regenerable coreagent or catalyst [8].

LABORATORY STUDIES ON DYE REDUCTIONS

Laboratory studies show that the bisulfite-catalyzed borohydride reduction of various water-soluble direct, acid, and reactive dyes containing azo linkages, at pH 6 and ambient temperatures, give color reduction results that compare favorably to those obtained with dithionite on an election equivalent basis (cf. Table 2.1). Under the same laboratory test conditions, FAS acid completely failed to decolorize these dye solutions. Chemical reduction of the azo linkages in these types of acid, direct, reactive (or even disperse) dyes results in the formation of lower-molecular-weight aromatic amine entities, concurrent with the elimination of color in the reduced solutions of these dyes (cf. Figure 2.1).

TABLE 2.1   Lab Studies on Bisulfite-Catalyzed Borohydride Reduction of Azo Dyes (12)

Dye Type[a]	Treatment[b]	Color[c]		
		Initial	Final	% Reduction
Direct Red 254	250 mg $Na_2S_2O_5$/L 50 mg $NaBH_4$/L or 4600 mg $Na_2S_2O_4$/L 2850 mg formamidine sulfinic acid	84.6	7.3 No color reduction	91
Direct Yellow 4	500 mg $Na_2S_2O_5$/L 40 mg $NaBH_4$/L or 600 mg $Na_2S_2O_4$/L	132.8	0.7	99
Acid Red 1	500 mg $Na_2S_2O_5$/L 25 mg $NaBH_4$/L	101.4	5.5	95
Acid Red 361	500 mg $Na_2S_2O_5$/L 25 mg $NaBH_4$/L	86.4	14.6	83
Reactive Red 24[d]	250 mg $Na_2S_2O_5$/L 17 mg $NaBH_4$/L	102.8	10.7	90
Reactive Red 120[d]	500 mg $Na_2S_2O_5$/L 25 mg $NaBH_4$/L	97.2	1.0	99

[a]0.01% aqueous solution of each dye.
[b]Mixed for 15 min at ambient temperature. Maintained at pH 5–6.
[c]Measured using the CIE L^*, a^*, b^* scale and expressed as ΔE or the total color difference between the sample and distilled water reference solution.
[d]Cationic agent, flocculant, and filtration required for separation of reduced species and complete color reduction.

DYE	REDUCTION PRODUCTS

ACID RED 1
(C. I. #18050)

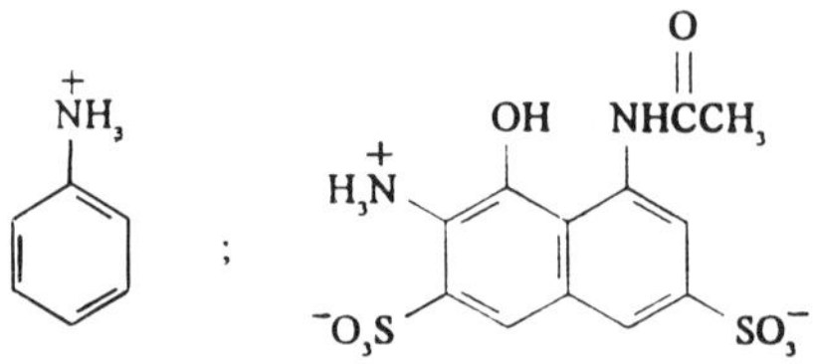

DIRECT YELLOW 4
(C. I. #25890)

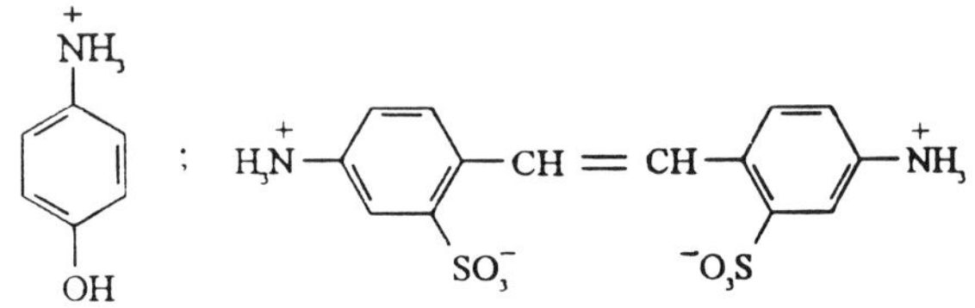

DIRECT BLUE 86
(C. I. #74180)

Cu° :

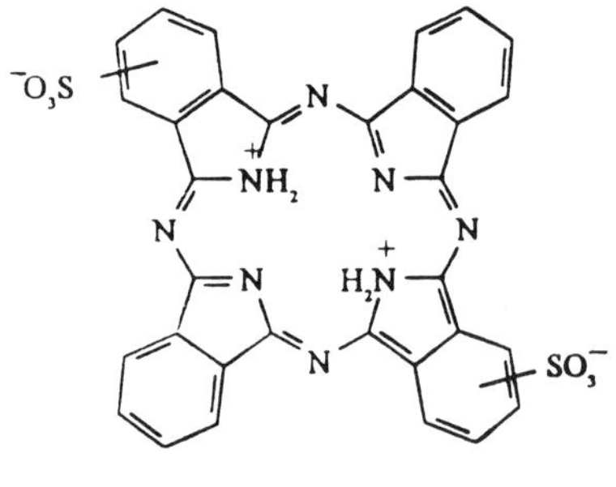

Cu° ;

DIRECT BLUE 218
(C. I. #24401)

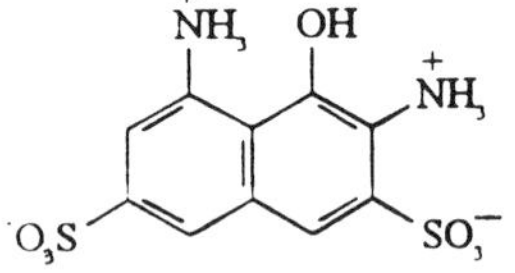

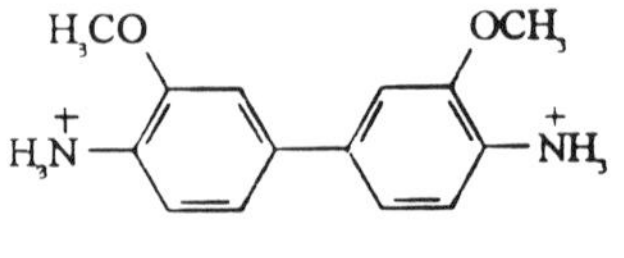

FIGURE 2.1 Anticipated products from chemical reduction of various azo or copper complex dyes.

Reduction of copper-metallized dyes can also be achieved by a bisulfite-catalyzed borohydride treatment (cf. Table 2.2). In these studies, various levels of the copper–phthalocyanine dyes were employed to provide an initial 1-mmol/L concentration of copper (as the metallized dye) in solution. In the cases involving the copper–phthalocyanine dyes, a cationic coagulant followed by an anionic polymeric flocculant was employed during or following reductive treatments to afford complete color reduction. In the case of the copper–azo dye, a concentration of 0.01% Direct Blue 218 was employed. The amounts of bisulfite employed were in the 250–1000 mg $Na_2S_2O_5$/L range, and the amount of borohydride employed was determined empirically to maximize the degree of color reduction.

Colors in Tables 2.1 and 2.2 were measured on the ColorQUEST spectrocolorimeter manufactured by Hunter Associates Laboratory, Inc. (Reston, VA 22090). The color of each solution was measured by transmittance using the American Dye Manufacturers Institute (ADMI) color scale. The method is described in ASTM standard E450-82, Standard Method for Measurement of Color of Low-Colored Clear Liquids Using the Hunterlab Color Difference Meter. In our studies, color is expressed as ΔE, or the total color difference between the sample and a distilled water reference solution, using the CIE (Commission Internationale de l'Eclairage) L^*, a^*, b^* scale. ΔE is calculated according to the formula

$$\Delta E = \sqrt{(\Delta L^*)^2 + (\Delta a^*)^2 + (\Delta b^*)^2}$$

TABLE 2.2 Bisulfite-Catalyzed Borohydride Reduction of Copper-Metallized Dyes

Dye	Type	Treatment[a]	Color[b] Initial	Final	% Reduction
Direct Blue 218 (0.01%)	Cu–Azo	250 mg $Na_2S_2O_5$/L 17 mg $NaBH_4$/L	70.6	18.6	74
Direct Blue 189 (0.23%)	Cu–phthalocyanine	500 mg $Na_2S_2O_5$/L 34 mg $NaBH_4$/L Cationic Agent Anionic agent/filter	95.1	0.5	>99
Direct Blue 86 (0.16%)	Cu–phthalocyanine	Cationic polymer 800 mg $Na_2S_2O_5$/L 20 mg $NaBH_4$/L Anionic polymer	95.0	0.5	99.5
Direct Blue 199 (0.6%)	Cu–phthalocyanine	Cationic polymer 500 mg $Na_2S_2O_5$/L 42 mg $NaBH_4$/L Anionic polymer	96.1	0.1	99.9

[a]Mixed for 15 min at ambient temperature after addition of $NaBH_4$. Maintained at pH 6. Solids removed via filtration through 0.8-μm pore filter.
[b]Measured using the CIE L^*, a^*, b^* scale and expressed as ΔE, or the total color difference between the sample and the distilled water reference solution.

where ΔL^* measures the differences in the lightness or grayness component of the sample color vs. the reference, Δa^* measures the difference in the red–green component, and Δb^* measures the difference in the yellow–blue component.

RESULTS ON INDUSTRIAL DYE WASTEWATERS

Field results on industrial effluents containing a mixture of azo dyes (cf. Table 2.3) confirm the benefits of using a bisulfite-catalyzed borohydride reduction followed by precipitation with a cationic coagulant. Using this process color reductions of >90% were possible in all cases. In these field studies, effective treatment was

TABLE 2.3 Bisulfite-Catalyzed Borohydride Reduction of Industrial Dye Manufacturing or Textile Effluents (12)

		Color		
Sample	Treatment[a]	Initial	Final	% Reduction
A[b]	500 mg $Na_2S_2O_5$/L $NaBH_4$ soln. to -500 mV (ca. 12 mg $NaBH_4$/L)	7825[c]	184[c]	98%
B[b]	500 mg $Na_2S_2O_4$/L $NaBH_4$ soln. to -600 mV (ca. 30 mg $NaBH_4$/L) 200 mg cationic agent	20255[c]	1329[d]	93%
C[d]	400 mg $Na_2S_2O_5$/L 300 mg cationic agent $NaBH_4$ soln. to -600 mV (ca. 15 mg $NaBH_4$/L)	3.160[e]	0.007[e]	>99%
D[d]	400 mg $Na_2S_2O_5$ $NaBH_4$ soln. to -590 mV (ca. 25 mg $NaBH_4$/L) 150 mg cationic agent	3.160[e]	0.021[e]	99%
E[d]	400 mg $Na_2S_2O_5$/L $NaBH_4$ soln. to -595 mV (ca. 25 mg $NaBH_4$/L) 100 mg cationic agent	3.980[e]	0.029[e]	99%
F[d]	400 mg $Na_2S_2O_5$/L $NaBH_4$ soln. to -600 mV (ca. 15 mg $NaBH_4$/L) 100 mg cationic agent	3.180[e]	0.029[e]	99%

[a]All treatments done at pH 5, and any solids formed were removed by filtration before final color treatment.
[b]Primarily contains spent reactive azo black dyes.
[c]ADMI color scale.
[d]Primarily contains disperse azo dyes.
[e]Fram color scale.

obtained at pH 5–6 with preaddition of bisulfite (ca. 200–500 mg $Na_2S_2O_5$/L) followed by addition of a caustic solution of sodium borohydride. The reaction pH stabilized in the pH 5.5–7.0 range. The borohydride solution is added to a slight excess relative to the requirements of the solution. This can be conveniently monitored in the field using an oxidation reduction potential (ORP) probe. At this pH range, the presence of bisulfite ion (HSO_3^-) and dithionite ion ($S_2O_4^{2-}$) (formed by the reduction of bisulfite with the slight excess of borohydride, cf. [4]) gives a stable negative ORP reading at approximately −500 to −600 mV. (The ADMI and Fram color scales were employed by the respective client companies at which this testing was conducted.)

Currently a major azo dye manufacturer has successfully implemented bisulfite-catalyzed borohydride reductive technology along with a cationic polymer, to remove color and stabilize resulting solids in their wastewater and tank washes at one of its manufacturing plants. This azo dye manufacturer had previously been employing only a coagulation procedure (with no solids separation). On several occasions, the color reappeared at the subsequent publically operated treatment works (POTW), resulting in added discharge costs to the manufacturer. Attempts by the manufacturer at using a reductive treatment based on dithionite were not successful. After extended field evaluations, the bisulfite-catalyzed borohydride reduction technology was effectively developed to decolorize the various wastestreams from the plant. As this plant campaigns the manufacture of various azo dyes, its effluent varies considerably. Clearly the treatment technology needs to be flexible enough to accommodate these production changes. Bisulfite-catalyzed borohydride reductive technology has met this need for the past several years. The decolorized solids currently discharged by this manufacturing site now present no problems at the POTW's activated sludge treatment facility and impart no color to the discharged waters. Evaluations on color removal from several other effluents containing dyes have also yielded positive technical results, but this process has not as yet been commercially implemented at these other sites, due to several other site-specific circumstances.

Although actual treatment procedures vary depending on the specific species in the wastewater, the general process steps are as follows:

1. pH adjustment to 5–6
2. Bisulfite addition (typically 200–500 mg $Na_2S_2O_5$/L)
3. Borohydride solution addition to an ORP of −500 to −600 mV (or until the desired color reduction is achieved)
4. (Optional) The addition of a cationic agent (typically a polyamine or aluminum-based coagulant) followed by filtration
5. Carbon adsorption, peroxide oxidation, or biological posttreatment

In reference to step 3, a 1.2% $NaBH_4$ solution was obtained on site dilution of a commercial solution (containing 12% $NaBH_4$ and 40% NaOH).

TABLE 2.4 **Summary of Effluent Batch Reactor Characteristics for Each Treatment over a 48-hr Period**[a] **(12)**

Treatment	pH 0 hr	pH 48 hr	Oxygen[2] 0 hr	Oxygen[2] 48 hr	COD[2] 24 hr	COD[2] 48 hr	Conductivity 0 hr	Conductivity 48 hr	BOD$_5$[b] 48 hr	Color Removal 48 hr
Control	7.6	6.2	5.6	6.7	175	250	320	410	89.9	NA
Untreated dye	7.6	6.4	5.5	6.1	180	440	345	420	79.9	Slight
Reduced dye	7.1	6.4	5.5	6.3	325	625	780	1100	86.6	Extensive

[a]Study done at Biological Monitoring Inc., Blacksburg, VA, using an activated sludge inhibition test with 30% sewage, 20% of a 0.1% Direct Red 254 dye solution (untreated and reduced with borohydride), 40% activated sludge and 10% dilution water. Aeration was maintained to achieve a dissolved oxygen level of ca. 5 μg/l over the 48 hr test period.
[b]Oxygen values in μg O_2/L; BOD$_5$ values in mg/L; conductivity values in μmho/cm.

COMPARISON OF EFFECTS ON BIOLOGICAL TREATMENTS

Since this reductive technology affords more biologically active amine species, the effects of such a treatment on activated sludge were studied using a Direct Red 254 dye solution in a 48-hr activated sludge respiration inhibition test (cf. Table 2.4). These results indicated no adverse effect of the reduction treatment on the subsequent activated sludge process. The control, the sample containing the untreated dye, and the sample containing the reduced dye had similar pHs, oxygen, and BOD$_5$ levels after 48 hr.

COMMERCIAL BENEFITS OF A SODIUM BOROHYDRIDE DECOLORIZATION PROCESS

This process employs chemically stable, commercially available solutions and produces only boric acid and sodium bisulfite as by-products. Bisulfite, employed at catalytic levels, can typically be discharged without further treatment (although bisulfite is readily air oxidized to sulfate). This treatment process can be utilized for either batch or continuous applications. Automated borohydride feed can be readily accomplished with the use of ORP controllers. This process chemically reduces the complex azo dyes to smaller nonchromophoric amines, or in the case of a copper complex reduces the copper ion to metallic copper). The reduced species can be separated and removed as solids, or, in the case of the by-product amines, be adsorbed on carbon, oxidized, or safely treated biologically.

REFERENCES

1. T. F. Jula, *Proc. Eng.* 123 (1975); *Chem. Abstr.* **83**, 151770R (1975).
2. M. M. Cook and J. A. Lander, *Poll. Eng.* **13**, No. 12, 36 (1981) and references therein.

3. J. A. Ulman, *Spec. Publ.—R. Soc. Chem.* **61**, 173 (1986); *Chem. Abstr.* **106**, 72317p (1987) and reference therein.

4. N. Lopez and J. T. Regan, *Chem. Proc.* **47**, 58 (1984).

5. M. M. Cook and J. A. Lander, *J. Appl. Photographic Eng.* **5**, 144 (1979), and references therein.

6. E. C. Chou, B. W. Wiegere, D. K. Huggins, and E. I. Wiewiorowski, U.S. Pat. 4,544,541, 1984.

7. L. B. Friar and K. M. Smith, U.S. Pat. 4,279,644, 1980; *Chem. Abstr.* **95**, 223522j (1981).

8. C. Lores and R. B. Moore, U.S. Pat. 3,770,423, 1973; *Chem. Abstr.* **80**, 137055t (1974).

9. E. L. Cadmus, U.S. Pat. 3,764,528, 1971; *Chem. Abstr.* **80**, 30452t (1974).

10. Anon., *Res. Discl.* **51**, 136 (1975); *Chem. Abstr.* **83**, 153022x (1975).

11. R. S. Stennick and M. M. Cook, *Abstracts of Papers*, 204th National Meeting of the American Chemical Society, Washington, D.C., 1992; Environmental Chemistry Division.

12. M. M. Cook, J. A. Ulman, A. Iraclidis, R. S. Stennick, and D. A. Barratt, *Abstracts of Papers*, 203rd National Meeting of the American Chemical Society, San Francisco, CA, 1992; Environmental Chemistry Division.

13. J. A. Ulman and M. M. Cook, *Abstracts of Papers*, 204th National Meeting of the American Chemical Society, Washington, DC, 1992; Environmental Chemistry Division.

14. M. Akiba and H. Sakamoto, Jpn. Kokai Tokkyo Koho 04 11 991, 1992; *Chem. Abstr.* **116**, 220901y (1992).

15. A. Reife, *Kirk-Othmer Encyclopedia of Chemical Technology*, 4th ed., Vol. 8, 1993, p. 753, and references therein.

16. M. Kolb, P. Korger, and B. Funke, *Melliand Textilber.* **69**, 286 (1988); *Chem. Abstr.* **108**, 226293h (1988).

17. Y. Tanto, H. Takagi, and S. Hayashi, Jpn. Kokai 77 116 644, 1977; *Chem. Astr.* **89**, 30358c (1978).

18. M. C. Cheek, *TAPPI Proceedings Book*, TAPPI Papermakers Conference, 1991.

19. R. D. Voyksner, R. Straub, J. T. Keever, H. S. Freeman, and W. N. Hsu, *Environ. Sci. Technol.* **27**, 1665 (1993).

20. W. H. Stockmayer, D. W. Rice, and C. C. Stephanson, *J. Am. Chem. Soc.* **77**, 1980 (1955); *Chem. Abstr.* **49**, 9361d (1956).

21. P. R. Sanglet, U.S. Patent 4,859,448, 1989.

22. J. Ko, D. C. Munroe, and S. H. Levis, U.S. Pat. 5,094,833, 1992.

CHAPTER 3

OZONATION

MASAKI MATSUI
Department of Chemistry, Faculty of Engineering, Gifu University, Gifu, Japan

The reaction of dyes with ozone is important from the viewpoint of ozone fading of coloring matter and dye wastewater treatment. Even though the structures of dyes are often complicated, ozone fading can be explained in terms of ozone attack at the most reactive site in the molecule. This chapter describes 1) the generation of ozone, 2) toxicity of ozone, 3) some general reactions of ozone, 4) the reaction of ozone with dyes, 5) reactivity of ozone with dyes, and 6) ozone treatment of dye wastewater.

GENERATION OF OZONE

Ozone (O_3) is generated from oxygen by methods such as corona discharge and ultraviolet (UV) irradiation. In order to generate highly concentrated ozone, the corona discharge process has been generally used. The principle is shown in Figure 3.1.

The half-life of ozone in the vapor phase at room temperature is known to be about 16 hr. The stability of dissolved ozone in water depends on the pH value of the solution. Under acidic conditions, ozone is stable, while under alkaline conditions it is very unstable. The half-life of ozone in distilled water at room temperature is known to be about 20 min.

OZONE EXPOSURE AND TOXICITY

Ozone is one of the most active oxidants and exhibits strong toxicity toward animals. A large fraction of the photochemical oxidants in air pollution arises from ozone (1). Recently, the generation of highly concentrated ozone by copying machines, in an office environment, has been also reported (2). The toxicity of ozone is summarized in Table 3.1 (3).

Environmental Chemistry of Dyes and Pigments, Edited by Abraham Reife and Harold S. Freeman.
ISBN 0-471-58927-6 © 1996 John Wiley & Sons, Inc.

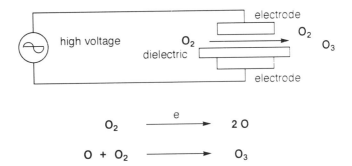

$$O_2 \xrightarrow{\quad e \quad} 2\,O$$

$$O + O_2 \longrightarrow O_3$$

FIGURE 3.1 Generation of ozone.

TABLE 3.1 Toxicity of Ozone

Ozone (ppm)	Action
0.01–0.02	An offensive smell
0.1	A strong smell, sharp irritation to the nose and throat
0.2–0.5	Decrease of the power of vision in 3–6 hr
1–2	Distinct irritation on the upper part of throat, headache, pain in the chest, chronicity by repeated exposure, cough and drying of throat
5–10	Increase in pulse rate, edema of lungs
15–20	Decease of infant animals in 2 hr
50	Potentially fatal

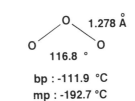

bp : -111.9 °C
mp : -192.7 °C

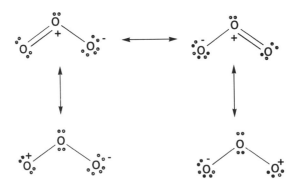

FIGURE 3.2 Resonance in the ozone molecule.

REACTIVITY OF OZONE

Since ozone has the resonance structures depicted in Figure 3.2, its reaction with organic compounds can be classified into 1,3-addition, 1,3-insertion, electrophilic, nucleophilic, and electron-transfer reactions (Scheme 1).

1,3-Addition Reaction

This is the most widely known and studied reaction of ozone. Ozone reacts with olefins to give a primary ozonide, which decomposes into carbonyl oxide and carbonyl compounds (4). In the case of unsymmetric olefins, the primary ozonide can decompose in two ways. The primary ozonide decomposes in a manner that stabilizes the positive charge on the carbon atom of the carbonyl oxide. Since the carbonyl oxide is unstable, it will recombine with the carbonyl compound to afford ozonide, dimerize, and react with water to give a second carbonyl compound. The reactivity of carbonyl oxides depends on the reaction conditions, that is, solvent and reaction temperature.

1,3-Insertion Reaction

Ozone reacts with aldehydes and acetals to give unstable hydrotrioxides, which can be identified by nuclear magnetic resonance (NMR) spectroscopy at low temperature (5). The hydrotrioxides easily decompose to give the corresponding carboxylic acid and singlet oxygen.

Electrophilic Attack of Ozone

Ozone easily reacts with amines, sulfides, and ethers, each of which contains a lone pair of electrons on the heteroatoms (6). The adducts are converted into oxidation products of the alkyl side chains.

Nucleophilic Attack by Ozone

An example is the reaction of ozone with isatin to afford isatoic anhydride (7). In this reaction, ozone has been reported to act as a nucleophile to attack the carbonyl carbon in the 3-position of isatin.

Electron-Transfer Reaction of Ozone

Organic compounds having low oxidation potential, such as nitrogen-containing substrates, transfer a single electron to ozone to form the radical cation of the substrate and radical anion of ozone (8). See reaction 5 in Scheme 1.

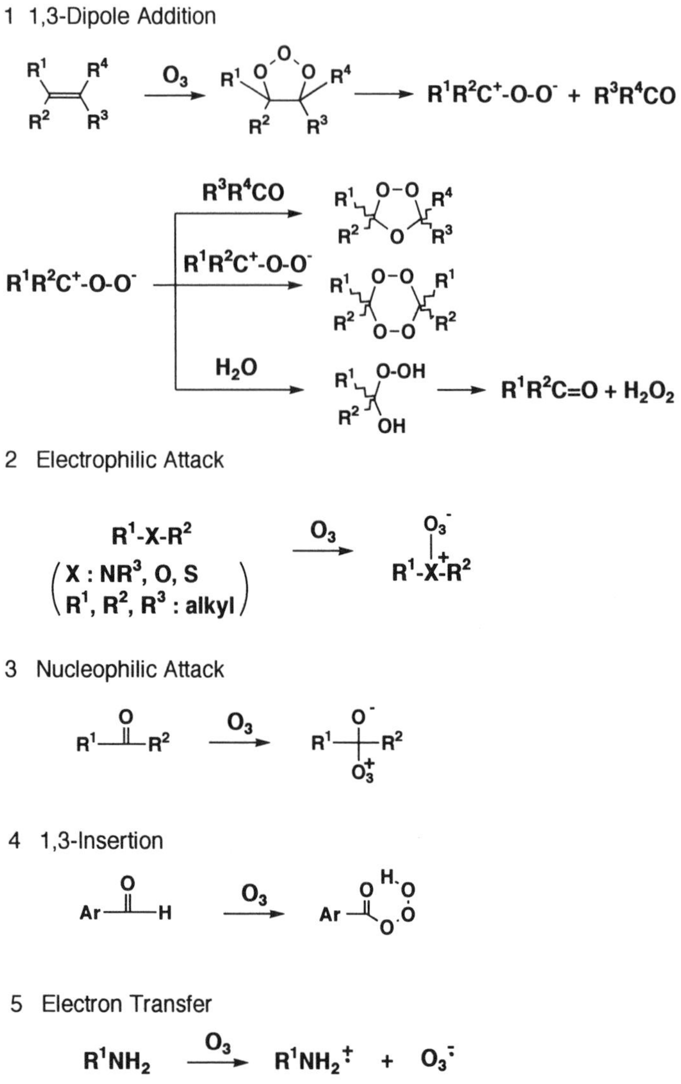

1 1,3-Dipole Addition

2 Electrophilic Attack

$$R^1\text{-}X\text{-}R^2 \quad \left(\begin{array}{l} X : NR^3, O, S \\ R^1, R^2, R^3 : \text{alkyl} \end{array}\right) \xrightarrow{O_3}$$

3 Nucleophilic Attack

4 1,3-Insertion

5 Electron Transfer

$$R^1NH_2 \xrightarrow{O_3} R^1NH_2^{\ddagger} + O_3^{\overline{\cdot}}$$

SCHEME 1 Reactions of ozone.

OZONE–DYE CHEMISTRY

Ozone fading of dyes occurs by the oxidative cleavage of the conjugated system of the molecule. The following is a summary of the chemistry that takes place when ozone reacts with specific dye classes.

Stilbene Dyes

Crysophenine G (**1**) reacts quantitatively with ozone to give the corresponding aldehyde **2** (9). In this case, ozone attacks the central olefinic bond of **1** by a 1,3-addition reaction (Scheme 2). The reactivity of the olefinic bond toward ozone is higher than that of the azo bond. As a result, the azo linkages of dye **1** remain intact under the conditions of the reaction.

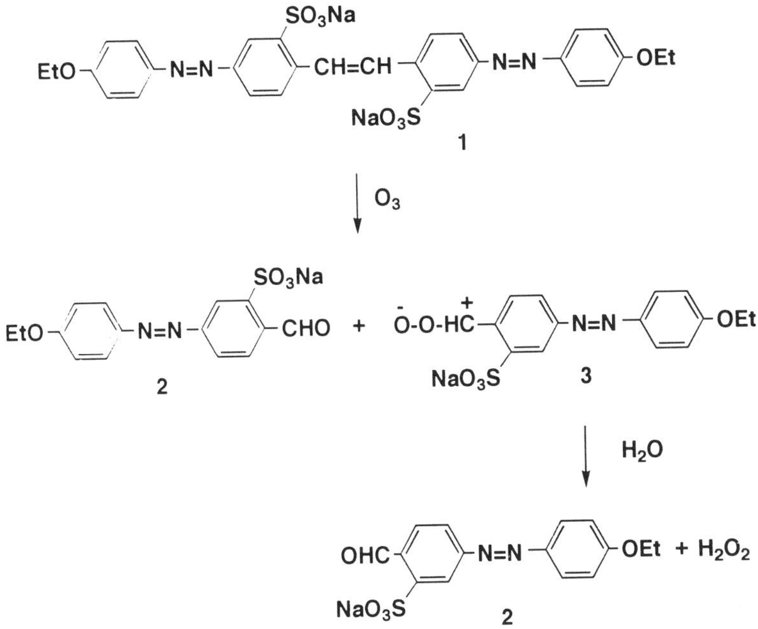

SCHEME 2   Ozone fading of crysophenine G.

Indigoid Dyes

Indigo (**4**) reacts with ozone to give isatin (**6**) (10). This reaction is a useful method for the quantitative analysis of ozone. Ozone attacks the central olefinic bond of **4** by a 1,3-addition reaction to generate two molecules of isatin per molecule of indigo (Scheme 3).

Azomethine Dyes

Ozone attacks both the azomethine and dimethylamino moieties of dye **8** to afford the corresponding oxaziridine and *N*-dealkylated derivatives **9–11** (Scheme 4) (11). The mechanism for the formation of the oxaziridine ring is explained by an electrophilic reaction of ozone at the azomethine nitrogen atom followed by loss of molecular oxygen (Scheme 5).

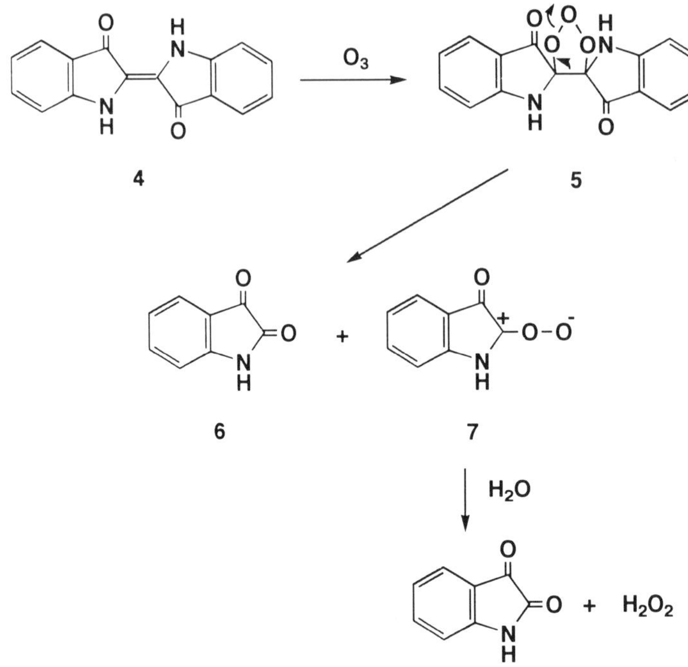

SCHEME 3 Ozone fading of indigo.

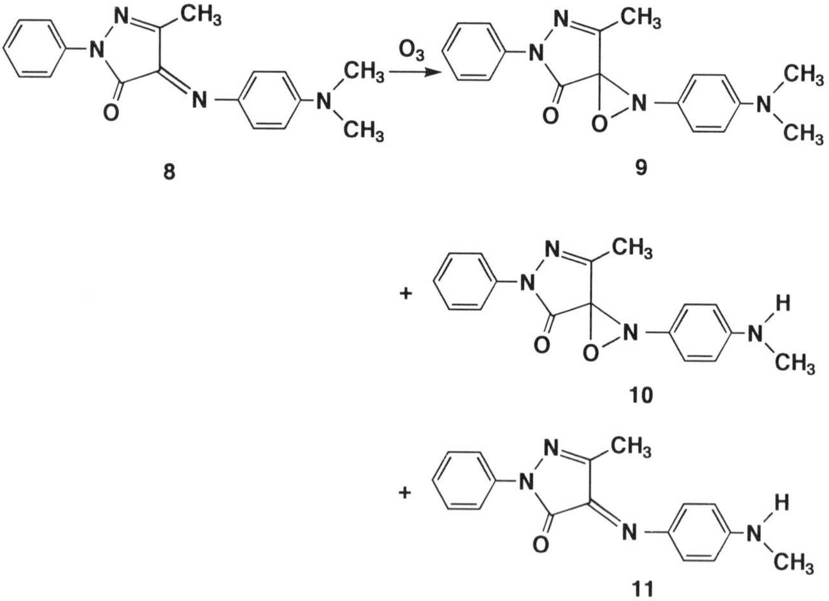

SCHEME 4 Ozone fading of an azomethine dye.

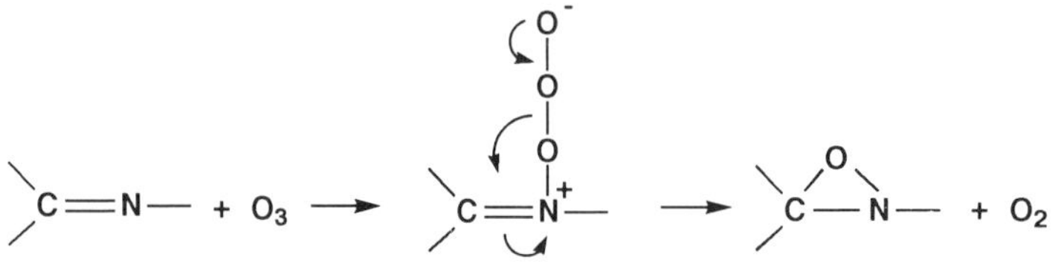

12 13 14

SCHEME 5 Formation of oxaziridine ring.

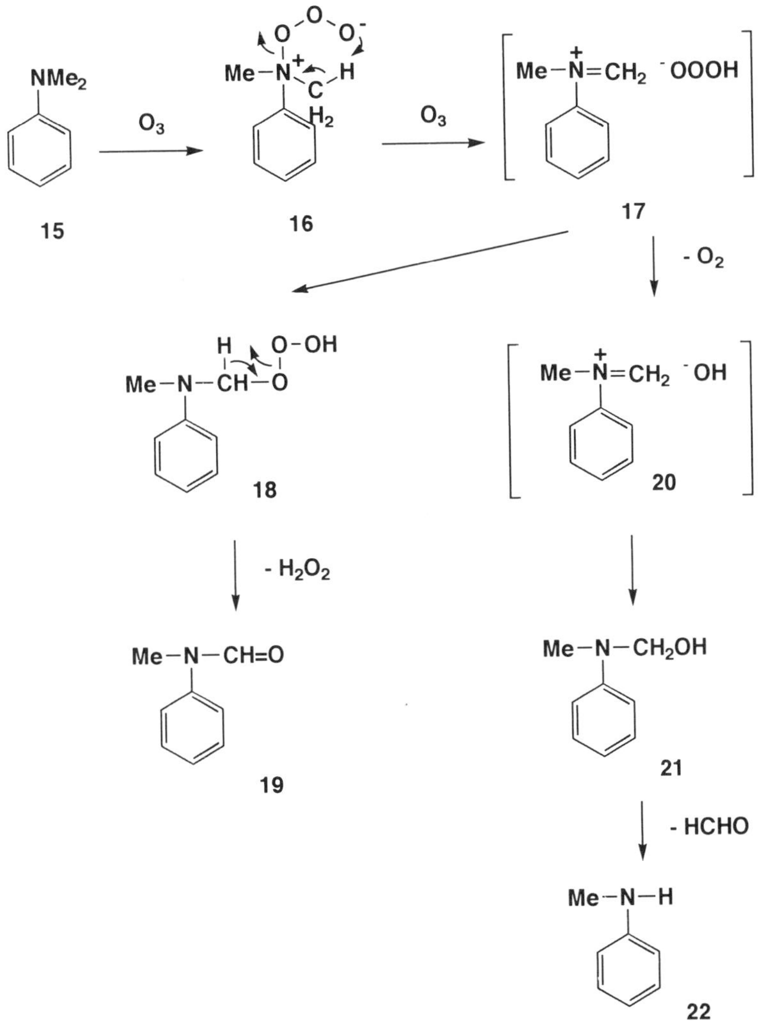

SCHEME 6 Ozone attack at an *N*-dimethylamino moiety.

The formation of the *N*-methylamino moiety by the ozone attack of the *N,N*-dimethylamino group is explained by the initial electrophilic attack of ozone at the nitrogen atom (Scheme 6) (12). From the identical intermediate (**16**), the formylmethylamino derivative **19** is also obtained (Scheme 6).

Azo Dyes

The reactivity of an azo linkage with ozone is very low compared to an olefinic or azomethine group. In the ozonization of azobenzene, both azoxybenzene and glyoxal are formed (13). However, the ozonization of phenylazonaphthalene and azonaphthalene is so complicated that the products have not been identified. These results suggest that aromatic rings are more reactive toward ozone than an azo linkage.

Many commercial azo dyes contain a hydroxyl group *ortho* to the azo linkage. The ozonization of azopyrazolone dye (**23**) in a chlorinated solvent gives ketone (**27**), benzene (**28**), phenol (**29**), and chlorobenzene (**30**) (14). Since dye **23** exists predominantly as hydrazone tautomer **24** (>90%) in solution and the Hammett's plots of the ozonization reaction exhibit a negative slope, electrophilic attack of ozone on the hydrazone isomer has been proposed for this reaction (Scheme 7).

Similarly, ozonization of 4-phenylazo-1-naphthol has been reported to give phenol, phthalic acid, and nitrogen gas (15). The main ozonization product of structurally similar azo dyes, such as Orange II, Congo Red, Amaranth, Tartrazine, and Direct Brown 95, has been reported to be nitrogen gas (16).

Diphenylmethane Dyes

Auramine O (**31**) reacts with ozone to give Michler's ketone (**32**) in good yield, accompanied by monoformylmethylamino derivative **33** (Scheme 8) (17). Formation of the latter product is consistent with the chemistry outlined in Scheme 6.

Triphenylmethane Dyes

The reaction of Malachite Green (**34**) with ozone gives the formylmethylamino derivative **35**, diarylketones **36** and **37**, and *p*-dimethylaminophenol (**38**) (Scheme 9) (18). The selectivity of ozone attack at an *N*-dimethylamino group (formation of **35**) versus carbon skeleton (formation of **36**) depends on the substituent R. Electron-donating groups such as dimethylamino (Crystal Violet) and methoxy groups accelerate ozone attack at the dimethylamino groups, while an electron-withdrawing nitro group activates the carbon skeleton. A probable mechanism of ozone attack at the carbon skeleton is shown in Scheme 10. The mechanisms of ozone attack at the dimethylamino moiety was depicted in Scheme 6.

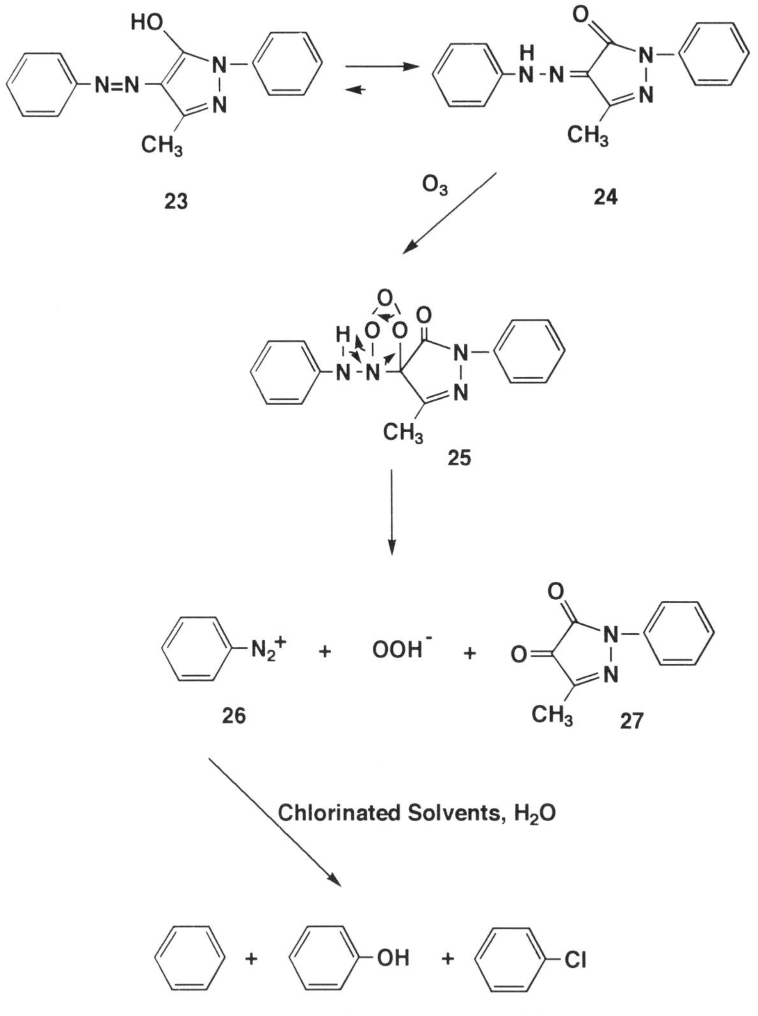

SCHEME 7   Ozone fading of azopyrazolone dye.

The ozonization of phenolphthalein (**43**) depends on the pH of the solution (19). Under acidic conditions, the reactivity of **43** is very low, because of the presence of the lactone structure. However, under alkaline conditions, the dye reacts readily to give ketone (**46**), phthalic acid (**50**), and hydroquinone (**52**). In alkaline solution, the carbon skeleton of quinonoid structure **44** is easily attacked by ozone, as shown

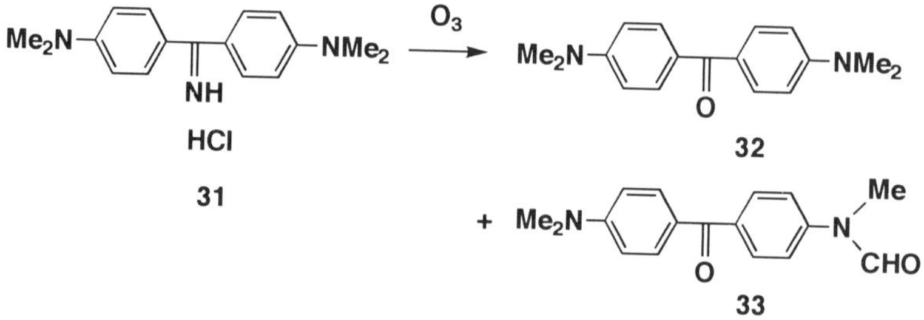

SCHEME 8   Ozone fading of Auramine O.

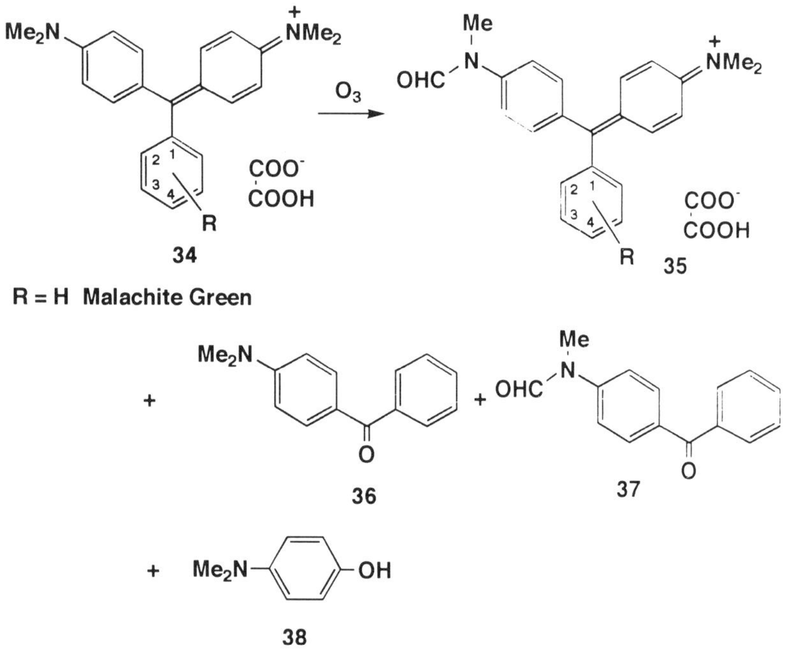

R = H **Malachite Green**

SCHEME 9   Ozone fading of Malachite Green.

in Scheme 11. Nucleophilic attack of ozone at the carbonyl carbon of ketone (**46**) affords ester (**49**), and hydrolysis of the ester linkage gives **50** and **52**.

Anthraquinone Dyes

Ozone fading of Disperse Blue 3 on nylon carpet has been reported to give phthalic acid (**20**). The ozonization of Alizarin and Alizarin Crimson also gives phthalic

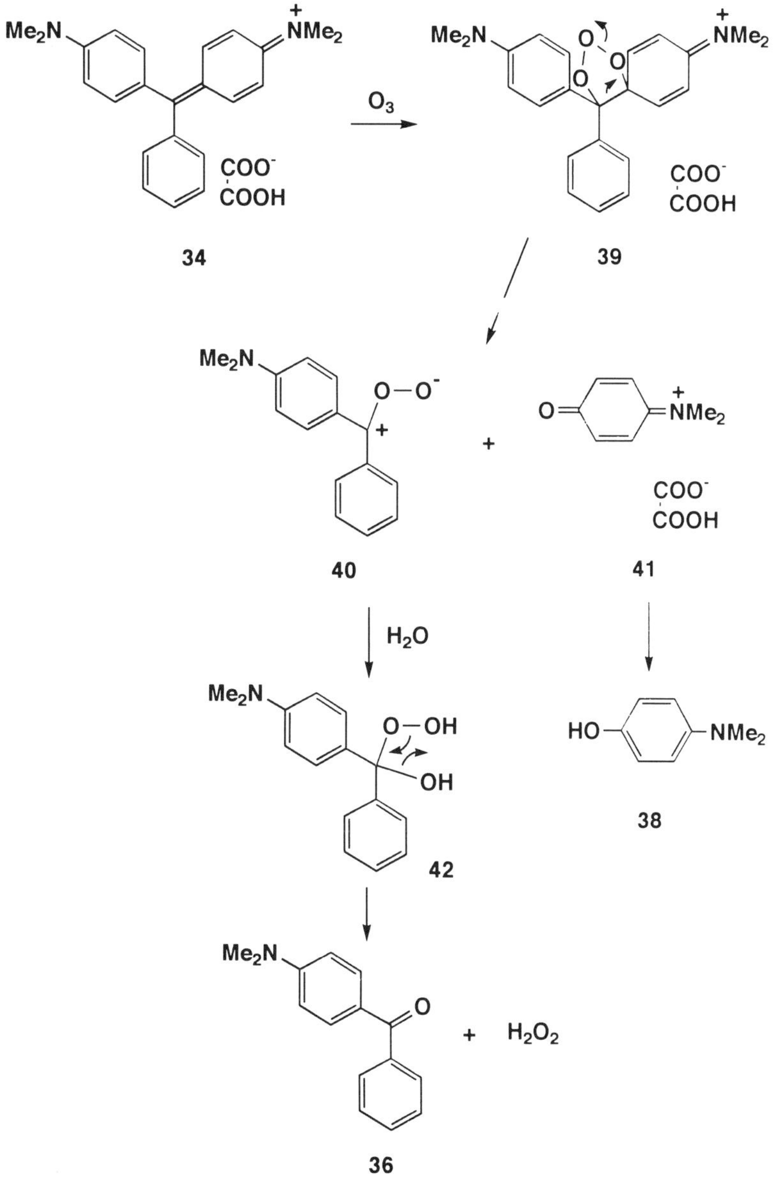

SCHEME 10   Ozone attack at the carbon skeleton of Malachite Green.

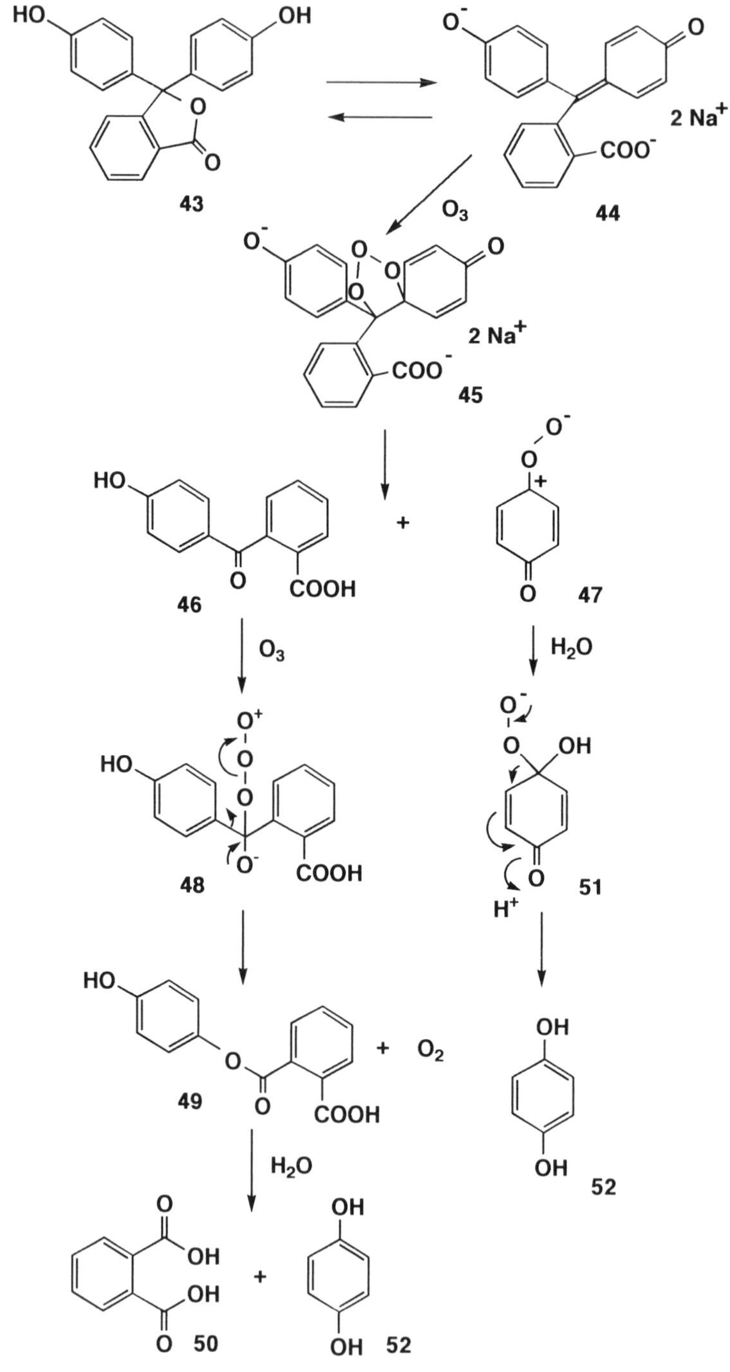

SCHEME 11   Ozone fading of phenolphthalein.

acid (**21**). The chemistry is no doubt the same as that shown in Scheme 11 for the degradation of **46**.

Metal Containing Dye

Copper-complex dye **53** reacts with ozone to give phenol (**54**), 2-naphthol (**55**), and phthalic anhydride (**56**) (Scheme 12) (22).

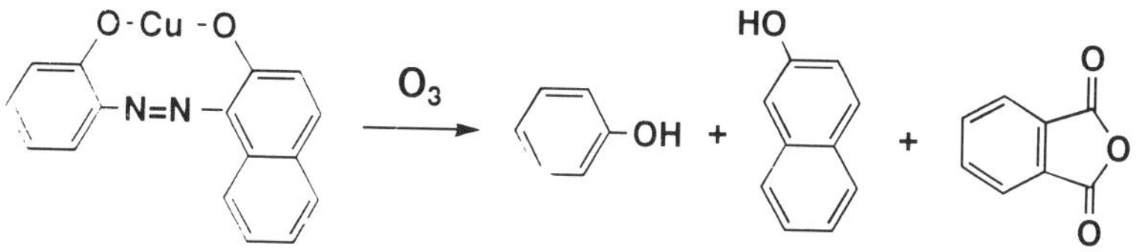

SCHEME 12 Ozone fading of a metal complexed dye.

Xanthenes

Xanthene itself (**57**) reacts with ozone to give xanthone (**59**), by way of xanthydrol (**58**), accompanied by the formation of singlet oxygen (Scheme 13) (23). This suggests a 1,3-insertion reaction at the C–H bond of the 9-position of xanthene and xanthydrol to afford the corresponding hydrotrioxide.

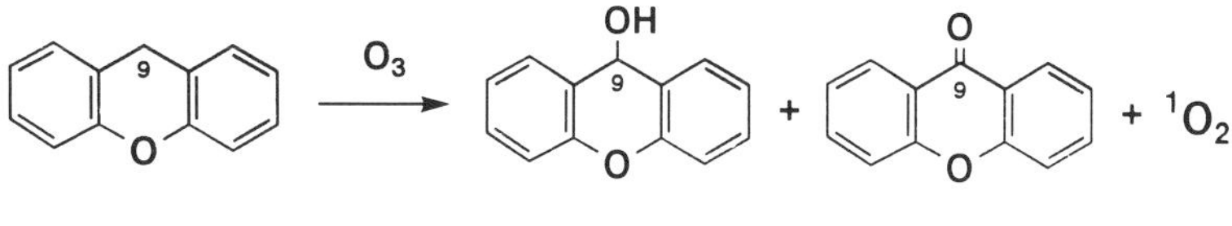

SCHEME 13 Ozonization of xanthene.

Phenothiazines and Phenoxazines

The reactions of phenothiazines and phenoxazines having a low ionization potential with ozone in dichloromethane has been reported to exhibit single-electron transfer from the dye to ozone to give the radical cation of the dye and the radical anion of ozone (Table 3.2) (24). The radical cations are observed by UV spectroscopy. The radical anion of ozone, produced *in situ*, is trapped by reaction with triphenylmethane and cumene. In a nonpolar solvent such as carbon tetrachloride, no single-electron transfer is observed. Compounds having a high ionization potential,

TABLE 3.2 Ozonization of Phenothiazines

X	Y	IP / V [a)]	Electron-Transfer
NH	S	0.59	yes
NH	O	0.60	yes
S	O	1.17	no
S	S	1.21	no

a) vs. SCE in acetonitrile

for example, thianthrene and phenoxathiin, do not exhibit single-electron transfer in dichloromethane.

REACTIVITY OF OZONE WITH DYES

The reactivity of ozone with dyes, in solution, is evaluated by subjecting the decrease in the absorption maxima to first-order rate kinetics (Table 3.3) (22). Dyes having olefinic and hydrazone groups react readily with ozone. Metal-containing and anthraquinone dyes are rather stable toward ozone. When an ozonization product is still colored, the reactivity is judged to be low, as in the case of Malachite Green, where the main product is **35**.

In dye structure **60** (Figure 3.3), the introduction of an electron-withdrawing or bulky substituent (R) in the 2-position of the diazo component depresses ozonization (25).

OZONE TREATMENT OF DYE WASTEWATER

Ozone has been used for the treatment of drinking water. There are more than 900 plants in Europe, 40 in America, 35 in Canada, and 12 in Japan utilizing ozonation. In Japan, ozone treatment has usually been employed to inhibit the formation of halogenated products and to remove the unpleasant odor in drinking water. In order to decompose colored components, ozone has also been used for the treatment of

TABLE 3.3   Reactivity of Ozone with Dyes

Structure	Name	$k\ /\ s^{-1}$
	Indigo Carmine (CI Food Blue 1, CI Acid Blue 74)	0.0130
	Azo dye	0.0123
	Chrysophenine (CI Direct Yellow 12)	0.0096
	Orange II (CI Acid Orange 7)	0.0066
	Cu Complexed Dye	0.0057
	Sodium Alizarinsulfonate (CI Mordant Red 3)	0.0052
	Malachite Green (CI Basic Green 4)	0.0046

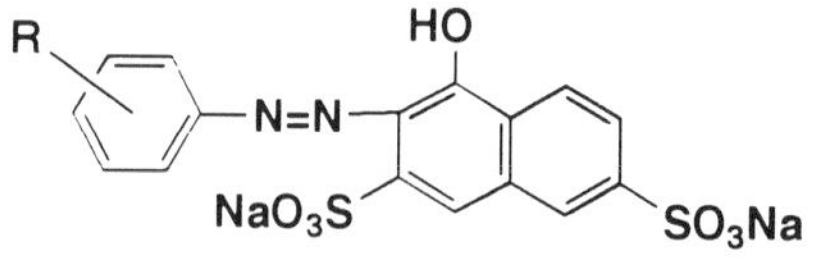

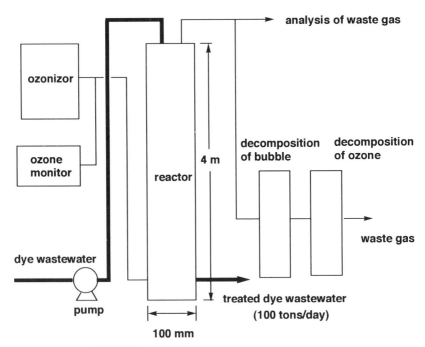

FIGURE 3.3 Azo dye **60**.

night soil. In general, ozone treatment accompanied by activated carbon adsorption is more effective.

Although no practical commercial method for the ozone treatment of dye wastewater has been reported, there have been interesting results from pilot plant studies (26,27). An example of ozone treatment at Bisai, Aichi, Japan, is now illustrated. In this area, there are about 120 dyehouses. More than 95% of the wastewater consists of dye effluent. The composition and amount of wastewater changes day by day, but its color is always black-purple. The average amount of wastewater is about 100,000 tons/day.

In order to evaluate the ozonation process, ozone treatment of practical dye wastewater from a pilot plant (100 tons/day), shown in Figure 3.4, was examined. The results are depicted in Table 3.4. The absorption at longer wavelength (580

FIGURE 3.4 Ozone treatment of dye wastewater.

TABLE 3.4   Ozone Treatment of Dye Wastewater

	Concentration of Ozone (mg/L)		
	0	9.2	9.1
TOC (ppm)	30.7	24.5	20.8
COD (ppm)	28.6	22.9	19.6
BOD (ppm)	2.2	5.3	6.9
A (580 nm)	0.146	0.024	0.008
A (410 nm)	0.284	0.125	0.076

nm) was easily reduced, whereas that at the shorter one (410 nm) was still observed, indicating that yellow organic components in the wastewater were not decomposed completely. The values of total organic carbon (TOC) and chemical oxygen demand (COD) were decreased by ozone treatment, whereas that of biological oxygen demand (BOD) increased. This suggests that the chemical substances underwent biodegradation after ozone treatment. Addition of a higher concentration of ozone to the wastewater was more effective. It was concluded that ozone could be used for dye wastewater treatment if this process were to be made cost effective. The cost of the ozone treatment process (calculated for 100,000 tons/day) accompanied by active sludge is ¥147.14/ton (approx. $1.57/ton). This cost is about twice that of active sludge process by ¥64.65/ton (approx. $0.72/ton).

CONCLUSION

Although, the chemistry of ozone has been widely studied, information on the ozonation of dyes is quite sparse. In this chapter, ozonation has been described as it relates to dye chemistry. A few issues need to be addressed (e.g., how to make the operating cost of the plant reasonable, and information pertaining to the toxicity of the final ozonation products) before ozonation is employed as a practical method for dye wastewater treatment. It is hoped that this chapter will stimulate further basic and pilot-scale research regarding the ozonation of commercial dyes.

REFERENCES

1. American Chemical Society, Ozone Chemistry and Technology, Advanced Chemistry Series, Vol. 21, 1959.
2. International Ozone Association, *Ozone News* **20**, 15 (1992).
3. O. Tada, Yuugai Bussitsu Kanri No Tameno Sokutei Houhou (Measurement of Harmful Materials), Roudoukagaku Kenkyuusho Shuppan Bu (Institute of Labor Science), p. 102, 1969 (in Japanese).

4. P. S. Bailey, *Ozonation in Organic Chemistry, Vol. I*, Academic Press, New York, 1978.

5. F. E. Stary, D. M. Emge, and R. W. Murray, *J. Am. Chem. Soc.* **98**, 1880 (1976); F. Kovac and B. Plesnicar, *J. Am. Chem. Soc.* **101**, 2677 (1979).

6. P. S. Bailey, *Ozonation in Organic Chemistry, Vol. II*, Academic Press, New York, 1982.

7. J. van Alphen, *Recl. Trav. Chim. Pay-Bas* **57**, 911 (1938).

8. P. S. Bailey and A. Y. Khashab, *J. Org. Chem.* **43**, 675 (1978).

9. M. Matsui, T. Hibino, K. Shibata, and Y. Takase, *Am. Dyest. Reptr.* **72**, 40 (1983).

10. M. Matsui, M. Morita, K. Shibata, and Y. Takase, *Nippon Kagaku Kai Shi* 1268 (1982).

11. D. P. Harnish, H. J. Osborn, and B. W. Rossiter, *J. Org. Chem.* **34**, 1687 (1969).

12. G. H. Kerr and O. Meth-Cohn, *J. Chem. Soc. (C)* 1369 (1971).

13. M. Matsui, Y. Iwata, T. Kato, and K. Shibata, *Dyes Pigments* **9**, 109 (1988).

14. M. Matsui, A. Konda, K. Shibata, and Y. Takase, *Bull. Chem. Soc. Jpn.* **58**, 2829 (1985).

15. M. Matsui, K. Kobayashi, K. Shibata, and Y. Takase, *J. Soc. Dyers and Colour* **97**, 210 (1978).

16. G. Ishizaki, K. Sakata, N. Shinriki, and A. Ikehata, *Nippon Kagaku Kai Shi* 1124 (1979).

17. M. Matsui, H. Nakabayashi, K. Shibata, and Y. Takase, *Sen'i Gakkai Shi* **37**, T381 (1981).

18. M. Matsui, H. Nakabayashi, K. Shibata, and Y. Takase, *Bull. Chem. Soc. Jpn.* **57**, 3312 (1984).

19. M. Matsui, T. Koike, and K. Shibata, *J. Soc. Dyers Colour* **104**, 425 (1988).

20. J. C. Haylock and J. L. Rush, *Textile Res. J.* **46**, 1 (1976).

21. D. Grosjean and P. M. Whitmore, *Environ. Sci. Technol.* **21**, 635 (1987).

22. M. Matsui, K. Tsubota, K. Shibata, and H. Muramatsu, *Bull. Chem. Soc. Jpn.* **64**, 2961 (1991).

23. M. Matsui, Y. Miyamoto, K. Shibata, and Y. Takase, *Bull. Chem. Soc. Jpn.* **57**, 603 (1984).

24. M. Matsui, Y. Miyamoto, K. Shibata, and Y. Takase, *Bull. Chem. Soc. Jpn.* **57**, 2526 (1984).

25. M. Matsui, N. Midzui, K. Shibata, and H. Muramatsu, *Dyes Pigments* **20**, 67 (1992).

26. Y. Kunishima and Y. Sakai, *28th Gesuido Kenkyuu Happyou Kai Abstr.*, 1991, p. 547 (in Japanese); Y. Hori, Y. Kunishima, and K. Honda, *29th Gesuido Kenkyuu Hapyyou Kai Abstr.*, 1992, p. 453 (in Japanese).

27. F. L. Evans, *Ozone in Water and Wastewater Treatment*, Ann Arbor Science, Ann Arbor, MI, 1975.

CHAPTER 4

USE OF ELECTROCHEMICAL TECHNOLOGY TO REMOVE COLOR AND OTHER CONTAMINANTS FROM TEXTILE MILL EFFLUENTS

ANNE E. WILCOCK,[1] MICHAEL BREWSTER,[2] and G. PECK[2]
[1]University of Guelph, Guelph, Ontario, Canada
[2]Andco Environmental Processes, Buffalo, New York

Methods used to electrolytically treat industrial wastewater were first developed at the turn of the twentieth century. At that time, the technology was primitive, electric power was expensive, and wastewater did not appear to be a major threat to the environment. Numerous methods for electrolytic treatment have been developed since then and are described in the literature. However, this chapter will focus on only a single commercial process for electrochemical wastewater purification that is widely used in North America.

BACKGROUND

The commercial development of electrochemical technology for treating contaminated water dates back only to the early 1970s, when Lee (1) first patented this technology. Since then, several patent modifications and improvements have been made (2).

The electrochemical process was applied initially to the chemical reduction and removal of hexavalent chromium from cooling tower blowdown water. However, toxicity concerns led to a movement away from the use of chromates in cooling tower water. The electrochemical process was then applied to the treatment of wastewater from electroplating, surface finishing, and printed circuit board operations. This work has led to methods for the removal of heavy metals from the wastewater of various industries and for the treatment of groundwater.

Environmental Chemistry of Dyes and Pigments, Edited by Abraham Reife and Harold S. Freeman.
ISBN 0-471-58927-6 © 1996 John Wiley & Sons, Inc.

It was this expertise, exhibited in the treatment of heavy metals in wastewater, that led a researcher at a major American textile institute to suggest the potential for treatment of metallized dyestuffs. A trial was initiated and, to everyone's delight, not only the heavy metals but also the color was removed from the dye liquor! Thus, research into the application of this technology to the treatment of various textile effluents, primarily those generated by dyeing processes, began.

The first patent (3) on the application of the process to the treatment of textile wastewater was issued to Andco Environmental Processes in 1989. Since that date, there have been several other patent applications filed.

LITERATURE SURVEY

One of the earliest studies reported on the use of the Andco electrochemical system to treat textile wastewater was by Demmin and Uhrich (4). Their studies, which focused on the treatment of effluent from dyebaths containing a variety of classes of dyestuffs using an iron cell, achieved reductions in biochemical oxygen demand (BOD), chemical oxygen demand (COD), total suspended solids (TSS), heavy metals, and color, with efficiencies as high as 90%.

Tincher, Weinberg, and Stephens (5) and Weinberg (6) used the same technology to study decontamination of simulated acid dyebaths, simulated solutions of stain-blocking agents (in which a "phenolic resin" was the main active ingredient), influent to a treatment plant, effluent from a carpet dyeing facility, and effluent from a wastewater treatment plant. An iron cell, which was shown to remove acid dyes at an efficiency of more than 80%, was used. A greater percent color removal occurred following treatment of the mill effluent than following treatment of the influent, although it was not clear whether that difference was real or the result of experimental error. The authors reported reductions in COD and organic contaminants, as well as in stain-blocking agents, from the carpet dyeing facility.

Following a study on the potential for recycling electrochemically treated disperse dye effluent, Wilcock and Hay (7) reported substantial decolorization and detoxification of disperse dye effluent following treatment with iron electrodes. Reuse of the treated effluent (following up to eight consecutive laboratory dyeings) resulted in first-quality dyeings.

A paper by Wilcock, Brewster, and Tincher (8) reported results from three case studies in which the Andco system effectively decolorized effluents containing metal complex, disperse, and acid dyestuffs. The case study that involved the removal of metal complex dyestuffs confirmed the removal of chromium ions as well as the dyestuffs. The studies that focused on disperse and acid dyes demonstrated the success of reusing the treated dyebaths, a strategy that can significantly minimize the quantity of effluent discharged into the environment without compromising dyeing quality.

A study by McClung and Lemley (9) considered the mechanism by which electrochemical treatment might occur and attempted to identify the breakdown products that could be released into the environment or be contained within the recycled decolorized effluent. Using high-performance liquid chromatography (HPLC) techniques, these researchers determined that the electrochemical removal of pure dye from an aqueous solution resulted from adsorption and/or degradation of the dyestuff itself, following interaction with iron electrodes. In the same study, degradation of one of the two azo dyes that were studied resulted in the formation of aniline. The degradation products of the second azo dye studied were not characterized. No breakdown products were found in the decolorized effluent that originally contained an anthraquinone dyestuff. No attempt was made to examine the efficacy of electrochemical treatment of the nondyestuff components normally found in a commercial dyestuff formulation.

Wilcock and co-workers (10) treated disperse dye effluent with an aluminum cell instead of the more common iron cell and reported that the mechanism by which decolorization occurred involved physical adsorption rather than dye degradation. The difference between the integrity of the colorants in this study and the dye degradation reported by McClung and Lemley (9) was explained by the choice of electrode in the two studies. The aluminum electrodes cannot act as a reducing agent since aluminum enters solution in a single valence state. In contrast, the iron electrodes used by McClung and Lemley cause the release of ferrous iron, which is capable of acting as a reducing agent to cause dye degradation. The results of the aluminum study further indicated that quinolone dyes were easier to decolorize than either anthraquinone or azo dyes, and that dyes with similar chemical structures reacted in a similar manner to electrochemical treatment. Dispersing agents of the lignin sulfonate type were shown to be more completely removed from the effluent along with the disperse dyes than were those of the naphthalene sulfonate type. The authors concluded that colorant, dispersing agent, and other components of commercial dyebath formulations probably all influence the efficacy of electrochemical treatment.

THE TECHNOLOGY

Most textile wastewaters contain dyes, heavy metals, and/or organic contaminants that can be removed from the aqueous phase through physical means such as adsorption and coprecipitation. However, electrochemical ion generation is a superior technology for physical treatment of dyes, including metal complex dyes. This technology will also remove BOD, COD, total organic carbon (TOC), total dissolved solids (TDS), TSS, and heavy metals such as chromium (Cr), copper (Cu), molybdenum (Mo), and zinc (Zn).

The system most commonly utilizes an electrochemical cell to generate ferrous hydroxide directly from steel electrodes. In a full-scale system (Figure 4.1), the electrochemical cell consists of a fiberglass body containing a number of electrodes separated from each other by small gaps. Wastewater flows through the gaps and is in contact with the electrodes. A direct current (dc) power supply is connected between the two end electrodes of the cell, and the treatment level is controlled by the power provided to the cell. As current flows from one electrode to another

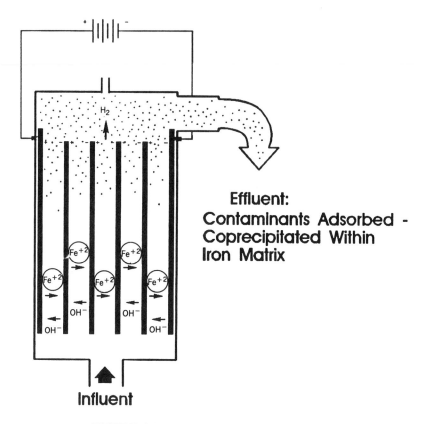

FIGURE 4.1   Illustration of electrochemical cell.

through the process water, the positively charged sides of the electrodes (the anodes) give off ferrous ions. At the negative sides (the cathodes), water decomposes into hydrogen gas and hydroxyl ions. Using iron electrodes as a model system, the overall reaction results in the formation of a combination of one or more of the following hydrous iron oxides: ferrous hydroxide, ferric oxyhydroxide, and ferric oxyhydroxide. If aluminum electrodes are used, the end products are a combination of aluminum hydroxide and aluminum oxyhydroxide.

The electrodes are slowly consumed as the metal hydroxide matrix is generated. After exiting the electrochemical cell, the process water enters a degassing tank where dissolved gases are allowed to dissipate. In liquors in which the pH is less than 7 or greater than 11 (the range at which satisfactory Fe^{2+} precipitation occurs), the pH may be adjusted at this stage. The water is then pumped from the reactor tank to a clarifier where the newly forming solids settle preparatory to subsequent solids handling processes. Before the solution enters the clarifier, a small amount of polymer is added to improve the flocculation and settling of the precipitated solids. The clarifier supernatant flows to a polishing filter before leaving the system.

When iron and/or aluminum matrix production occurs electrochemically, there is no simultaneous addition of anions such as sulfate or chloride. This is in contrast to conventional chemical precipitation methods that may introduce either chloride or sulfate anions, both of which interfere with contaminant removal and increase the TDS in the effluent stream. This latter factor may be significant in effluent reuse. When implemented properly, electrochemical ion generation removes contaminants while providing optimal reuse potential.

TREATABILITY STUDIES

Electrochemical processes are designed with the characteristics of the sample being treated in mind. Treatability studies have been used to identify which electrochemical cell configuration will best achieve treatment objectives while being the most economical to implement and operate. Electrochemical treatment provides physical removal capabilities through the generation of an iron, aluminum, or combined iron–aluminum matrix.

The full-scale process can readily be simulated on either a bench-scale batch or a semicontinuous system (Figure 4.2). A known volume of the effluent is circulated between the electrodes while a dc power supply provides the current for the reaction. Faraday's law is used to determine the generation time for a specific sample volume at a controlled amperage. Since hydrogen is formed during the operation of either test cell, a 10-min degassing time is allowed. During degassing, pH is

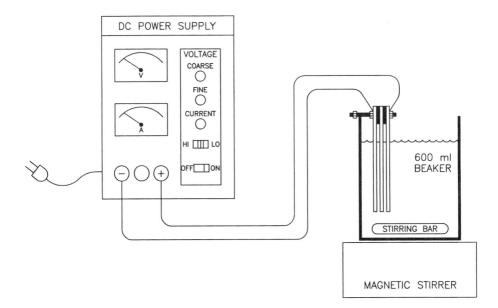

FIGURE 4.2   Bench-scale electrochemical treatment system.

monitored and adjusted, if necessary, to maximize precipitation and dye removal by adsorption. A polymer flocculant is added to assist clarification. The polymer dosage needed to achieve optimal settling characteristics ranges from 1 to 5 ppm by weight. After settling, the sample is filtered to remove suspended particulate materials.

Treatability tests are tailored to meet the specific objectives established by textile mills. The combinations and concentrations of contaminants present in most textile mill wastewater usually result in some or all of the following client objectives:

1. To reduce all heavy metals and organic contaminants either to below required discharge limits or to concentrations required for reuse in some wet process within the textile mill

2. To evaluate iron and/or aluminum coprecipitation using electrochemical ion generation

3. To determine the optimum treatment dosage and clarification pH

4. To quantify sludge generation and relate it to the amount collected during operation of a full-scale treatment system

5. To quantify chemical consumption for bench-scale experiments and the projected full-scale process

6. To reduce TDS concentration or minimize contribution to TDS

7. To obtain a clear and colorless effluent

In the bench-scale studies reported in this section, textile mill clients were responsible for collecting, packing, and shipping representative composite samples to a central facility that conducted treatability studies. Upon arrival at the laboratory, each wastewater composite was subjected to several standard electrochemical treatments, with a specific amount of iron (II) or aluminum being delivered to each 500-mL aliquot.

The measured pH of the untreated composite sample dictated whether or not pH adjustment was required. To minimize the electrochemical cell contribution to the TDS of the treated sample, pH adjustment was performed when metal ion solubility would prevent water reuse or when the adsorbing matrix produced was inadequate to achieve all treatment objectives. When pH adjustment was necessary, NaOH, $Ca(OH)_2$, or H_2SO_4 was used to achieve the desired pH for maximum precipitation and efficient clarification. A small amount of gas was formed during the operation of the cell and, therefore, a 10-min degassing period was included as part of the treatment protocol.

Once pH stabilization was obtained, a polymer flocculent was added to assist floc formation and improve clarification. The addition rate of the polymer was between 1 and 5 mg/L by weight. Both the dosage and type of polymer (cationic, anionic, and/or nonionic) employed were dependent on the nature and quantity of the dyes present, nature of the dyebath auxiliaries, and the treatment ion addition rate. Following floc formation, settling was allowed to occur until phase separation was completed. Finally, the supernatant was decanted off and filtered through What-

TABLE 4.1   Details of Treatability Studies

Client	Location	Dyes Utilized
A	Massachusetts	Direct, disperse, sulfur, acid, basic
B	Pennsylvania	Acid, disperse
C	Georgia	Acid
D	New Jersey	Direct, reactive
E	Tennessee	Acid, direct, disperse, reactive
F	North Carolina	Basic, direct, disperse, reactive

man No. 40 (8-μm) filter paper, which accurately simulates the results of multi-media filtration.

Of the numerous treatability studies that have been conducted on actual textile mill effluents using electrochemical ion generation, six were selected to represent a range of contaminant removal efficiencies. Plant locations and the dye classes contained within the treated effluents were as shown in Table 4.1.

Table 4.2 presents BOD, COD, and TOC removal data from three of the textile mill wastewaters. For clients A and B, only minimal ion generation was required to achieve a high level of BOD and COD removal. Successful treatment of acid and disperse dyes (client B) was easily accomplished. Charged acid dyes have a strong affinity for the iron (II) hydroxide matrix of the electrochemical cell, while

TABLE 4.2   BOD, COD, and TOC Removal Results

	EC–Fe Dosage (ppm)	EC–Al Dosage (ppm)	BOD (ppm)	COD (ppm)	TOC (ppm)
Client A	50	—			
Influent, ppm			230	920	—
Effluent, ppm			8	270	—
Removal, %			96	71	—
Client B	37.5	—			
Influent, ppm			—	587	—
Effluent, ppm			—	69	—
Removal, %			—	88	—
Client C	200	—			
Influent, ppm			230	910	280
Effluent, ppm			65	370	130
Removal, %			72	59	54
Client C	—	125			
Influent, ppm			230	910	280
Effluent, ppm			29	250	64
Removal, %			87	73	77

EC = electrochemical.

insoluble disperse dyes are coagulated as the iron matrix breaks the dispersion. The data presented for client C were included to compare and contrast the results of tests obtained using two different electrochemical cells. The use of an aluminum cell and clarification at a lower pH (7.25 vs 10.55) resulted in BOD, COD, and TOC levels that were 15, 14, and 23% lower (respectively) than when the same sample was treated with an iron cell. In addition, the lower aluminum dosage meant that less sludge was generated during treatment. Removal efficiencies relate to both the chemistry of the individual contaminants and different adsorption characteristics provided by the iron oxyhydroxide and aluminum oxyhydroxide matrices.

Removing heavy metals is complicated by both the broad range of possible constituents in textile mill wastewaters and the numerous metal forms (and valence states) that could exist. When extracting heavy metals from textile wastewater, the following considerations are necessary:

1. Is the heavy metal present as an integral part of the molecular structure of the unused dye? Is the dye soluble or insoluble?
2. Does the metal exist as a soluble inorganic ion?
3. Are chelating agents present that could interfere with the precipitation mechanisms?

If metal complex dyes are present, dye solubility and charge are the most important factors for successful removal of heavy metals. If dye solubility varies across the pH range, it may be important to maximize dye insolubility. Electrical charge of the dye is also affected by pH. Protonation/deprotonation reactions influence exposed charges, thereby modifying the dye's affinity for the electrochemically generated iron and aluminum species. Treatment by coagulation requires less iron or aluminum matrix than if removal relies solely on adsorption.

When metals are present as soluble inorganic ions, coprecipitation and adsorption will be the primary removal mechanisms. Precipitation is the most common method of removing dissolved metal ions from solution. Many heavy metals, including chromium, copper and zinc, are precipitated as metal hydroxides. Solubility curves for precipitation of cadmium hydroxide, lead hydroxide, chromium hydroxide, and copper hydroxide are illustrated in Figure 4.3.

Adsorption has also been identified as a viable mechanism for removal of contaminants via electrochemical processing. Removal efficiencies are dependent on the type (iron or aluminum) and quantity of adsorbing matrix introduced. Figure 4.4 illustrates how solubility curves are flat and broad when adsorption on $Fe(OH)_2$ is promoted.

Surface complexation and electrostatic attraction are the two most commonly used adsorption mechanisms. Soluble inorganic heavy-metal contaminants can exist as either cations (e.g.; Cr^{3+}, Cd^{2+}, Pb^{2+}, Cu^{2+}) or anions (e.g., MoO_4^{-2}). A schematic representation of surface complex formation on a suspended particle of $Fe(OH)_2$ is presented in Figure 4.5 (11). After forming all possible surface complexes, contaminants may be removed by simple electrostatic attraction. $Fe(OH)_2$ or $Al(OH)_3$, in combination with various surface complexes, contains areas of ap-

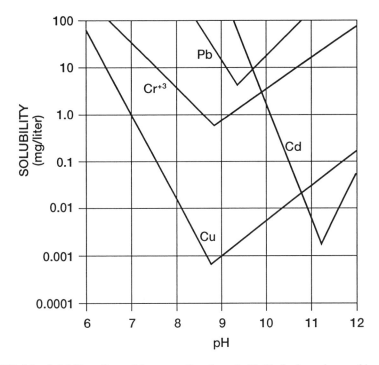

FIGURE 4.3 Solubility of metal ions as a function of pH. No hydrous iron oxide present.

parent positive and negative charge. Opposite charges attract and are capable of removing some dissolved species from the aqueous phase.

The two most important factors to consider when adsorptive removal of heavy metals is desired are pH and the adsorbing matrix-to-contaminant weight ratio. If adequate adsorbing matrix is provided for complete adsorption, careful control of pH is critical. Adsorption of cations typically improves as the pH is increased, while adsorption of anions increases as the pH is decreased.

Heavy-metal removal results for three case studies can be found in Table 4.3. Electrochemical iron dosage is dependent on types, forms, and concentrations of contaminants present. Wastewater from client D was the most difficult to treat. The combination of direct and reactive dyestuffs, their concentrations, and the presence of molybdenum required a substantial iron dosage and necessitated low pH (3.51) clarification for maximum molybdate adsorption. The data provide evidence that electrochemical iron generation is capable of removing heavy metals from textile effluents of widely varying characteristics.

Textile plant discharges are also regulated for color, TDS, and TSS. Table 4.4 contains removal results for three representative case studies. As color (dye concentration) increased, iron dosage had to be elevated to achieve maximum removal, and significant TDS reduction occurred because no pH adjustment was required.

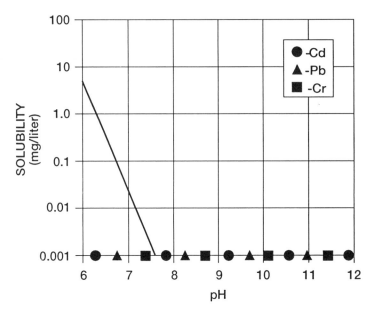

FIGURE 4.4   Solubility of metal ions as a function of pH. Hydrous iron oxide present.

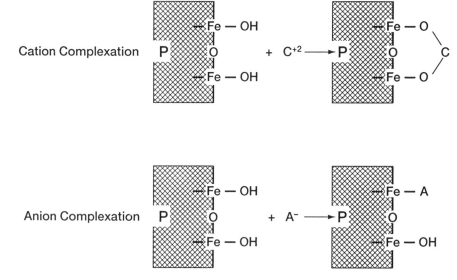

FIGURE 4.5   Surface complexation on hydrous iron oxide.

TABLE 4.3 Heavy Metals Removal Results

	EC–Fe Dosage (ppm)	Cr (ppm)	Cu (ppm)	Mo (ppm)	Zn (ppm)
Client A	37.5				
Influent, ppm		0.69	—	—	0.28
Effluent, ppm		0.03	—	—	0.02
Removal, %		96	—	—	93
Client D	400				
Influent, ppm		—	27.0	26.0	—
Effluent, ppm		—	0.21	1.4	—
Removal, %		—	99	95	—
Client E	25				
Influent, ppm		—	—	—	2.30
Effluent, ppm		—	—	—	0.02
Removal, %		—	—	—	99

EC = electrochemical.

The TSS, which consisted of insoluble dyes, precipitated inorganic compounds and lint, were easily removed through coagulation, coprecipitation, and adsorption.

Table 4.5 illustrates typical costs associated with treating textile mill wastewater using a system based on electrochemical iron generation. Included in these costs are the electrodes, electrical power for the system, chemicals for pH adjustment, and polymer for flocculation. In general, solids disposal costs depend on the classification of dyes and chemicals used in the textile facility. If they are nonhazardous, the solids should be considered nonhazardous. Dyes containing heavy metals

TABLE 4.4 Color, TDS, and TSS Removal Results

	EC–Fe Dosage (ppm)	Color (ADMI Units)	TDS (ppm)	TSS (ppm)
Client A	50			
Influent		—	—	260
Effluent		—	—	8
Removal, %		—	—	97
Client B	37.5			
Influent		108	530	230
Effluent		4	310	6
Removal, %		96	42	97
Client F	200			
Influent		1337	—	—
Effluent		107	—	—
Removal, %		92	—	—

EC = electrochemical.

TABLE 4.5   Operating Costs of Each of Six Treatability Trials

Client	Operating Costs ($/1000 gal)
A	0.53
B	0.53
C	1.20
D	2.10
E	0.40
F	1.04

are often complexed with the iron matrix and can be classified as nonhazardous in that form.

As an example of a full-scale system, a 700-gal/min system located in Kentucky will now be described. This facility generates wastewater containing vat and direct dyes. After removal of lint and other large solids, the colored water is processed through the electrochemical cell (Figure 4.6) for iron addition and dye removal. Subsequent processing for pH adjustment, solids flocculation, and separation produces treated water of which 70–80% is suitable for reuse. While the initial system was provided with a carbon absorption effluent polishing unit for the total flow, it is not needed for the recycled stream because of the high-quality water produced by the electrochemical process. Operating costs are approximately $0.80/1,000 gal, with solids disposed of as nonhazardous materials.

CONCLUSIONS

Electrochemical ion generation is a proven technology for removing color, BOD, COD, TOC, solids (suspended and dissolved), and heavy metals from textile mill wastewater. Wastewater from textile mills can also be treated using electrochemically added aluminum ions. Many factors must be addressed when choosing which process to implement. Whenever electrochemical ion generation is used, it must be coupled with conventional processes such as clarification and filtration. Along with successful removal of contaminants, this technology is capable of reducing toxicity and producing clean water for reuse.

In the representative case studies discussed here, iron dosages ranged from 25–400 mg/L. This illustrates the importance of a comprehensive evaluation of needs, in addition to a treatability study. The chemistry and concentrations of dyes and contaminants in wastewater will dictate the best treatment alternative. Since only limited research using aluminum has been conducted, it is difficult to estimate the range of treatment dosages that may be necessary. When utilized properly, electrochemical iron and aluminum ion generation is a cost-effective process for treating textile effluents.

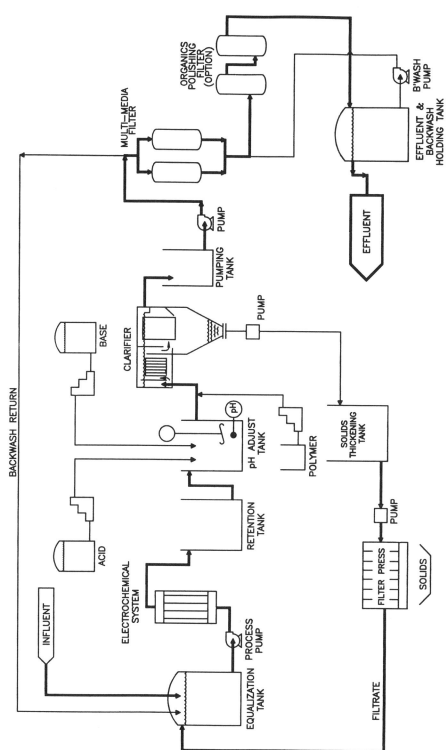

FIGURE 4.6 Process flow diagram.

73

ACKNOWLEDGMENTS

The authors would like to thank Jackie Tebbens, University of Guelph, for her patience and willingness to treat a seemingly endless number of dyebaths. They are also indebted to the textile companies that allowed them ready access to their laboratories and production facilities. The encouragement of Jim Frank and Bob Peters (Argonne National Laboratory) to complete this chapter is also most appreciated.

REFERENCES

1. Sung Ki Lee, Inventor, Andco Incorporation, Assignee, Electrochemical Contaminant Removal from Aqueous Media, U.S. Pat 3,926,754, Oct. 16, 1973.

2. S. B. Gale, P. P. O'Donnell, and S. Bruckenstein, Inventors, Andco Industries, Inc., Assignee, Method and Apparatus for Electrochemical Contaminant Removal from Liquid Media, U.S. Pat. 4,123,339, July 19, 1977.

3. K. D. Uhrich, Inventor, Andco Environmental Processes, Inc., Assignee, Method for Removing Dyestuffs from Wastewater; U.S. Pat. 4,880,510, Nov. 14, 1989.

4. T. Demmin and K. D. Uhrich, *Am. Dyest. Reptr.* **77**(6), 13–14, 17–18, 32 (1988).

5. W. C. Tincher, M. Weinberg, and S. Stephens, Removal of Dyes and Chemicals from Textile Wastewater, Book of Papers, International Conference and Exhibition, Nashville, TN:AATCC, 1988, p. 25.

6. M. K. Weinberg, Effectiveness of the Andco Electrochemical Treatment Process and Its Application in Textile Wastewater Treatment, Georgia Institute of Technology, M.S. Thesis, August 1989.

7. A. Wilcock and S. Hay, *Canad. Textile J.* **108**(4), 37 (1991).

8. A. Wilcock, M. Brewster, and W. Tincher, *Am. Dyest. Reptr.* **81**(8), 15 (1992).

9. S. McClung and A. Lemley, Electrochemical Treatment of Acid Dye Wastewater, Preprint Extended Abstract, American Chemical Society Division of Environmental Chemistry, Washington, D.C., August 23, 1992.

10. A. Wilcock, J. Tebbens, F. Fuss, J. Wagner, and M. Brewster, *Textile Chem. & Color.* **24**(11) (1992).

11. J. Buffle, *Complexation Reactions in Aquatic Systems*, Ellis Horwood Ltd., Chichester, 1988.

CHAPTER 5

CHEMICAL PRETREATMENT AND AEROBIC–ANAEROBIC DEGRADATION OF TEXTILE DYE WASTEWATER

SHARON F. DUBROW,[1] GREGORY D. BOARDMAN,[2] and DONALD L. MICHELSEN[3]
[1]Science Applications International Corporation
McLean, Virginia
Departments of [2]Civil Engineering and [3]Chemical Engineering
Virginia Polytechnic Institute and State University
Blacksburg, Virginia

INTRODUCTION

Optimizing procedures for the treatment of dye-containing wastewater is an enormous task, made extremely complex by the thousands of dyestuffs commercially available in the United States. Dyes among the various classes (e.g., azo, anthraquinone) exhibit different chemical characteristics and, consequently, respond differently to treatments aimed at their removal. This chapter presents an overview of the research conducted in four treatment areas and provides a discussion of results obtained from the pertinent studies. The four areas include: (1) chemical pretreatment, (2) aerobic biological treatment, (3) anaerobic biological treatment, and (4) anaerobic–aerobic biological treatment in tandem. Conclusions and observations that are consistent between related studies are outlined, as well as those that are inconsistent with earlier work. As more research is conducted with specific dyes and dye classes, more general conclusions and guidance will become available regarding the optimum treatment technology for a particular waste stream.

The section on chemical pretreatment primarily includes oxidizing and reducing agents targeted at removing color and preparing the wastewater for subsequent biological treatment. It also addresses the treatment practices of coagulation and flocculation. Because separate chapters in this book are dedicated to specific chemical oxidizing and reducing agents, the chemical pretreatment information in this chapter emphasizes recent work conducted at the Virginia Polytechnic Institute and

Environmental Chemistry of Dyes and Pigments, Edited by Abraham Reife and Harold S. Freeman.
ISBN 0-471-58927-6 © 1996 John Wiley & Sons, Inc.

State University (VPI&SU). In addition, fewer references specific to dye textile wastewater exist on chemical pretreatment than biological treatment.

In contrast, numerous studies have been conducted on the aerobic biological treatment of dye textile wastewater and are reviewed in this chapter. The presentation of aerobic and anaerobic biological treatments leads to the fourth topic of the chapter, the anaerobic–aerobic biological treatment sequence. The theory of reducing dyestuffs, either chemically or anaerobically prior to aerobic treatment, is currently of great interest to the dye community.

CHARACTERIZATION OF TEXTILE DYE WASTEWATER

Dye wastewaters are extremely variable in composition, due to the large number of dyes and other chemicals used in dyeing processes. In general, the composition of a particular wastewater, in addition to site-specific conditions, dictates the most appropriate treatment method. Detailed wastewater characterization is, therefore, an integral step in selecting wastewater treatment methodologies.

Dyes often receive the most attention from researchers interested in textile wastewater treatment processes because of their color, as well as the toxicity of some of the raw materials used to synthesize dyes (e.g., certain aromatic amines used to produce azo dyes). However, they are often not the largest contributor to the waste stream. For example, textile wastewater often contains high concentrations of inorganic salts. In addition, a 7-year research study conducted in the United Kingdom indicated that sources for over 90% of the chemical oxygen demand (COD) load from a typical plant that dyes and finishes woven cotton fabrics included desizing, scouring, reducing agents, and bleaches (1). The data from this study are presented in Table 5.1. Additional data presented by Park and Shore (1) indicate that dyeing operations involving cotton contribute 10–20% of the COD load.

Shriver and Dague (2) evaluated the characteristics of process wastewater generated from a textile mill in North Carolina in order to select a treatment process that would reduce 5-day biological oxygen demand (BOD_5) by 85% and achieve a high degree of organics and color removal. The authors reported that the major

TABLE 5.1 Typical Annual Pollution Loads from a Cotton Fabrics Textile Plant (1)

Waste Source	COD (tons/yr)	Percent
Desizing and scouring agents and bleaches	164	86.6
Reducing agents	8	4.2
Detergents and wetting agents	7	3.7
Finishing agents	5	2.6
Organic acids	3	1.6
Dyes and fluorescent brighteners	2.5	1.3

waste sources arise from sizing, dyes, soap, oils, and other products typical of the industry. The wastewater in this study was characterized as containing mainly soluble compounds, little suspended solids, colors that varied between reds and blues, and insufficient concentrations of nitrogen and phosphorus to sustain biological activity. Textile dye wastewater is generally nutrient deficient. Nutrients needed for biological treatment, however, can be provided by mixing the dye wastewater with municipal wastewater.

Additional wastewater characterizations are provided throughout this chapter, along with a description of the studies from which they were taken.

PRETREATMENT

Physical Pretreatment

Pretreatment processes in the textile dyeing industry are generally designed to remove color from the wastewater prior to discharge to an aerobic biological treatment system. Chemical coagulants can successfully be used to treat organic dyes, pigments, and other compounds that tend to adsorb to biomass or flocs and resist biodegradation. Textile plants often utilize coagulation and flocculation compounds such as lime, alum, ferric salts, or polyelectrolytes, followed by sedimentation or dissolved air flotation (DAF). The disadvantage of using coagulation and flocculation is the generation of large amounts of sludge that require disposal. Therefore, the coagulant dose required to achieve the desired color removal is an important consideration in evaluating chemical coagulants.

In a study conducted by the American Dye Manufacturers Institute (ADMI), physical, chemical, and biological treatments of 20 dyebaths from the textile industry, covering various dye classes and fibers, were investigated (3). Powdered activated carbon (PAC) was generally effective in removing color but required large doses relative to the amount of wastewater treated. For example, although color reductions to less than 100 ADMI units could be achieved, PAC doses from 400 to 2500 mg/L were required. The average percent color reduction for the 8 dyes evaluated with PAC treatment was 94%. Alum achieved average color reductions of 82% for 9 dyes in the study, with doses ranging from 8 to 80 mg/L. Neither PAC nor alum was as effective in reducing total organic carbon (TOC) as each was in reducing color. Powdered activated carbon achieved an average TOC reduction of 43%, while alum reduced TOC by only 33%. Based on results from the evaluation of 17 dyes, Horning (3) found that carbon treatment did not perform well with vat and disperse dyes but performed better than alum for reactive and basic dyes. Horning recommended a combination of PAC and alum as a potential cost-effective solution to the problem of high PAC dose requirements.

In his literature review on coagulation and flocculation in the textile dye industry, Boe (4) reported that Hatton and Simpson (5), using laboratory tests, found alum to be more effective in removing color from sewage containing textile waste than ferric sulfate or polyaluminum chloride. A subsequent plant trial involving alum was conducted on a waste stream containing 60% industrial (primarily textile) wastewater. Reductions in color, COD, and BOD of up to 37.8, 40.7, and 23.5%, respectively, were achieved using an alum dose of 400 mg/L.

Alspaugh (6) reported that alum, or alum combined with a cationic polymer, is a more effective treatment than the use of lime or iron salts. An effluent color of 15 on the platinum cobalt scale was achieved when a mixture of alum (150–250 mg/L), lime (10 mg/L), and cationic polymer (20 mg/L) was added to the mixed liquor of a biological treatment plant receiving textile dye wastewater. Alum alone, 200–400 mg/L, achieved color removals of greater than 95%. Lower doses of coagulant were generally needed when the addition was made to the mixed liquor rather than the effluent from biological clarification (6).

Abo-Elela and co-workers (7) found lime (CaO) and ferrous sulfate ($FeSO_4$) to be effective in reducing organic from textile wastewater. Lime doses of 750–2500 mg/L produced COD reductions of 80–82%, and BOD reductions of 80–92%. With ferrous sulfate, doses of 200–450 mg/L produced COD reductions of 78–79%, and BOD reductions of 67–84%. The authors found that a combination of lime and ferrous sulfate achieved higher removals than either coagulant alone. Specifically, doses of 500–700 mg/L CaO in combination with 125–150 mg/L $FeSO_4$ achieved COD reductions of 84–88% and BOD reductions of 86–91%, leaving residual COD and BOD concentrations of 99–143 and 19–23 mg/L, respectively. Although color was not evaluated in this study, the authors stated that alum and ferric or ferrous sulfate are capable of achieving up to 99% color removal. Other researchers reported that lime, either alone or with ferric salts, may achieve up to 90% color removal (8).

In another study, the pretreatment of textile wastes using alum and lime followed by biological treatment was evaluated (9). Lime doses of 4500 mg/L achieved color removals of 81.5%, while 600 mg/L of lime achieved 62.8% color removal. Subsequent biological treatment did not enhance color removal.

Poon and Virgadamo (10) installed and evaluated a pilot plant utilizing both activated carbon and biological treatment at a dyeing and finishing operation. The system consisted of an aeration basin, two activated carbon columns operating in parallel with an anaerobic biological regeneration phase, and a second pair of parallel activated carbon columns with an aerobic biological regeneration phase.

The pilot plant facility produced reductions in TOC of 53–67%, COD of 55–70%, and BOD_5 of 66–73%, at an average flow rate of 75,000 gal/day (gpd). The percent reductions were lowered by more than 10% when the flow rate was increased to its maximum of 125,000 gpd. With recommended modifications to the system (e.g., increased size of the aeration basin, three additional carbon columns containing additional amounts of carbon, and increased carbon regeneration times), the authors projected that the percent reductions would be raised to 91% for TOC, 92% for COD, and 94% for BOD_5, at a flow rate of 75,000 gpd.

Boe (4) reported results from another study in which activated carbon was effective in removing color from a textile plant wastewater. In this case, Davis (11) treated the plant effluent with organic polymer, DAF, sand filtration, and activated carbon filtration. The results of the study indicated that the sand filter removed

19% of the color, and the carbon filter removed an additional 78%. The polymer and DAF treatment steps did not contribute significantly to color removal.

In a recent study (12), various adsorbents, flocculents, and reducing agents were evaluated for their efficiency in removing color from Navy 106 slack washer water. Twenty-two different adsorbents (including rice grain, chitin, and crab meal) and dewatered sludge from a local publically owned treatment works (POTW) were employed in closed containers under anaerobic conditions. Color removal using these adsorbents was generally very effective; however, disintegrated adsorbent often contributed to increased TOC. In addition, large amounts of expensive adsorbent were required (15,000–20,000 ppm) to achieve a high degree of color removal. The authors concluded that the use of adsorbents does not appear to be a practical mechanism for the removal of reactive dyes from textile wastewater.

Fifteen polymers were also evaluated as adsorbents and flocculents for treating reactive dye wastewater (12). In general, very little color removal was achieved. Both the adsorbent and flocculent experiments were evaluated at pH levels of 4.6, 7.5, and 10.4. While the adsorbents appeared to perform better at lower pH, there was no observable trend in flocculent performance versus pH. The best performer was a strongly cationic, high-molecular-weight flocculent that achieved approximately 50% color removal. However, over 7600 mg/L of flocculent was required to achieve this reduction. Therefore, the use of a polymer flocculent is not a practical treatment method for decolorizing textile wastewater containing hydrolyzed reactive dyes.

Chemical Pretreatment

Oxidation   Many dyes may be effectively decolorized using chemical oxidizing agents. Chlorine—in the form of a liquid or gas, chlorine water, or hypochlorite—is often used to decolorize wastewater. The pretreated, decolorized wastewater may then be sent to activated sludge treatment. However, chlorine is viewed with increasing disfavor because it has the potential for generating toxic chlorinated organics that are harmful to both humans and the environment.

Chlorination was evaluated by Ghosh and co-workers (13) for its effectiveness in color removal. At a chlorine level of 150 mg/L, color was reduced by 77%, but 110 mg/L of total chlorine remained in the wastewater. At chlorine doses below 100 mg/L, no residual chlorine was detected, but only 57% of the color was removed.

Ozone is a more powerful oxidant than chlorine and offers a mechanism for oxidizing dye wastewater without the potential for generating chlorinated organics. As part of an ADMI study (3), ozone was evaluated as an oxidizing agent and was reported to be effective in decolorizing wastewater that contains reactive and basic dyes but ineffective in decolorizing wastewater that contains disperse dyes, using about 1 g/L ozone. The effectiveness of ozone treatment on direct dyes was variable. In addition, ozone treatment produced little reduction in TOC. Cost and efficiency are often barriers associated with ozonation, although researchers continue efforts to determine both the actual amount of ozone required to decolorize wastewater and the level of decolorization achievable. A separate chapter in this book is dedicated to ozone treatment and provides more information on this topic.

A study conducted at VPI&SU involved an evaluation of the effects of ozone and Fenton's reagent on effluent from a textile slack washer and jet dyebath (14). The dyes present in the effluent were reactive azo and anthraquinone dyes containing a vinylsulfone reactive group. Ozonation was evaluated using a semibatch method in which a reactor column was filled with a known amount of wastewater, and ozone was bubbled through the wastewater solution at a constant rate. The reaction was allowed to proceed until color change was no longer observed. Pure dye solutions were used as controls.

The experiments utilizing Fenton's reagent were conducted on a batch basis. Iron and peroxide were added at a specified rate and allowed to react until no further reaction was observed.

Laboratory-scale, sequential batch reactors (SBRs) were used to evaluate the amount of biodegradation following oxidation, for both the ozone and Fenton's reagent experiments. The laboratory units contained 2 liters of wastewater and utilized a 2-day retention time. The feed to the SBRs consisted of 1 part treated jet dyebath (untreated for the control reactor) to 9 parts primary effluent from a local wastewater treatment plant. The treatment wastewater from the slack washer was diluted at a ratio of 1:3.

The results of the ozone pretreatment indicate that ozone was effective in removing color from three dye waste streams. The streams showed approximate color removals of 86, 68, and 63% at ozone concentrations of 100 ppm (stream 1, pH 11), 250 ppm (stream 1, pH 3), 570 ppm (stream 2, pH 11 and 3), and 125 ppm (stream 3 pH 11 and 3). Stream 1, containing a low concentration of dissolved organic carbon (DOC, 200 ppm), of which only 10–15% was estimated to be dye, required a much greater amount of ozone at pH 3 than at pH 11 to achieve the same level of color removal. Variations in removal efficiencies between streams may be caused by competition and selective oxidation of chemicals (15). Ozone treatment did not produce any significant reductions in DOC.

Although Fenton's reagent reduced color to low levels, removal efficiency was dependent on the initial DOC of the waste streams, with higher initial DOC resulting in less color removal. Powell (15) recommended a ratio of peroxide to iron of 10 or 20 to 1, with the higher ratio used when solids generation would present a problem. Unlike ozone, Fenton's reagent was also effective in removing DOC from the waste stream (up to 60% reduction). In addition, the biological reactors were neither enhanced nor hampered by the oxidative pretreatment processes, compared to the control reactors.

Reduction   It is widely known that azo dyes undergo conversion to aromatic amines, via reductive cleavage of the azo bond, when subjected to a reducing environment (16). This chemistry has been the focus of many studies on anaerobic biological degradation, a process that will be discussed later in this chapter.

Chemical reduction of azo dyes also causes cleavage of the azo bond, generating aromatic amines that, in theory, are more amenable to subsequent aerobic biological treatment than the parent dye structures. Consequently, research has been conducted to assess the efficiency of the aerobic biological treatment, a step that is critical to avoiding a discharge of toxic aromatic amines into the environment. In addition, published results indicate that certain reactive dyes are adsorbed better onto activated carbon when pretreated with a reducing agent (17).

For many dyes, particularly azo dyes, chemical reduction is an effective decolorization technique. The most commonly used chemical reducing agent is sodium hydrosulfite (also known as sodium dithionite). Thiourea dioxide [also known as formamidine sulfinic (FAS) acid], sodium borohydride, sodium formaldehydesulfoxylate, and tin (II) chloride are additional reducing agents for treating dye wastewater. When evaluating the chemical decolorization of wastewater, it is important to investigate the potential reversal of the reaction upon exposure to oxygen, since color may reappear upon discharging the wastewater to the environment.

A study was recently conducted to evaluate the effectiveness of three reducing agents in decolorizing Navy 106 (a mixture of 3 azo dyes) washwater generated from a textile plant (18). The effect of chemical pretreatment on subsequent aerobic biological treatment was also investigated. Three reducing agents were evaluated: (1) 225 ppm sodium hydrosulfite, (2) 225 ppm thiourea dioxide, and (3) 150 ppm sodium borohydride. Oxidation employing hydrogen peroxide was also evaluated for color removal. In addition, experiments were conducted to evaluate the effects of reduction with sodium hydrosulfite followed by (1) pH adjustments to 7.5 using sulfuric acid, (2) phosphate addition, (3) ammonium, potassium, magnesium, calcium, and iron salt nutrient additions, (4) pH adjustment and nutrient addition, (5) lime addition, and (6) hydrogen peroxide addition. Subsequent aerobic biological treatment was evaluated using 8-liter SBRs containing mixed liquor suspended solids between 2000 and 2500 mg/L. The reaction time was approximately 21 hr and the settling time was 1 hr.

The results of preliminary tests from various hydrosulfite, thiourea, borohydride, and hydrogen peroxide treatments are presented in Table 5.2. Preliminary tests were conducted using either 17 or 100% Navy 106 wash water. For each pretreatment method (e.g., sodium hydrosulfite pretreatment on a 100% wash water solution), the percent removals due to chemical addition are provided on one line, and the percent removals due to the SBR step are provided on the next line. The percent removals are calculated based on the influent to the entire system such that the sum of all treatment steps would equal the total removal by the system.

Following chemical pretreatment and before SBR treatment, 25% municipal wastewater was added to the Navy 106 wash water to simulate the commingling of wastewaters prior to the POTW. Some dilution of the values occurred during this step, which would increase the overall reductions shown in Table 5.2 by the amount of the dilution.

The chemical pretreatments, except for hydrogen peroxide pretreatment, were effective in removing color. Only the hydrogen peroxide treatment was effective in removing BOD_5 (21% removal), though it did not enhance biodegradation relative to the control experiment. It would appear that chemical reduction, using the three reducing agents listed in Table 5.2, created a wastewater that was inhibitory

TABLE 5.2   Effect (% Reduction) of Various Treatments on Navy 106 Wash Water (19)

Pretreatment Method	Color	TOC	BOD$_5$[a]	COD
Sodium hydrosulfite[b] on 100% wash water				
Chemical addition	92	−3.7	NR	−2.6
SBR	0.2	1.8	1.3	3.2
Sodium hydrosulfite on 17% wash water				
Chemical addition	11	−1.4	0	−1.7
SBR	3.5	1.4	2.2	9.2
Sodium borohydride[c] on 17% wash water				
Chemical addition	11	−1.5	NR	−3.3
SBR	2.8	0.8	1.0	6.7
Thiourea dioxide[d] on 17% wash water				
Chemical addition	11	−1.3	NR	−0.8
SBR	4.4	0.9	0.8	6.2
Hydrosulfite and nutrients[e] on 17% wash water				
Chemical addition	13	−2.7	0	−0.9
SBR	2.8	2.1	2.5	4.3
Hydrosulfite/nutrients/pH on 17% wash water				
Chemical addition	12	1.4	0	0.4
SBR	1.7	2.9	3.8	4.5
Hydrosulfite/lime[f] on 17% wash water				
Chemical addition	12	−1.0	0	−0.3
SBR	2.0	2.4	2.1	8.6
Hydrogen peroxide[g] on 100% wash water				
Chemical addition	0	−1.0	NR	−0.7
SBR	14	27	21	25

[a]NR: Not reported.
[b]Sodium hydrosulfite: 225 ppm.
[c]Sodium borohydride: 150 ppm.
[d]Thiourea dioxide: 225 ppm.
[e]Nutrients: 1 ml per liter of feed of the following salt solutions: phosphate, ammonium, potassium, magnesium, calcium, and ferric salt per liter of feed.
[f]Lime: 100 ppm.
[g]Hydrogen peroxide: 150 ppm.

or toxic to the subsequent aerobic biological system. The authors suggested that either the reducing agents or the aromatic amines generated from reductive cleavage of the azo bonds may have been toxic to the microorganisms in the biological treatment plant. Overall, none of the chemical pretreatments evaluated was successful in removing both color and BOD_5, when used in tandem with aerobic biological treatment.

A second experiment was conducted to evaluate the effect of chemical reduction followed by hydrogen peroxide addition and aerobic biological treatment (18). Hydrogen peroxide addition was expected to create an oxidized environment that would be more conducive to aerobic biological degradation. This pretreatment sequence was performed on increasing percentages of the wastewater (i.e., 17, 33, 66, and 100% of the influent). As the amount of wastewater pretreated with chemical reduction increased, the amount of subsequent biological degradation required decreased, as would be expected (i.e., there was less residual color needing further treatment). In all cases, residual color (color not amenable to either chemical or aerobic biological degradation) was observed in the wastewater. Results from pretreatment of the undiluted textile wastewater are presented in Table 5.3, along with results from a control reactor in which chemical reduction was not performed.

McCurdy and co-workers (18) concluded that the addition of hydrogen peroxide following chemical reduction created an oxidized environment that was more conducive to aerobic biological treatment than the environment resulting from chemical reduction alone, although the oxidation step itself did not contribute toward additional reductions in color, COD, TOC, or BOD_5. In addition, the toxic aromatic amines created by chemical reduction may be oxidized to aromatic amine oxides or aromatic quinones that are no longer toxic to the bacteria used in the biological system (20). When evaluating a pilot- or plant-scale system, effluent analyses should be conducted in order to identify residual compounds that would be discharged to the environment.

A subsequent study by Loyd and co-workers (21) was initiated to refine the chemical reduction/aerobic biological treatment system established in the McCurdy study. When the optimized pretreatment protocol was used on a fresh batch of Navy 106 wash water, which was five times more concentrated than the original wash water evaluated, the color was initially removed but then reappeared within

TABLE 5.3 Percent Removals Arising from Chemical Reduction Followed by Oxidation and Aerobic Biological Treatment of 100% Navy 106 Wash Water (19)

Treatment Step	Color	TOC	BOD_5	COD
Reduction	62	−0.5	NR	0.4
Oxidation and dilution	9.4	11	16	17
SBR	5.1	51	49	44
Total reduction	77	62	65	61
Control reactor	48	38	50	43

24 hr as a dark, blue-gray color. It may be that some textile dye wastewater would be amenable to decolorization by chemical reduction, with subsequent organic removal by aerobic biological treatment; however, laboratory and pilot-scale evaluations of the particular wastewater would be required to ensure success.

Three reducing agents were recently evaluated by Michelsen (12) on Navy 106 slack wash water from cold pad/batch dyeings at a textile mill. The reducing agents were sodium hydrosulfite, thiourea dioxide, and sodium formaldehydesulfoxylate. Each reducing agent was evaluated at various pH levels. After considering color reduction efficiency, dosage required, and cost, it was found that at pH 4, sodium hydrosulfite was the most efficient (approximately 71% color removal at a dose of 400 mg/L), but at pH 10.7, thiourea dioxide was the most efficient agent (approximately 63% color removal at a dose of 100 mg/L). Due to the higher cost of thiourea dioxide, however, sodium hydrosulfite may be the reducing agent of choice, even at high pH.

A follow-up study to the earlier work of teams led by McCurdy (18), Powell (14), and Michelsen (12) was conducted in an attempt to design, fabricate, and test a 1-L/min pilot plant using a sequence of two continuously stirred tank reactors (22,23). The objective was to treat, decolorize, and remove metals and TOC from selected reactive dyebath concentrates (i.e., following dye application) using Fenton's reagent, thiourea dioxide, and sodium hydrosulfite.

The initial pilot plant studies were conducted using Navy 106 jet dye concentrate. Reductive pretreatment resulted in 92.2% color removal, with partial color regeneration upon aeration. This afforded a net color removal of 76.6%. Oxidative pretreatment resulted in 98.8% color removal, with a net color removal after aerobic treatment of 96.8%. Total organic carbon removal during aerobic treatment was greater for the oxidized wastewater than the reduced wastewater.

On-site operation of the pilot plant on other dye wastes showed color removals above 95%, and TOC removals of 38 and 19% for an azo and copper-phthalocyanine-based (CPC-based) dye concentrate, respectively. The copper concentration in the CPC concentrate was reduced from 19.22 to 4.5 ppm. This corresponded to a threefold increase in suspended solids from 0.575 to 1.505 g/L.

The results showed that continuous oxidative pretreatment, with 15-min residence hold-up, was controllable and more effective than reductive treatment, for color removal. Oxidative pretreatment also reduced copper in a copper-containing waste and did not hinder biological activity.

Pretreatment Overview

Overall, pretreatment processes using chemical coagulants and oxidizing and reducing reagents have afforded varying degrees of success in laboratory and field studies. As shown by the results of the studies presented in this section, determining the optimum combination and doses of chemical additives for a particular waste stream can determine the ultimate success of the project. In practice, many members of the textile industry currently employ pretreatment techniques to reduce the color

of the effluent. However, as effluent standards become more stringent, higher performance pretreatment systems at affordable costs will continue to be needed.

AEROBIC BIOLOGICAL TREATMENT

Aerobic biological treatment is one of the most commonly used treatment methods
for wastewater generated from textile dyeing operations. Historically, dye wastewater has often been discharged to POTWs where activated sludge is the most
commonly used system. However, the ineffectiveness of aerobic biological treatment in reducing color has caused aesthetic problems in the receiving waters and
encouraged dischargers to investigate alternatives. Dyes themselves are generally
resistant to oxidative biodegradation, since one of the most important properties
built into commercial dyes is resistance to fading caused by chemical- and light-
induced oxidation.

The issue of toxicity of the effluent is also a rising concern. As analytical methodologies improve, and more sophisticated instrumentation is used to analyze for
toxic compounds in water, sludge, and sediment, BOD_5 no longer remains the only
criteria for evaluating the effectiveness of waste treatment processes. Concentration
limits for specific toxic constituents in the waste stream, in addition to criteria for
general toxicity tests (e.g., *Ceriodaphnia dubia* test), already exist in some areas,
and may increasingly be required in the future.

Another problem with aerobic biological treatment of dye wastewater is the
difficulty in acclimating the organisms to the substrate. Acclimation presents a
problem with textile wastewater due to constant product changes and batch dyeing
operations. Porter and Snider (24) found that 30-day BOD tests demonstrated much
greater biodegradability of dye wastes than the standard 5-day BOD test (using
unacclimated cultures). The average BOD_5, using eight commercial textile dyes
that included direct, disperse, reactive, and vat dyes, was determined to be 20,000
mg/L, which is 38% of the 30-day BOD load of 53,000 mg/L. In addition, the
BOD curves for one half of the dyes tested continued to increase beyond 30 days.
These results indicate that dye wastes require long periods of acclimation and are
slow to degrade. The BOD_5 value for most readily biodegradable wastewater, such
as domestic wastewater, is 60–70% of the ultimate BOD. Porter and Snider (24)
concluded that BOD_5 is often not an accurate measurement of the ultimate biodegradability of a waste.

Shriver and Dague (2) conducted preliminary BOD tests and concluded that
color in textile wastewater would be expected to undergo biodegradation at a slower
rate than that from typical domestic wastewater. For example, after 10 days, biodegradation of the textile wastewater was 31% complete, whereas the biodegradation of colored domestic wastewater was 92% complete.

Other researchers have also shown dyes to be degraded slowly under aerobic
conditions (25). As a result, research has been conducted to identify treatment
technologies that will perform better than aerobic biological treatment in reducing
the organic loadings, color, and toxicity of wastewater effluent.

Application of This Technology to Textile Plant Effluent

Based on the characterization studies described earlier in this chapter, Shriver and Dague (2) constructed four pilot-scale, activated sludge units with detention times of 12, 24, 36, and 48 hr, respectively, and a mixed liquor suspended solids (MLSS) concentration of 2000 mg/L. The authors observed a substantial decrease in color and COD as the detention time increased from 12 to 48 hr. For example, with a 12-hr detention time, BOD, COD, and color reductions of 84, 41, and 22%, respectively, were achieved, whereas a 48-hr detention time allowed for reductions of 92, 74, and 50%. Based on these pilot studies, a plant was designed and constructed having two concrete, asphalt-lined aeration cells, six floating aerators in each cell, two final clarifiers, two aerobic digesters, and a nutrient feed system capable of feeding ammonia and phosphoric acid at ratios of 100:5:1 for BOD:N:P. Among the results obtained was a high COD-to-BOD$_5$ ratio (average of 13.5:1), indicating that a nonbiodegradable fraction of organics remained in the effluent. On average, over 85% of the influent BOD$_5$ was removed by the system after 90 days, reaching an effluent BOD$_5$ of 24 mg/L.

Shriver and Dague (2) also investigated treatment options for the secondary effluent, so that wastewater recycling might be possible. Activated carbon columns were recommended and designed based on wastewater loadings. The authors concluded that activated carbon treatment produces wastewater effluent of sufficient quality for some reuse applications but that its economic practicality depends on the cost and availability of fresh water. Laboratory carbon column tests achieved an effluent quality of no detectable BOD, 5 mg/L COD, no measurable color, and no increase in hardness. The authors recommended more extensive activated carbon pilot tests to confirm design criteria and feasibility.

Scientists at the Environmental Protection Agency's (EPA's) Water Engineering Research Laboratory in Cincinnati, Ohio, studied (25) the fate of water-soluble acid and direct azo dyes in the activated sludge process. Screened raw wastewater was used from a local sewage treatment plant as influent to three pilot-scale activated sludge biological treatment systems operated in parallel. The wastewater influent to the treatment system was spiked with commercial dye in concentrations of either 1 or 5 mg/L. Dye analyses were then conducted on the wastewater effluent and activated sludge.

As a quality control measure, the dyes were also spiked into laboratory-prepared organic free water and wastewater and sludge samples from the control activated sludge process, in order to measure the percent recoveries obtained by the analytical methods. Most of the values fell within the targeted range of 80–120%. Recoveries from four of the dyes fell outside the targeted range; however, the results were accepted as indicating that little or no chemical transformation of the dyes occurred as a result of contact with the wastewater or sludge matrices. This quality control methodology is not presented for other studies in this chapter but is described here because of its importance in evaluating analytical results.

Data obtained from this study demonstrated that 11 out of 18 dyes passed through the pilot-scale, activated sludge process untreated, 4 adsorbed onto the activated sludge, and 3 underwent biodegradation. Shaul and co-workers (25) conducted a literature review that indicated that adsorption to the sludge is the primary removal mechanism for dyes in a biological wastewater treatment system and that factors inhibiting permeation of the dye through the microbial cell membrane (e.g., increased water solubility and increased molecular weight) reduce the effectiveness of biological degradation (26–29). The results from that review support the conclusion that wastewater containing dyes of high water solubility are generally not decolorized effectively by the activated sludge process. The 4 dyes that adsorbed to the sludge had fewer sulfonic acid groups and were of higher molecular weight; both features tend to reduce water solubility.

In the study conducted by Dohanyos and co-workers (26), 20 textile dyes were evaluated in a laboratory-scale, activated sludge system. The dyes evaluated included azo acid, direct, and reactive dyes. The results of this study led to the conclusion that the dyes were removed from wastewater via physical and physicochemical processes such as diffusion, adsorption, and chemical reaction. It was also noted that the amount of dye removed was increased by the presence of hydroxyl, nitro, and azo groups in the dye structure, and the size of the dye molecule, but was decreased by the presence of sulfonic acid groups. These results are consistent with the conclusions reached in the EPA study (25), indicating that dyes having high water solubility are less likely to be removed by the activated sludge process. Following the batch experiments, continuous-flow activated sludge treatment tests were conducted that indicated that: (1) dye removal occurred via the same mechanisms as in batch systems; (2) the concentration of dye ($10-240$ g/m^3 day) did not affect treatment efficiency; and (3) the amount of dye removed (7–88%) was directly proportional to sludge production.

In a second study conducted at the EPA's Water Engineering Research Laboratory (30), scientists investigated the fate of Disperse Blue 79, a water-insoluble monoazo dye, in a continuous feed pilot-scale wastewater treatment system. The system included a primary clarifier, plug-flow aeration basin, secondary clarifier, and an anaerobic digester. Screened, raw wastewater from the local municipal wastewater treatment plant was spiked with 5 mg/L of the commercial formulation of Disperse Blue 79. A second, identical system served as a control and received unspiked municipal wastewater. Following the investigation of the fate of the dye in the effluent, primary and waste activated sludge, digester supernatant, and digester effluent, the fate of the dye degradation products from the anaerobic digester was evaluated in a laboratory-scale, activated sludge system. This process was designed to simulate the recycle of digester supernatant to the front of the activated sludge system.

The results of this study showed, on average, that 20.4% of the dye remained in the effluent from the activated sludge system, 3.6% was found in the primary sludge, and 62.3% was present in the waste activated sludge. A mass balance around the entire system accounted for 86.3% of the dye present in the influent. However, the authors did not detect any compounds related to Disperse Blue 79 that would provide evidence of biodegradation in the activated sludge process and account for the remaining 13.7%. The results indicated that although 62.3% or more of this water-insoluble dye was removed by the activated sludge treatment, the primary removal process was adsorption to the sludge. The authors recommended coagulant aids and a longer retention time in the secondary clarifier as steps to improve the removal of dyestuffs from the sludge, though specific coagulants and retention times were not an outcome of this study.

In contrast to the amount of biodegradation that occurred in the waste activated sludge process, anaerobic sludge digestion degraded greater than 97% of the Disperse Blue 79 fed to the digester. Identification of the anaerobic metabolites was not performed in this study.

The feed to the bench-scale activated sludge unit consisted of primary effluent mixed with supernatant from the sludge thickening operation, and "centrate" from centrifuging the digester effluent. This system was designed to simulate the return of supernatant from sludge lagoons or other sludge-thickening operations to the front of the wastewater treatment system. The influent, effluent, and waste sludge were analyzed for total suspended solids (TSS), total biological oxygen demand (TBOD), total chemical oxygen demand (TCOD), Disperse Blue 79, and the presence of any degradation products. The results showed that 62% of the dye was removed by the lab-scale activated sludge unit. The mass balance around the unit accounted for 73.2% of the dye. In addition, the concentration of degradation products decreased across the unit, although the compounds produced were not positively identified or quantified.

The Ecological and Toxicological Association of the Dyestuffs Manufacturing Industry (ETAD), the trade association representing dyestuff manufacturers, developed a test method to evaluate the tendency of a particular dye to be treated by the activated sludge process (27). The test consists of adding dye solution to an activated sludge suspension, centrifuging the treated sludge solution, and measuring the absorbance of the supernatant using a colorimeter. The amount of sludge adsorption is determined by comparing the absorbance of the supernatant to that of a blank solution (i.e., prepared in the absence of dye) and a standard dye solution (of known concentration). ETAD applied the test method to six classes of dyes and concluded that the suitability of the dye to activated sludge treatment is determined by the class of dye and structural differences within each class. The test results were consistent among six laboratories using two types of activated sludge.

The ETAD report is consistent with earlier studies in that adsorption to the sludge was the primary removal mechanism for dyes treated by activated sludge and that any aerobic biodegradation of the dye was unusual. ETAD found that the presence of a sufficient amount of activated sludge is critical to effective adsorption and successful dye removal. The test results indicated that low adsorption occurs with acid and reactive dyes, high adsorption occurs with basic and direct dyes, and high to medium adsorption occurs with disperse dyes. Within the acid dye class, the degree of adsorption was partly related to the number of sulfonic acid groups in the dye molecule; however, a substituent-related adsorption effect was not observed with the other dye classes.

A study was conducted on wastewater generated from a textile mill in Guilford, Maine, where plant operations consisted of carding, spinning, weaving, knitting, dyeing, and finishing fabrics (13). The objectives of the study were to evaluate the treatability of the wastewater alone and with municipal wastewater and to design a full-scale treatment facility. The wastewater varied from gray to black, with reds and greens dominating on some days, and was characterized by moderate BOD (180 mg/L), high COD (1300 mg/L), low suspended solids (38 mg/L), high color (350 APHA (American Public Health Association) units), and high turbidity (330 JTU (Jackson Turbidity Units)). The investigators utilized four continuous-flow, laboratory-scale, complete-mix activated sludge units. Color removal was reported as essentially negligible following hydraulic retention times from 8 to 20 hr. The required removal, as specified in the 1977 Best Practical Control Technology Guidelines (BPCTG), for BOD_5 and suspended solids (SS) was achieved by the activated sludge units; however, 30-day COD limits were not met. None of the criteria were met for the 1983 BPCTG without additional treatment units. Such units were not evaluated in this study.

Although results of the studies cited indicate that little degradation of dyestuffs occurs in aerobic biological treatment systems, other studies have shown that some dyes may be biodegraded under aerobic conditions. Some researchers are presently studying the isolation, identification, and characterization of aerobes capable of degrading dyes. For example, a study involving the degradation of two disperse dyes showed that Disperse Red 5 was degraded under aerobic conditions, whereas Disperse Orange 5 required anaerobic conditions for degradation (31). This experiment was conducted using a 13-liter laboratory-scale activated sludge system with a stream of air entering from the bottom of the tank to keep the sludge in motion. Domestic sludge from a local treatment plant was gravity fed to the laboratory unit at a flow rate of 1 L/hr, providing for a 13-hr retention time before the effluent flowed over the weirs. Analysis of the dyes and their metabolites were conducted using thin-layer chromatography and infrared spectrophotometry. The results indicated that both dyes were degraded to the aromatic amines arising from reductive cleavage of the azo bond.

A study involving aerobic degradation of azo dyes, conducted at the Institute of Environmental Health at the University of Cincinnati, revealed that whether degradation is abiotic or biotic depends on the particular dye (32). Two azo dyes, Acid Orange 7 and Acid Red 151 (see Figure 5.1), were investigated utilizing activated sludge from a local municipal sewage treatment works that receives discharges from several azo dye manufacturers.

The degradation pathway for Acid Orange 7 was biotic and produced aromatic amines, whereas the pathway for Acid Red 151 was abiotic and oxidized products were generated. Degradation of Acid Red 151 occurred in both the biologically active and the control reactor, indicating that the degradation is an abiotic process. The reaction mechanism was further characterized with the aid of high-pressure liquid chromatography (HPLC) and was determined to involve a photoinduced, metal-cation-catalyzed, free-radical oxidation reaction. This effort has been supplemented with research conducted by P. Bishop (33), who studied the aerobic degradation of azo dyes on biofilms. Biofilms accommodate microorganisms that effect the aerobic biodegradation of specific dyes by al-

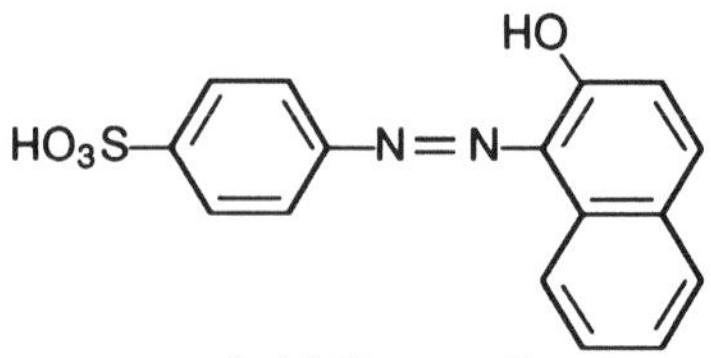

Acid Orange 7

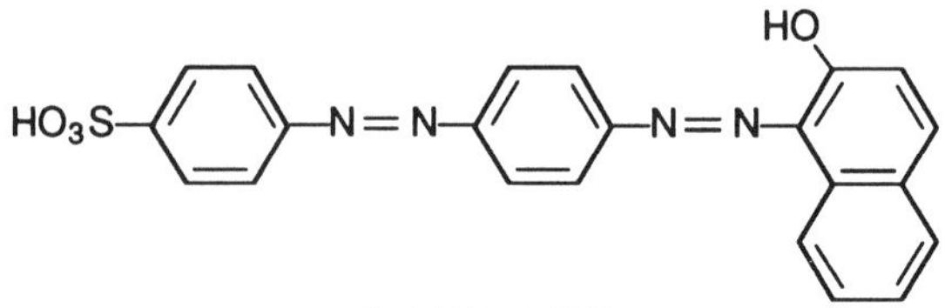

Acid Red 151

FIGURE 5.1   Acid Orange 7 and Acid Red 151.

lowing for longer retention times, necessary acclimation, and the development of specialized microorganisms.

Aerobic Treatment Overview

The studies presented in the prior section on aerobic biological treatment of textile dye wastewater indicate that little biodegradation of dyes actually occurs and that the primary removal mechanism involves adsorption to activated sludge. In addition, textile dye wastewater was shown to undergo biodegradation more slowly than domestic wastewater, under aerobic conditions. High residual COD-to-BOD$_5$ ratios indicate that difficult to degrade and/or nonbiodegradable materials remain in the effluent from aerobic systems. Ghosh and co-workers (13) concluded that carbon adsorption, with or without chemical coagulation, and other sorption processes are more effective in removing color than aerobic biological treatment. Although work has been published (32) pertaining to the isolation of specific aerobes that degrade one or more dyes, it may be difficult to apply the technology to a textile dye wastewater that contains many dyes and varies daily in composition.

Significant research has been and continues to be conducted in the areas of chemical pretreatment, anaerobic biological treatment, and anaerobic–aerobic treatment sequences, in order to improve the biodegradation of dyestuffs. Representative studies from the latter two areas are presented in the section that follows.

ANAEROBIC BIOLOGICAL TREATMENT

As described in the ''Pretreatment'' section of this chapter, azo dyes undergo reductive cleavage when subjected to certain chemical or biological reducing environments. The anaerobic biological reduction of azo dyes has been investigated from many perspectives. Some researchers have been concerned primarily with color removal and chemical degradation (i.e., treatment), while others have studied the potential generation of toxic aromatic amines and the consequences of their release to the environment.

Examples of anaerobic reducing conditions found in the environment include sediments at the bottom of streams and in certain sections of landfills. Aromatic amines could be released into the ground and surface waters at such locations. Research addressing these concerns has been directed toward the degradation of the dyestuffs in anaerobic sediment/water systems, identification of the resulting compounds, and subsequent biodegradation of the anaerobic metabolites (34).

Reduction Pathways

Many studies concerning dye degradation are laboratory studies involving prepared solutions of commercial dyestuffs. This approach introduces fewer variables than would an experiment utilizing a "real-world" complex wastewater. However, the more carefully controlled laboratory studies may be useful in identifying reduction pathways and in providing a theoretical potential for dye degradation.

In a study sponsored by ETAD, 14 solutions of commercial azo dyes in distilled water were analyzed following anaerobic digestion (16). Nearly all the azo dyes evaluated, which included acid, direct, and mordant dyes, were greater than 90% degraded. Subsequent analysis and quantitation of the aromatic amines expected from the dye structures indicated that cleavage of the azo bond was the initial degradation step. Structures for two of the dyestuffs, Acid Yellow 36 and Direct Yellow 12, and the corresponding aromatic amine reduction products are presented in Figures 5.2 and 5.3, respectively.

Results from the ETAD study also indicate that other pathways contributing to dye degradation may include adsorption of the dyestuff to the sludge and reduction of functional groups other than the azo bond. Sludge adsorption and reduction of functional groups, other than the azo bond, may result in color reduction; however, they do not contribute toward the elimination of dyestuff from the waste. In addition, the degradation of polyazo dyes varied from 1 to 100% among the participating laboratories, whereas the monoazo and disazo dyes were consistently degraded (35).

A laboratory study conducted on Acid Red 88 under anaerobic conditions resulted in 90% of the dye being removed within 8 hr, approximating a first-order

Acid Red 88

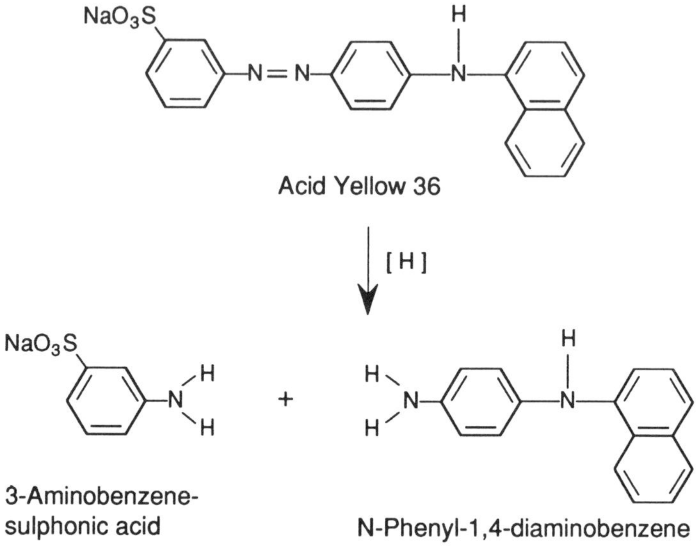

Acid Yellow 36

3-Aminobenzene-
sulphonic acid

N-Phenyl-1,4-diaminobenzene

FIGURE 5.2 Anaerobic reduction of Acid Yellow 36.

reaction rate with a half-life of 6 hr (36). Among the reduction metabolites iden-tified were naphthionic acid and 1-amino-2-naphthol, which are the aromatic amines generated from cleaving the azo bond. Several other compounds identified, including 1,2-naphthoquinone, 2-naphthol, isoquinoline, and quinacridone, are be-lieved to have been generated from subsequent reactions of the initial reduction products. The results of this study also suggested that a carbon source other than the dye itself was important for enhanced degradation rates. This was demonstrated

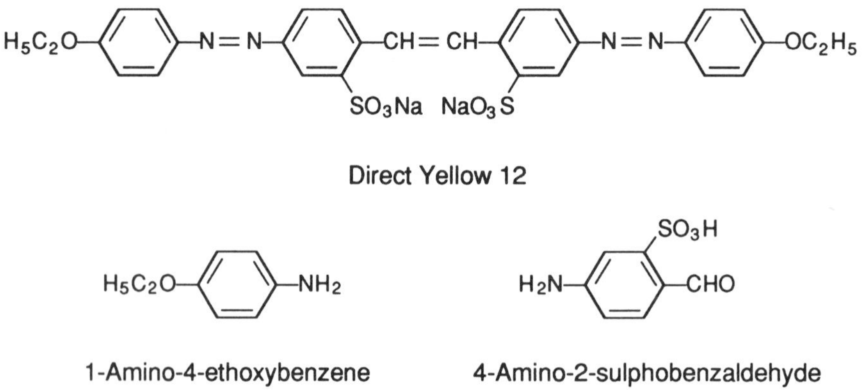

Direct Yellow 12

1-Amino-4-ethoxybenzene

4-Amino-2-sulphobenzaldehyde

FIGURE 5.3 Anaerobic reduction of Direct Yellow 12.

by increased rates of mineralization and metabolite formation occurring with the addition of acetate and propionate as supplemental carbon and energy sources.

Application of Technology to Textile Mill Effluent

Although anaerobic biological treatment alone does not completely degrade organic dyestuffs present in wastewater, the aromatic amines generated by anaerobic reduction may be effectively treated using aerobic biological treatment. This makes the anaerobic–aerobic sequence a plausible solution to the treatment problem.

Historically, the reduction of azo dyes as a treatment step has been contemplated by industry members with great reluctance. The intentional generation of aromatic amines that are generally more toxic than the dye itself, was not appealing from an environmental or regulatory perspective. However, the breakdown of azo dyes to their corresponding amines has been shown to accomplish two goals: (1) decolorization of the wastewater and (2) preparation of the wastewater for subsequent aerobic biological treatment. With this in mind, researchers are now evaluating the efficiency of aerobic biological treatment in degrading the anaerobic biological degradation products. Due to the potential toxicity of certain aromatic amines generated from the anaerobic process, a high efficiency of subsequent aerobic treatments must be achieved in order to prevent the release of toxic compounds into the environment.

Few anaerobic biological treatment studies have been conducted on actual textile dye wastewater. One such study was conducted on a cotton yarn and fabric dyeing and finishing plant effluent using an upflow anaerobic filter (37). The results of the study demonstrated 66–80% removal of COD at a daily loading of up to 0.8 kg COD/m^3 day. The COD and BOD_5 removals of 50% were obtained at higher loadings.

More recently, a study was conducted by Loyd and co-workers (21) to evaluate an anaerobic–aerobic treatment sequence on a textile wastewater. The textile company involved in the study dyes mostly cotton fleece for apparel and primarily uses reactive azo dyes. In this study, a solution of 75% Navy 106 wash water and sanitary primary effluent was used as influent to the system, in order to simulate the influent to the POTW receiving the textile dye wastewater. Anaerobic seed, acclimated to the Navy 106/sanitary effluent solution, was obtained from laboratory-scale, continuous-flow biological reactors (CFBRs). A preliminary experiment was conducted in which the acclimated seed and varying ratios of Navy 106/sanitary effluent solution were added to 300-mL BOD bottles, and color measurements were taken on each of seven intermittent days, including the twenty-first day. Some reappearance of color was observed upon exposure to the atmosphere. To ensure accurate color readings, the solution was exposed to oxygen for 15 min prior to color measurement.

Based on ADMI color measurements, the dyestuff degraded from 31 to 86%, depending on the concentration of dyestuff present in the solution. The more concentrated solutions gave rise to less color degradation overall and afforded a slower rate of color loss than the more dilute solutions. Though the slower rate was expected, due to a higher dye–microorganism ratio, the lower overall degradation may be attributed, according to Loyd and co-workers (21), to inhibition caused by the dye molecules or the reduction metabolites. A similar effect was observed by Kremer (36) where metabolites inhibited the reduction process, though the trend did not hold for all concentrations. Other explanations for the difference in the extent of color reduction include sensitivities of the ADMI measurement, nutrient availability, and insufficient microbial degradative capacity (21).

The same preliminary test was also conducted by Loyd and co-workers (21) on a sample of influent to the Martinsville, Virginia, POTW that receives a high percentage of textile wastewater (approximately 70–75%). Color reductions of 73 and 74% were observed in both seeded and unseeded samples. Of course, wastewater will readily become anaerobic, which may explain the color reduction achieved in the unseeded sample.

ANAEROBIC–AEROBIC SEQUENCE

Following the preliminary experiments of Loyd and co-workers (21), two laboratory-scale treatment systems were constructed to simulate (1) aerobic biological treatment and (2) anaerobic biological pretreatment followed by aerobic treatment. Anaerobic pretreatment was conducted in 26-liter Nalgene carboys. Following pretreatment, the wastewater was sent to a CFBR containing approximately 9.7 liters of liquid and divided by a baffle into two compartments. The smaller compartment served as a settling tank where biomass collected on the bottom and recycled through an opening at the bottom to the larger, aerated compartment. Among the analyses performed after each treatment step were color, TOC, BOD_5, COD, pH, and toxicity.

The anaerobic–aerobic sequence produced significantly greater color reduction than aerobic treatment alone (88 vs. 28%). In the anaerobic–aerobic sequence, 46% of the color reduction occurred during seeding and dilution with the sanitary primary effluent, 44% occurred during anaerobic treatment, and a 2% *increase* in color occurred during aerobic treatment. The aerobic treatment alone produced a 28% color reduction during dilution and seeding and a slight color increase from the aerobic treatment. The experiment was repeated with the addition of a pH adjustment step used to lower the pH to between 8 and 8.5, which is more conducive to biological activity. Using the pH-adjusted wastewater, aerobic treatment produced 30% color reduction, whereas anaerobic–aerobic treatment produced 89% color reduction. The contribution from each of the process steps was similar to that observed in the non-pH-adjusted experiment. Table 5.4 presents the results for color, COD, BOD_5, and TOC reductions from both treatment schemes. The results in Table 5.4 represent the average of results from experiment with and without pH adjustment.

In addition to the improved color reduction resulting from the anaerobic–aerobic sequence, TOC reduction (79%, without pH adjustment) also improved, when compared to aerobic treatment alone (69%). Nearly all of the TOC reduction occurred

TABLE 5.4   Percent Reductions from Laboratory-Scale Treatment Processes Involving Navy 106 (38)

Treatment Process	Treatment Step	Color	TOC	BOD$_5$	COD
Anaerobic–aerobic	Dilution	40	41	13	38
	Anaerobic	50	3	−2	1
	Aerobic	−2	38	80	38
	Total reduction	88	82	91	77
Aerobic	Dilution	28	38	15	39
	Aerobic	1	36	75	30
	Total reduction	29	74	90	69

during aerobic biological treatment (44–47%) and dilution (25–30%). A statistically insignificant ($p < 0.05$) amount of TOC reduction occurred during anaerobic treatment. The pH-adjusted experiment produced overall TOC reductions of 80 and 85% for the aerobic and anaerobic–aerobic processes, respectively, with the largest reduction again resulting from the aerobic treatment step. Overall, the anaerobic–aerobic sequence was approximately 10% more effective in reducing TOC than the aerobic treatment alone.

Similar results were observed for BOD$_5$, where 96% reduction was obtained through the aerobic treatment process, and 98% reduction was obtained from anaerobic–aerobic treatment when the influent pH was adjusted. Without pH adjustment, 83 and 84% reductions were obtained, respectively.

Analyses of COD produced trends similar to those observed for TOC and BOD$_5$. Aerobic treatment produced a COD reduction of 78% with pH adjustment and 60% without pH adjustment, while the anaerobic–aerobic sequence produced a COD reduction of 88% with pH adjustment and 63% without pH adjustment.

Following the experiments involving Navy 106, the same processes were used to treat the Martinsville POTW influent. The results are presented in Table 5.5.

Overall, the idea that anaerobic reduction of dyestuffs generates organic amines that are more amenable to aerobic biodegradation than the original dyestuffs ap-

TABLE 5.5   Percent Reductions from Laboratory-Scale Treatment Processes with Martinsville POTW (38)

Treatment Process	Treatment Step	Color	TOC
Anaerobic–aerobic	Anaerobic	39	−10
	Aerobic	20	83
	Polymer addition	28	2
	Total reduction	87	75
Aerobic	Aerobic	−13	73
	Total reduction	−13	73

pears to be supported by these results. Color reductions achieved from the anaerobic reduction step (i.e., 50% reduction for Navy 106) were significantly greater than those obtained from aerobic reduction, which were actually color increases in some cases. Little, if any, reduction in COD, BOD_5, and TOC was observed during the anaerobic step; however, anaerobic treatment did appear to enhance the ability of the aerobic system to degrade organic matter. The reduction in COD, BOD_5, and TOC was greater for the anaerobic–aerobic sequence than for aerobic treatment, by up to 8%.

Loyd and co-workers (21) also calculated the BOD_5–COD ratio at several points, as a measure of the biodegradation potential of the wastewater. The ratio was determined to be 0.19 in the raw Navy 106 wash water, 0.21–0.22 following dilution and seeding, and 0.08–0.09 following aerobic treatment. The anaerobic pretreatment step increased this ratio slightly, and subsequent aerobic treatment decreased it to below 0.10. Ratios as low as 0.20 in the raw wastewater indicate that a significant amount of organics, as measured by COD, was present in the raw wastewater that was not amenable to aerobic biodegradation.

A second, laboratory-scale experiment was designed (21) based on SBRs constructed from 2-liter Pyrex Erlenmeyer flasks. The anaerobic SBRs operated on a repetitive cycle of fill, react, settle, and withdraw. The aerobic SBRs were constructed from 4-liter Pyrex beakers and operated using the same cycle as the anaerobic reactor. The cycle times were less than 5 min for filling and withdrawing and varied from 5 min to 24 hr for reacting, and from 30 min to 1 hr for settling, depending on the detention time desired for the experiment.

A single, timed, anaerobic treatment study was conducted on Navy 106, with pH adjustment, using the SBR with a 5-min reaction time and 30-min settling time (21). During this experiment, color was reduced 75% by hour 8, and then increased slightly through the remainder of the experiment (48 hr). Based on both color and TOC reduction (approximately 20%), it appeared that the biomass floc initially adsorbed the dyestuff, but that the dyestuff was subsequently degraded and released. This was observed as a continuous darkening and lightening of the floc throughout the experiment.

Next, Loyd performed a timed anaerobic–aerobic experiment on the Martinsville POTW influent. As with the Navy 106, significant decolorization (35% reduction in color) was measured immediately at the start of the experiment ($t = 0$). These results indicated that color was adsorbed onto the bacterial floc. In addition, subsequent lightening and darkening of the floc without release of color to the water column indicated that the sorbed dye was undergoing degradation. Reduction in TOC was insignificant during the anaerobic process, and the reduction value of 59% measured at $t = 0$ of the aerobic process was attributed to dilution and sorption by the floc.

The above work (21) was followed by an evaluation of an anaerobic–aerobic pilot plant system by Boe and co-workers (39). Effluent from a primary clarifier of a local POTW receiving 75% textile dye wastewater was used as influent to the pilot plant. Two systems were evaluated: the first consisted of anaerobic reduction followed by aerobic treatment, and the second consisted of anaerobic reduction, chemical reduction using thiourea dioxide, followed by aerobic treatment. The anaerobic step had hydraulic retention times of 6 and 12 hr with 200 mg/L biomass, the aerobic system utilized a 30-day solids retention time with 2000–3000 mg/L biomass, and thiourea dioxide was added at a concentration of approximately 25 mg/L. On a batch basis, additions of polymeric flocculent and chlorine (10 mg/L) were also evaluated. Polymer and chlorine are currently used at the POTW in order to achieve their discharge limits for color and toxicity, which are 2000 ADMI units, and 100% for a 48-hr LC_{50} using *C. dubia*, respectively. This toxicity standard specified that the 48-hr LC_{50}, or concentration at which 50% of the organisms survive after 48 hr, must be 100% wastewater (i.e., undiluted wastewater).

Treatment of the wastewater using a 12-hr anaerobic retention time achieved average color reductions of 55–60%. For the total system (i.e., anaerobic plus aerobic treatment), color reductions of 65–70% were obtained, achieving an effluent of 560–590 ADMI color units. The 6-hr anaerobic retention time achieved only 15–20% color reduction, and the system eventually failed completely. The reducing agent was evaluated in order to determine whether the shorter detention time could be used and achieve the same results; however, little benefit was obtained from the addition of thiourea dioxide. Reductions of TOC of 70–75% were achieved at both 6- and 12-hr detention times, with essentially all of the organics removal occurring during the aerobic treatment step.

In order to achieve effluent color values below 200 ADMI units, as specified in the National Pollutant Discharge Elimination System (NPDES) permit, Boe and co-workers (39) determined that polymer treatment and chlorination would still be required to supplement the combined system. Addition of 60 mg/L cationic polymer achieved color reduction to 200–250 ADMI units, and chlorination (10 mg/L) further reduced the color to 160 ADMI units.

Toxicity, as measured by a general toxicity test, was significantly reduced by the addition of anaerobic pretreatment to the aerobic treatment process evaluated by Loyd (38). Toxicity testing was conducted with *C. dubia* and LC_{50} values were reported. The results of this testing, and the toxicity testing from the pilot plant evaluation (39) are provided in Tables 5.6 and 5.7, respectively.

TABLE 5.6 *Ceriodaphnia dubia* LC_{50} **Concentrations from Laboratory-Scale Anaerobic/Aerobic Treatment (38)**

Wastewater Description	Treatment Process	LC_{50} (%)
Navy 106 wash water	Dilution/seeding	7
	Anaerobic alone	7
	Aerobic alone	26
	Anaerobic–aerobic	56
POTW influent	Dilution/seeding	8
	Anaerobic alone	50
	Anaerobic–aerobic	70

TABLE 5.7 *Ceriodaphnia dubia* LC_{50} **Concentrations from Pilot Plant Aerobic Treatment (39)**

Wastewater Description[a]	Polymer Treatment	LC_{50} (%)
12-hr Anaerobic Hydraulic Retention Time		
Aerobic effluent (NR)	No	49
Aerobic effluent (NR)	Yes	22
Aerobic effluent (NR)	Yes	33
Aerobic effluent (R)	No	71
Aerobic effluent (R)	Yes	41
Aerobic effluent (R)	Yes	37
6-hr Anaerobic Hydraulic Retention Time		
Aerobic effluent (NR)	No	73
Aerobic effluent (NR)	Yes	84
Aerobic effluent (R)	No	84
Aerobic effluent (R)	Yes	77

[a]NR: not reduced; R: reduced (using 25 mg/L thiourea dioxide).

Ceriodaphnia dubia is now commonly used as an indicator of aquatic toxicity. *Ceriodaphnia dubia* is known to be a sensitive organism and is used, in some areas, by regulatory agencies in setting limits for NPDES discharge permits. Because this type of test is specific to one type of organism, conclusions about overall toxicity to human health and the environment can be tenuous. Prior to initiating a pilot-scale process, it would be important to analyze and determine the concentration of expected inorganic and organic constituents, because even a low concentration of toxic compounds may bioaccumulate and pose an environmental risk.

Results from the study by Boe and co-workers (39) indicate that the toxicity of the pilot plant effluent from both the 6- and 12-hr conditions was reduced to an LC_{50} greater than 100%, as measured by the fathead minnow toxicity test. However, effluent LC_{50} values less than or equal to 84% for *C. dubia* were obtained, which do not meet NPDES permit limitations.

Degradation Mechanisms

Research has also been conducted (21) to define the degradation mechanism occurring during the anaerobic phase. Oxidation-reduction potentials (ORPs) and color reductions were measured on autoclaved and nonautoclaved samples. The results of the experiment indicated that the presence of viable microbes was necessary for anaerobic degradation of dyestuffs. This conclusion was reached based on the following two observations: (1) the autoclaved samples exhibited lower color reductions than the nonautoclaved samples and (2) the ORP values for the non-autoclaved samples were much lower than those of the autoclaved samples, indicating the presence of facultative organisms and/or anaerobes. Other observations that led to this conclusion include the increase in decolorization over time (consistent with an increased number of microorganisms), and the continued decolorization after repetitive seeding using wastewater from the previous experimental batch as feed. If the reduction were chemical or enzymatic in nature, repetitive seeding would cause dilution of the reducing agent and slow decolorization. Instead, the microorganisms increased in number, replenished the diluted batch, and continued to achieve decolorization. Lastly, the anaerobic seed was observed under a microscope next to the Navy 106 raw wastewater and was found to contain much higher numbers and varieties of organisms. The exact mechanism of reduction, other than the requirement that organisms be present, was not investigated in this study. It was concluded, however, from the timed, anaerobic experiments, that sorption of dye to the floc preceded dye degradation.

Other researchers have concluded that the anaerobic degradation of azo dyes may be abiotic in nature and that sorption to sludge or sediment may be the primary removal mechanism (34). Weber and Wolfe (34) identified sorption as the first step in the anaerobic degradation of dyestuffs and developed a kinetic model based on abiotic anaerobic degradation of soluble azo dyes in sediment–water systems where transport to the sediment site is the rate-determining step. It may be, as with aerobic biological degradation of dyestuffs, that the degradation mechanism is dependent on the specific dye.

Anaerobic Treatment Overview

The results obtained from the studies presented in this section indicate that anaerobic–aerobic treatment of textile dye wastewater shows significant improvement over aerobic treatment alone. Anaerobic treatment achieves much greater color reductions than aerobic treatment and appears to improve organics removal as well. The pilot plant study (39) achieved color reductions of 65–70% and TOC reductions of 70–75% using an anaerobic system with only 200 mg/L biomass. The authors suggest that increased biomass could significantly improve the effluent color and organics concentrations.

INHIBITORY FACTORS

This section presents a number of factors that have been identified in laboratory and field experiments that are inhibitory to the biological treatment of dye-containing wastewater. For example, microorganisms that are not acclimated to the substrate are likely to inhibit the biodegradation process. In some cases, acclimation may be a requirement. As mentioned earlier, textile mills may not have the luxury of acclimating and maintaining the system to treat specific components of the waste, due to the batch nature of the dyeing process and the large number of dyes used in the dyeing operations. Kremer (36) suggested future research into the use of packed-bed reactors, which would provide an increased surface area for biomass contact and allow for longer growth times of the microorganism population. As mentioned earlier in the chapter, aerobic biofilms are also being investigated as a way to provide longer retention times (33).

Several studies indicate that sulfonic acid groups, which are widely used to increase the water solubility of dyes, inhibit aerobic biological degradation of dye wastewater (25,40). Other studies have shown that nitrites, nitrates, and sulfonic acid groups inhibit azo dye reduction by facultative or obligate aerobes (41) and that dyes themselves are often inhibitory to the activated sludge process (3,42). It has also been reported that the concentration of dyes found in wastewater is not sufficient to be toxic to the microorganisms unless toxic metals are also present (43).

Brown and Laboureur (35) studied four solutions of commercial anthraquinone dyes in distilled water, using an anaerobic sludge inoculum, and found that three of the four were significantly degraded. In this study, Acid Blue 80 was only slightly degraded, though the source of inhibition was not identified. Additional results, which were later confirmed by Brown and Hamburger (16), indicated that one anthraquinone dye, Acid Blue 25, may be structurally altered under anaerobic conditions to produce an insoluble anthraquinone pigment. Thus, the types of dyes and substituent groups found in the wastewater greatly influence the treatment selection. Dyes or other compounds that are inhibitory to biological treatment may be pretreated or segregated and treated in a separate treatment process. In some cases, a high concentration of inhibitory compounds may preclude the use of biological treatment.

Several researchers have investigated rate-determining factors in the anaerobic reduction of azo dyes. The results of one study suggest that the optimum pH range for the azo reductase enzyme is 5–8 and that the reduction rate increases exponentially as the pH is lowered within that range (44). In addition, transport of the dye through the cell wall may be a rate-determining factor. Increased cell age and reduced food–microorganism ratio increase cell permeability and improve the rate of dye reduction (45,46). Other researchers (47–50) have studied the influence of enzyme-generated flavins, and cofactors $NADP^+$, FAD^+, and glucose-6-phosphate on the rate of azo bond cleavage initiated by the azo reductase enzyme. The positions and types of substituent groups (e.g., amino, hydroxy, methyl, and sulfonic acid) and their relative positions on the aromatic rings of the dye molecule affect the availability of the substrate and, consequently, the rate of biological anaerobic reduction (44,51). Studies such as these are important in establishing laboratory conditions for optimum biodegradation and for identifying factors that may influence the degree and rate of biodegradation. For more details on these and other studies on factors inhibitory to the anaerobic biodegradation of azo dyes, refer to the literature review published by Kremer (36).

CONCLUSIONS

Aerobic biological degradation, the traditional method of treating textile dye waste-water, is increasingly being viewed as inadequate in meeting today's standards for acceptable effluent. The most obvious improvement is needed in the area of color reduction. Consequently, many creative approaches using physicochemical methods have been developed to address this problem. As toxicity standards become more common and stringent, the development of technological systems for minimizing the concentration of dyes and their breakdown products in wastewater also becomes necessary. Studies conducted on the aerobic biological degradation of dyes have revealed that dyes are slow to degrade using this method, and that a nonbiodegradable fraction remains in the dye-containing wastewater following treatment. This position is supported by results from several studies on BOD_5 versus the ultimate biodegradability of the waste, which indicate that (1) the BOD_5 for textile dye wastewater is a relatively small fraction of the 30-day BOD and (2) textile dye degradation is much slower than the degradation of biodegradable wastes found in domestic wastewater. Accounts of aerobic biological treatment systems achieving large reductions in BOD_5, while significant amounts of TOC and COD remain in the waste, is explained by the fraction of organics present in the waste stream that is either nonbiodegradable or difficult to biodegrade, and, consequently, is not measured by BOD_5. In addition, through biological processes, difficult to degrade COD or TOC are often generated as by-products of metabolism. Adding this new portion of COD or TOC to the already significant fraction of textile dye wastewater that is nonbiodegradable or difficult to biodegrade results in a significant organic fraction that is not detected by BOD_5.

The research results presented in this chapter concerning chemical pretreatment and anaerobic biological treatment of textile dye wastewater are promising. Chemical coagulants, powdered activated carbon, and oxidizing and reducing agents have demonstrated the capacity to effect significant color and organics removals. However, the required dosages are often too great to be economically feasible. Several researchers were able to lower the dosage requirements by selecting a specific combination of chemical additives. Optimized systems for specific wastewater, therefore, may accomplish the desired treatment at a reasonable cost.

Anaerobic biological treatment studies on textile wastewater have also yielded encouraging results. Significant color reductions and breakdown of the azo dye molecule have been demonstrated under anaerobic conditions. In addition, less biomass was needed in anaerobic–aerobic systems than standard aerobic biological systems, to achieve comparable reductions in color and organics.

As stated in the beginning of this chapter, the large number of dyes used commercially, combined with the complexity of textile dye wastewater, appears to make the development of a universal treatment system for all textile facilities impossible. It is necessary to evaluate the effectiveness of treatment systems at individual textile mills on a case-by-case basis, in order to address the influence of specific dye classes and auxiliary chemicals present at a particular plant. However, as treatment facilities are successfully implemented, there will undoubtedly be some common threads of information transferred that will facilitate a more rational design of the systems.

REFERENCES

1. J. Park and J. Shore, *JSDC* **100**, 383 (1984).
2. L. E. Shriver and R. R. Dague, *Am. Dyest. Reptr.* **67**, 34 (1978).
3. R. H. Horning, Crompton and Knowles Corporation for the American Dye Manufacturers Institute (ADMI), *Textile Chem. Color.* **9**(3), 24 (1977).
4. R. W. Boe, Pilot-Scale Study on Anaerobic/Aerobic Treatment of a Textile Dye Wastewater, thesis submitted in partial fulfillment of the M.S. Degree, VPI&SU, October, 1993.
5. W. Hatton and A. M. Simpson, *Environ. Tech. Lett.* **7**, 413 (1986); as quoted in Reference 4.
6. T. A. Alspaugh, *Text. Chem. Color.* **5**(11), 44 (1973).
7. S. I. Abo-Elela, F. A. El-Gohary, H. I. Ali, and R. A. Wahaab, *Environ. Tech. Lett.* **9**, 101 (1988).
8. R. Casanova and M. Jaggli, *Textilveredlung*, **11**, 532 (1976); as quoted in Reference 1.
9. M. L. Shelley, C. W. Randall, and P. H. King, *J. WPCF*, **48**(4), 753 (1976); as quoted in Reference 4.
10. C. P. C. Poon and P. P. Virgadamo, Anaerobic-Aerobic Treatment of Textile Wastes with Activated Carbon, prepared for the U.S. EPA Office of Research and Monitoring, Washington, D.C., EPA-R2-73-248, May, 1973.
11. J. Davis, *Am. Dyest. Reptr.* **80**(3), 19 (1991); as quoted in Reference 4.
12. D. L. Michelsen, L. L. Fulk, R. M. Woodby, and G. D. Boardman, Adsorptive and Chemical Pretreatment of Reactive Dye Discharges, ACS Symposium Series *Emerging Technologies in Hazardous Waste Management, Vol. III*, Washington, D.C., 1992.
13. M. M. Ghosh, F. E. Woodard, O. J. Sproul, P. B. Knowlton, and P. D. Guertin, *J. WPCF*, **50**, 1776 (1978).
14. W. W. Powell, D. L. Michelsen, G. D. Boardman, A. H. Dietrich, and R. M. Woodby, Removal of Color and TOC From Segregated Dye Discharges Using Ozone and Fenton's Reagent, preprinted and presented at Chemical Oxidation Technologies for the Nineties Second International Symposium, Feb., 1992.
15. W. W. Powell, The Removal of Color and DOC from Segregated Dye Waste Streams Using Ozone and Fenton's Reagent Followed by Biotreatment, thesis submitted in partial fulfillment of the M.S. Degree, VPI&SU, February, 1992.
16. D. Brown and B. Hamburger, *Chemosphere* **16**(7), 1539 (1987).
17. A. Reife, Waste Treatment of Soluble Azo, Acid, Direct, and Reactive Dyes Using a Sodium Hydrosulfite Reduction Treatment Followed by Carbon Adsorption, AATCC 1990 International Conference, Boston, October 1–3, Book of Papers, AATCC, Research Triangle, NC, 1991, as quoted in Reference 12.
18. M. W. McCurdy, G. D. Boardman, D. L. Michelsen, and R. M. Woodby, Chemical Reduction and Oxidation Combined with Biodegradation for the Treatment of a Textile Dye Wastewater, Proceedings of the 46th Purdue Industrial Waste Conference, Purdue University, W. Lafayette, Indiana, May, 1991.

19. M. W. McCurdy, Chemical Reduction and Oxidation Combined with Biodegradation for the Treatment of a Textile Dye Wastewater, thesis submitted in partial fulfillment of the M.S. Degree, VPI&SU, May, 1991.

20. K. L. Rhinehart, Jr., *Oxidation and Reduction of Organic Compounds*, Prentice-Hall, Englewood Cliffs, NJ, 1973, p. 148, as quoted in Reference 19.

21. K. C. Loyd, G. D. Boardman, and D. L. Michelsen, Anaerobic/Aerobic Treatment of a Textile Dye Wastewater, preprinted for the Mid-Atlantic Industrial Waste Conference, Morgantown, WV, July 15–17, 1992.

22. V. Price, III, Continuous Color Removal From Concentrated Dye-Waste Discharges Using Reducing and Oxidizing Chemicals—A Pilot Plant Study, thesis submitted in partial fulfillment of the M.S. Degree, VPI&SU, 1993.

23. V. Price, III, D. L. Michelsen, J. W. Balko, and R. T. Chan, Continuous Color Removal From Concentrated Dye Waste Discharges Using Reducing and Oxidizing Chemicals—A Pilot Plant Study, preprinted and presented at Chemical Oxidation Technology for the Nineties, Fourth International Conference, February, 1994.

24. J. J. Porter and E. H. Snider, *J. WPCF*, **91**(9), 40 (1976).

25. G. M. Shaul, C. R. Dempsey, and K. A. Dostal, Fate of Water Soluble Azo Dyes in the Activated Sludge Process, U.S. EPA Water Engineering Research Laboratory, Cincinnati, Ohio, August, 1987.

26. M. Dohanyos, V. Medera, and M. Sedlacek, *Prog. Wat. Tech.* **10**, (5–6), 559 (1978).

27. H. R. Hitz, W. Huber, and R. H. Reed, *JSDC*, **94**, 71 (1978).

28. W. C. Tincher, Survey of the Coosa Basin for Organic Contaminants from Carpet Processing, Contract No. 3-27-630 for the Environmental Protection Division, Department of Natural Resources, State of Georgia, 1978; as quoted in Reference 25.

29. K. Wuhrmann, K. Mechsner, and T. Kappeler, *Europ. J. Appl. Microbiol. Biotech.*, **9**, 325 (1980); as quoted in Reference 25.

30. D. A. Gardner, T. J. Holdsworth, G. M. Shaul, K. A. Dostal, and D. L. Betowski, Aerobic and Anaerobic Treatment of C.I. Disperse Blue 79, prepared by Radian Corporation under EPA Contract No. 68-03-3371 for the U.S. EPA Office of Research and Development, Cincinnati, Ohio. Document No. EPA/600/2-89/051A&B, 1989.

31. H. D. Pratt, Jr., A Study of the Degradation of Some Azo Disperse Dyes in Waste Disposal Systems, Georgia Institute of Technology, Master of Science Thesis, September, 1968.

32. J. S. Eberhard, S. T. Giolando, and M. W. Tabor, Biotic and Abiotic Degradation of Azo Dyes in Aerobic Bioreactors, preprinted extended abstract, presented before the Division of Environmental Chemistry, American Chemical Society, August 27–31, 1990.

33. P. L. Bishop, University of Cincinnati, Department of Civil and Environmental Engineering, presentation to the Ecological and Toxicological Association of the Dyestuffs Manufacturing Industry (ETAD), Pittsburgh, Pennsylvania, July, 1993.

34. E. J. Weber and N. L. Wolfe, *Environ. Sci. Tech.* **6**, 911 (1987).

35. D. Brown and P. Laboureur, *Chemosphere* **12**, (1–6), 397 (1983).

36. F. V. Kremer, Anaerobic Degradation of Monoazo Dyes, University of Cincinnati Department of Civil and Environmental Engineering, Dissertation, 1987.

37. N. Athanasopoulos and T. Karadimitris, *Biotechn. Lett.* **10**(6), 443 (1990); as quoted in Reference 38.

38. K. C. Loyd, Anaerobic/Aerobic Degradation of a Textile Dye Wastewater, thesis submitted in partial fulfillment of the M.S. Degree, VPI&SU, March, 1992.

39. R. W. Boe, G. D. Boardman, A. M. Dietrich, D. L. Michelsen, and M. Pudaki, Pilot Scale Study on Anaerobic Treatment of a Textile Wastewater, prepared for the 25th Mid-Atlantic Industrial and Hazardous Waste Conference, University of Maryland, College Park, July, 1993.

40. Y. Yonezawa and Y. Urushigawa, *Bull. Environ. Contam. Toxicol.* **17**(2), 208 (1977); as quoted in Reference 36.

41. K. Wuhrmann, K. Mechsner, and T. Kappeler, *Europ. J. Appl. Microbiol. Biotech.*, **9**, 325 (1980); as quoted in Reference 38.

42. A. Netzer and H. K. Miyamoto, The Biotreatability of Industrial Dye Wastes Before and After Ozonation and Hypochlorination-Dechlorination, Proceedings 30th Purdue Industrial Waste Conference, Purdue University, W. Lafayette, IN, 1975, pp. 804–814; as quoted in Reference 38.

43. D. K. Gardiner and B. J. Borne, *JSDC* **94**, 339 (1978).

44. T. Zimmerman, F. Glasser, H. G. Kulla, and T. Leismer, *Arch. Microbio.* **138**, 37 (1984); as quoted in Reference 36.

45. J. J. Roxon, A. J. Ryan, and S. E. Wright, *Food Cosmet. Toxicol.* **5** 645 (1967); as quoted in Reference 36.

46. King-Thom Chung, G. Fulk, and M. Egan, *Appl. Environ., Microbio.* **35**(3), 558 (1978); as quoted in Reference 36.

47. R. Gingell and R. Walker, *Xenobiotica.* **1**(3), 231 (1971); as quoted in Reference 36.

48. J. R. Fouts, J. J. Kamm, and B. B. Bordie, *J. Pharmacol. Exp. Therap.* **120**, 291 (1957); as quoted in Reference 36.

49. A. K. Khan, G. Guthrie, H. H. Johnston, S. C. Trulove, and D. H. Williamson, *Clin. Sci.*, **64**(3), 349 (1983); as quoted in Reference 36.

50. M. Huang, G. Miwa, N. Cronheim, and A. Lu, *J. Biol. Chem.* **254**(22), 11223 (1979); as quoted in Reference 36.

51. R. Walker and J. Ryan, Some Molecular Parameters Influencing Rate of Reduction of Azo Compounds by Intestinal Microflora, in J. W. Bridges and D. V. Parke, eds., *Biological Oxidation of Nitrogen for Organic Molecules*, Halsted Press, New York; as quoted in Reference 36.

CHAPTER 6

THE PACT® SYSTEM FOR WASTEWATER TREATMENT

DAVID G. HUTTON, JOHN A. MEIDL, and G. J. O'BRIEN

D. G. Hutton, Inc., Newark, Delaware
Zimpro Environmental, Inc., Rothschild, Wisconsin
E. I. du Pont de Nemours and Company, Deepwater, New Jersey

INTRODUCTION

Biological treatment, whether by the activated sludge process, trickling filters, or other process, is the standard method for treating dilute wastewaters in both municipalities and industry. Although biological treatment can treat many of the solvents or contaminants in a dye manufacturing, or dye use operation, most dyes are made to be nonbiodegradable and are, therefore, not well treated by conventional biological treatment processes. In addition, some relatively nonbiodegradable priority pollutants are found in waste streams containing dyes. For these reasons, an improvement in the biological treatment is needed for wastes containing dyes.

History

The combined powdered activated carbon–activated sludge system (PACT® system)* is a classic case of ''Necessity is the mother of invention.'' In the 1960s, there was a problem of fish kills in the Delaware River caused by low dissolved oxygen concentrations in the summer. The regulatory body for the Delaware River, the Delaware River Basin Commission, concluded from a river model study that all dischargers to the Delaware River would have to reduce their biochemical oxygen demand (BOD_5) load to the river by 85%. One of the facilities affected by the 85% BOD_5 removal regulation was the Chambers Works of E. I. du Pont de Nemours and Company. Chambers Works personnel evaluated wastewater treatment alternatives and concluded that the best alternative was biological treatment of a concentrated portion of their waste stream; so efforts were made to segregate

*Registered trademark of Zimpro Environmental, Inc.

Environmental Chemistry of Dyes and Pigments, Edited by Abraham Reife and Harold S. Freeman.
ISBN 0-471-58927-6   © 1996 John Wiley & Sons, Inc.

noncontact cooling water from process water. At that time, dye wastes from the large synthetic dye manufacturing facility at Chambers Works contributed perhaps half of the waste load at the plant. A laboratory activated sludge system was installed to develop design parameters for the envisioned activated sludge plant; but, months of operation led to frustration when the 85% BOD removal goal could not be met. Alternatives to activated sludge treatment were examined, all without success in attaining the 85% BOD removal requirement. A pair of engineers conceived the idea of adding powdered activated carbon to the activated sludge system. An experiment was tried in which 400 ppm of Darco® KB powdered activated carbon (Atlas Chemical Co., Wilmington, DE) was added to the laboratory activated sludge testing unit. Within a week the effects of the carbon addition were dramatically observed, as the BOD removal increased from 79 to 96% (Table 6.1) (1).

Concerns were raised about the proposed combined activated sludge–activated carbon process because conventional wisdom at the time was that bacteria would grow on the surface of the activated carbon and plug the pores, thereby rendering it ineffective; a concern that proved to be invalid.

General Process Description

The PACT system is a process for the treatment of organic materials in wastewater streams. Zimpro Environmental, Inc., who purchased the patent rights to the process from du Pont and expanded the technology, believes that the process fits into the overall scheme of technologies used for treating organics, as indicated in Figure 6.1 (2). The PACT system is used to treat aqueous waste streams that are too dilute for oxidation technologies, such as wet air oxidation or incineration, and for waste streams that are too strong to be economically treated with granular activated carbon. Thus, the treatable aqueous waste stream strength for the PACT system is between about 50 and 50,000 mg chemical oxygen demand (COD) per liter of wastewater (2).

The original flow sheet for the PACT system is illustrated in Figure 6.2, where it is seen that the PACT system is simply the addition of powdered activated carbon

TABLE 6.1   Effect of Adding Carbon to the Activated Sludge Process

	With Carbon	Without Carbon
Carbon dose, ppm	400	0
Influent BOD (unfiltered), mg/L	168	168
Effluent BOD (unfiltered), mg/L	5.4	73
BOD removal, %	96	79
Influent COD (unfiltered), mg/L	381	381
Effluent COD (unfiltered), mg/L	63	182
COD removal, %	86	56
Influent color, APHA units	800	800
Effluent color, APHA units	40	800
Color removal, %	95	0

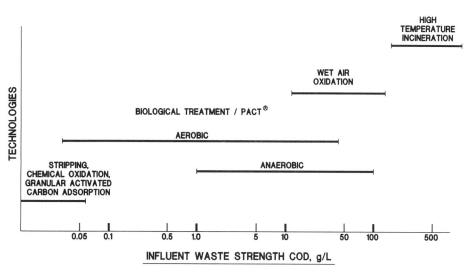

FIGURE 6.1 Industrial waste treatment technologies: general systems application.

to the activated sludge process. Depending on waste characteristics, the carbon dosage may range from 10 to over 5000 mg of powdered carbon added per liter of wastewater flow (10–5000 ppm). The powdered activated carbon may be added to the waste stream before treatment, to the recycle sludge, or to the aeration tank itself, as the point of addition generally makes little difference in overall results. As the carbon is recycled in the system, the mixed liquor carbon levels in the

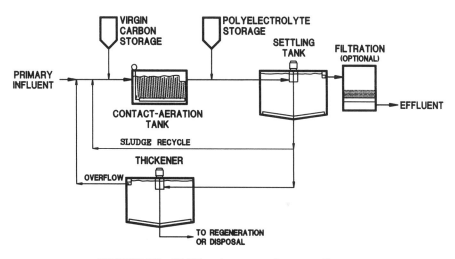

FIGURE 6.2 PACT system: general process diagram.

aerator will range from approximately 10 mg/L to as high as 50,000 mg/L, but usually between 1000 and 10,000 mg/L. Polyelectrolyte is often, but not always, added to the overflow from the aeration tank to aid coagulation of fine particles in the clarifier where the solids are separated out of the treated water. The clarifier overflow, depending on requirements, can go to additional treatment such as sand filtration, as illustrated in Figure 6.2. Underflow clarifier solids are pumped, in concentrations ranging from 3 to 5% solids, back to the inlet of the aeration tank. Excess solids are usually removed from the PACT system by withdrawal of a small stream directly from the sludge recycle stream. Waste solids, being a mixture of powdered activated carbon, activated sludge, adsorbed organic material, and inert material, are then directed to solids processing, illustrated in Figure 6.2 as a sludge thickener. Thereafter, solids handling can take the form of dewatering in a filter press or carbon regeneration. If carbon regeneration is employed, recovered carbon will go directly back to the aeration tank. In installations where regeneration is employed in the PACT system, ash removal systems have been included to prevent ash buildup in the mixed liquor.

Process Variations

Two-Stage Process   Since the origination of the PACT system, many variations of the basic process in Figure 6.2 have come to be regarded as part of the technology. The Chambers Works in Deepwater, New Jersey, has upgraded its original single-stage process to the two-stage process illustrated in Figure 6.3. This was done in 1991, mainly to improve the removal of priority pollutants, as specified in their discharge permit. The two-stage process works very much like two single-stage processes in series, but with options of where (1) the virgin carbon is added and (2) the waste sludge is removed. As illustrated in Figure 6.3, some virgin carbon is added to both the first and second stages. Waste sludge from the second stage is cycled back to the first stage so that the carbon used to treat the dilute effluent can be fully saturated before it is discarded, and waste sludge from the first stage is sent for carbon regeneration or landfill.

As shown in Table 6.2 from pilot testing results (3), the addition of the second stage did improve the priority pollutant removal. The full-scale second-stage PACT unit has been added to the 40-million-gallon per day (mgd) Chambers Works wastewater treatment plant and has performed extremely well in removing priority pollutants.

Treating a pesticide waste using a two-stage aerobic PACT system having a specially designed "concentrating aerator" but lacking the intermediate clarifier is illustrated in Figure 6.4. The performance of the two-stage system for removing pesticides was shown to be much better than that for the one-stage system (Table 6.3).

Batch Process   A batch aerobic PACT system was developed by Zimpro Environmental to handle relatively small (less than 150,000 gpd) wastewater flows

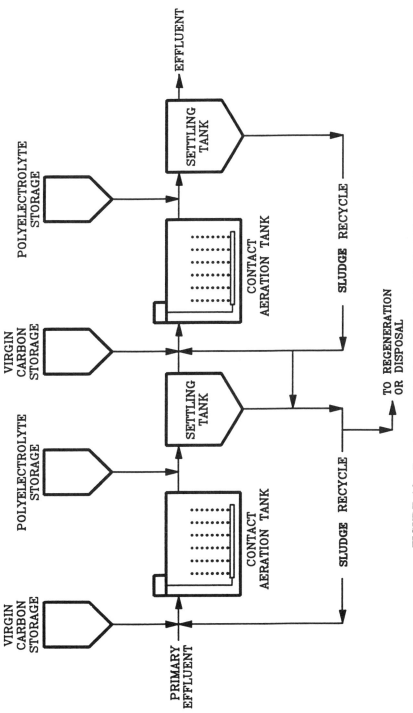

FIGURE 6.3 Two-stage PACT system as used at the DuPont Chambers Works.

(Figure 6.5). In this system the same tank is used to aerate and to settle, with operation "sequencing" as shown in Figure 6.6. Wastewater is first pumped into an aeration tank where it comes in contact with a mixture of biological solids and powdered activated carbon, and the flow is stopped. The contents of the aeration tank are then aerated for a time appropriate for the waste being treated. During aeration, the biodegradable portion of the waste degrades, while the nonbiodegradable components are adsorbed on the activated carbon particles. After aeration ceases, the aeration tank contents are allowed to settle, so that the solids fall to the bottom of the aeration tank. Then the clarified, treated portion of the aeration tank contents is drawn off for discharge to the receiving stream. The solids are retained in the aeration tank for use in the next batch of wastewater. Periodically, a portion of the aeration tank solids is removed as waste sludge, and fresh activated carbon is added as needed.

TABLE 6.2   Comparison of Single- and Two-Stage PACT for Priority Pollutant Removal at DuPont's Chambers Works

Compound	Primary Effluent Feed	Single-Stage Effluent	First-Stage Effluent	Second-Stage Effluent
Volatiles (μg/L)				
Chloroethane	2200	60	70	ND
Methylene chloride	730	72	180	11
Chloroform	260	16	20	ND
1,2-Dichloroethane	110	12	57	ND
Tetrachloroethane	80	7	5	ND
Toluene	3200	ND[a]	ND	ND
Chlorobenzene	350	ND	ND	ND
Ethylbenzene	73	ND	ND	ND
Benzene	500	ND	ND	ND
Trichloroethane	51	6	ND	ND
Acid extractables (μg/L)				
Phenol	1600	ND	ND	ND
2-Nitrophenol	720	ND	ND	ND
2-Methyl-4,6-dinitrophenol	313	ND	ND	ND
Base-neutral extractables (μ/L)				
N-Nitrosodimethylamine	230	ND	ND	ND
1,2-Dichlorobenzene	2400	30	10	ND
Nitrobenzene	3200	ND	ND	ND
2,6-Dinitrotoluene	350	60	60	40
2,4-Dinitrotoluene	1310	ND	ND	ND
bis(2-Ethylhexyl) phthalate	380	30	10	10

[a]ND = not detected.

Reprinted by permission of Lewis Publishers, a subsidiary of CRC Press, Boca Raton, Florida, from 44th *Annual Purdue Industrial Waste Conference Proceedings*, Table I, p. 330.

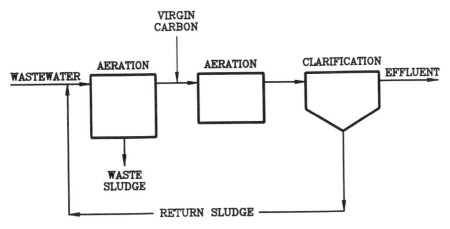

FIGURE 6.4 Two-stage PACT system without intermediate clarification, used at a pesticide manufacturer.

Other Processes

Anaerobic Adams (4) showed that the addition of powdered activated carbon to an anaerobic digester could stimulate methane production, thicken the sludge, improve sludge filterability, and improve supernatant clarity. Furthermore, Meidl and Vollstedt (5) found that powdered carbon stabilized an upflow anaerobic filter unit that was not performing when operated as a straight biological process.

Since then, Zimpro Environmental has developed a hybrid anaerobic PACT reactor for treating high-strength (10,000–20,000 mg/L COD) wastes (6). The treatment reactor consists of an upper filter zone and a lower suspended growth zone contained in a single vessel (Figure 6.7). This system operates in an upflow mode in which feed is to the bottom suspended growth zone, where most of the biological conversion takes place. As the waste stream passes through the upper filter zone, most of the solids are trapped and fall back into the lower suspended growth zone. All solids, whether inert solids from the feed or biosolids created in the reactor, are removed through a controlled removal of solids from the bottom of the reactor.

TABLE 6.3 Comparative Treatment of Pesticide Wastewater by Single-Stage and Two-Stage PACT Systems

	Single-Stage	Two-Stage
Carbon dose, ppm	1,480	600
Influent total pesticides, ppb	39,045	39,045
Effluent total pesticides, ppb	1,290	11

FIGURE 6.5 Batch-operated PACT system.

Pure Oxygen The PACT system is compatible with pure oxygen aeration systems; in fact, these processes may be complimentary. The two would be most useful together in cases where a high-strength waste (favoring oxygen) also contains nonbiodegradable materials or color. Other applications include problems where (1) space is limited or (2) odor cannot be controlled by carbon addition (a situation favoring covered aerators).

PACT System Vs. Activated Sludge Plus Carbon Columns

One alternative to the PACT system for treating dye wastes is to use granular activated carbon columns in conjunction with biological treatment, usually activated sludge. The carbon columns can be used either in front of or behind the biological treatment. Several investigators who evaluated activated sludge plus carbon columns versus the PACT system concluded that the PACT system was more effective (3,7–9). One problem is the *chromatographic effect* that is observed with granular carbon columns, wherein the effluent concentration of a compound can actually be higher than that fed to the carbon column. The chromatographic effect occurs after a contaminant is adsorbed on the carbon and is later displaced from the carbon column by a more strongly held contaminant. The chromatographic effect is of great concern in that it jeopardizes permit compliance. An influent

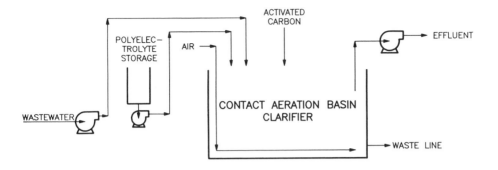

OPERATING SEQUENCE

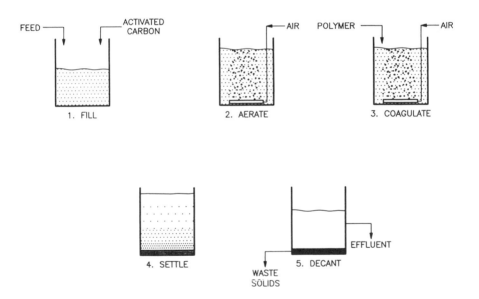

FIGURE 6.6 Batch PACT system diagram.

contaminant may be initially adsorbed to a degree that meets permit conditions, only to later be detected at a concentration causing a permit violation, because the contaminant is displaced from the carbon column by a more strongly adsorbed contaminant. This was observed by DiGiano (10) in pure compound studies and by Hutton (8) and O'Brien (3) using an actual plant effluent.

Other problems found when using granular activated carbon downstream of biological treatment are:

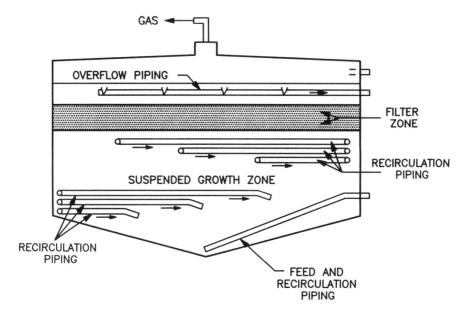

FIGURE 6.7 Multizone anaerobic reactor.

- Affords no toxicity protection to active biomass.
- Generates a hazardous waste carbon stream [waste carbon will not pass Toxicity Characteristic Leaching Procedure (TCLP)].
- Volatile organic carbon (VOC) stripping will occur in the biological treatment.
- Carbon columns are susceptible to plugging by suspended solids from the activated sludge process and microbial growth. Sand filters are required between the activated sludge process and the carbon columns, which add additional investment and complexity.

When granular carbon is used upstream of biological treatment:

- It removes biodegradable materials that are more efficiently treated by a biological process.
- Requires far more carbon than when used downstream.
- Generally requires upstream filtration.
- Generates a hazardous waste carbon stream (waste carbon will not pass TCLP).

A major advantage of the PACT system over the combination of activated sludge and carbon columns is the reduced investment cost. Even in retrofit situations, an activated sludge plant can be converted to a PACT system for considerably less cost than the installation of carbon columns and associated equipment. In addition, the PACT system is cheaper to operate than the comparable granular carbon system (3,7,11–12).

PACT SYSTEM PERFORMANCE

The advantage of the aerobic PACT system over the activated sludge process are improvements in:

- Organics removal (both conventional and nonconventional material, including priority pollutants)
- Volatile organic chemical control (odors)
- Color removal
- Metals removal
- Nitrification
- Process stability
- Resistance to shock loads
- Hydraulic residence time
- Low-temperature stability
- Sludge settleability and thickening
- Sludge handling and disposal
- Effluent toxicity
- Foam control
- Other
    - Protection from toxic upsets
    - Compatible with activated sludge processes
    - Increased operational flexibility
    - Carbon can be regenerated

An example of each of these advantages to using the PACT system is now discussed.

Organics Removal (Both Conventional and Nonconventional Material, Including Priority Pollutants)

Organic Chemical Wastewater   Even the earliest experiments with what is now the PACT system (Figure 6.2) showed the benefits of adding powdered activated carbon to an activated sludge process treating a complex organic chemical process wastewater (1). Dramatic improvements in the quality of a raw waste stream containing 168 mg/L of BOD and 381 mg/L COD was observed with the PACT system versus activated sludge (Table 6.1).

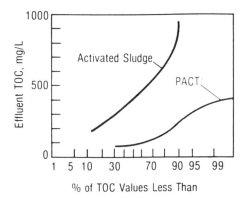

FIGURE 6.8   Effect of powdered carbon on effluent total organic carbon. (Reprinted by permission from *Hydrocarbon Processing*, October 1975, p. 107, copyright 1975 by Gulf Publishing Co., all rights reserved.)

Many nonbiodegradable or difficult to degrade molecules, such as substituted aromatics, are readily adsorbed on activated carbon. As a result, the PACT system can consistently produce better effluent quality than activated sludge in terms of total organic carbon for waste streams containing significant amounts of nonbio-degradable organics. Results shown in Figure 6.8 are typical of the performance that has been seen (13).

In a laboratory comparison made at DuPont's Chambers Works, a single batch of raw chemical plant wastewater was used to determine the effectiveness of the PACT and activated sludge processes for priority pollutant removal. All units operated in parallel, with an 8-hr hydraulic detention time and 10-day-old sludge. Carbon was used at two dosages, 100 and 300 ppm. For the base-neutral and acid-extractable organic compounds such as dinitrotoluene and chlorinated phenols, the PACT system gave significantly better results than activated sludge. Effluent concentration was generally inversely related to carbon dose (Table 6.4). Though effluent quality for volatiles appears to be similar, it must be pointed out that stripping would have been a factor in the activated sludge unit's performance; however, it would have been much less of a factor in the PACT system (see discussion on VOC control later in this chapter).

Full-scale Chambers Works plant data confirmed PACT's ability to control priority pollutant organics. Table 6.5 shows the priority pollutant removal, arranged from highest to the least, when operating at the conditions shown in Table 6.6. Of 36 organics tabulated, 30 are removed at levels higher than average soluble total organic carbon (TOC). Volatiles are generally removed well by PACT. The acid extractables are generally well removed, but phenolics removability decreases with increasing ring chlorination or substitution. For example, nitrobenzene is removed better than dinitrobenzene; dinitrobenzene is removed better than trinitrobenzene,

TABLE 6.4 Comparison of PACT and Activated Sludge for Organics Removal

Compound	Activated Sludge			PACT, 100 ppm[a]		PACT, 300 ppm[a]	
	Feed (µg/L)	Eff. Conc. (µg/L)	Removal (%)	Eff. Conc. (µg/L)	Removal (%)	Eff. Conc. (µg/L)	Removal (%)
Volatiles							
Benzene	81	1.2	98.5	0.3	99.6	0.3	99.6
Chlorobenzene	3660	34	99.1	12	99.7	8.2	99.8
Chloroethane	667	1.3	99.8	<0.6	>99.9	<0.7	>99.9
Chloroform	72	2.4	96.7	2.0	97.2	2.2	96.9
Methylene chloride	138	2.1	98.5	<0.4	>99.7	<0.4	>99.7
Tetrachloroethylene	33	<0.1	>99.5	<0.1	>99.5	<0.1	>99.5
Base-neutral extractables							
1,2-Dichlorobenzene	18	1.7	90.6	0.4	97.8	<0.1	>99
2,6-Dinitrotoluene	1000	690	31	400	60	100	90
2,4-Dinitrotoluene	1000	950	14	70	94	53	95
Nitrobenzene	330	18	94.5	1.7	99.5	<0.1	>99.9
1,2,4-Trichlorobenzene	210	<0.1	>99.9	<0.1	>99.9	<0.1	>99.9
Acid extractables							
2,4-Dichlorophenol	19	22	<0	3.1	84	1.3	93
2,4-Dinitrophenol	140	86	39	3.9	97	<0.1	>99
4-Nitrophenol	1100	830	25	100	91	29	97

[a]Carbon dose expressed as ppm, parts carbon added per million parts of wastewater treated.

TABLE 6.5   Relative Removability of Compounds by Chambers Works PACT System

Removal (%)[a]	Average Feed Conc. (ppb)[a]	Average PACT Effluent Conc. (ppb)[a]	Compound	Class
>99	1770	Nil[c]	Methyl chloride	Volatile
>99	33	Nil	Naphthalene	Base neutral
>99	454	2.1	Nitrobenzene	Base neutral
>99	28	Nil	N-nitrosodiphenylamine	Base neutral
99	519	4.7	Toluene	Volatile
99	19	Nil	1,2-Dichloroethane	Volatile
99	3.6	Nil	1,2-*trans*-Dichloroethylene	Volatile
98	105	0.85	Benzene	Volatile
98	1720	30	Chlorobenzene	Volatile
98	161	5.0	2,4-Dinitrophenol	Acid extractable
97	1020	10	4-Nitrophenol	Acid extractable
>95	18	Nil	N-nitrosodi-n-propylamine	Base neutral
>95	3.6	Nil	Methyl bromide	Volatile
95	11.4	0.6	2-Chlorophenol	Acid extractable
95	94	1.4	Carbon tetrachloride	Volatile
95	155	3.0	Trichlorofluoromethane	Volatile
94.9	174[d]	6.3[d]	BOD_5	
94	611	38	Phenol	Acid extractable
94	192	13	2-Nitrophenol	Acid extractable
94	41	1.7	Ethylbenzene	Volatile
94	41	1.9	Trichloroethylene	Volatile
94	280	12.3	Chloroethane	Volatile
93	24	1.7	Tetrachloroethylene	Volatile
>90	2	Nil	2,4-Dimethylphenol	Acid extractable
>90	1.6	Nil	Acenaphthalene	Base neutral
>90	0.6	Nil	Anthracene	Base neutral
>90	1	Nil	Fluoranthene	Base neutral
>90	0.8	Nil	Phenanthrene	Base neutral
89	13	0.6	1,1,1-Trichloroethane	Volatile
81	2.1	0.4	Pentachlorophenol	Acid extractable
81	201	20.5	Chloroform	Volatile
80.8	174[d]	32.9[d]	Soluble TOC	
73	370	100	1,3- and 1,4-Dichlorobenzenes	Base neutral
67	0.3	0.1	2,4,6-Trichlorophenol	Base neutral
66	523	169	1,2,4-Trichlorobenzene	Base neutral
65	1900	243	2,4-Dinitrotoluene	Base neutral
64	1640	575	2,6-Dinitrotoluene	Base neutral
63.7	1440[b]	484[b]	Color	
44	214	120	1,2-Dichlorobenzene	Base neutral

[a]All are average values for several analyses. They, therefore, do not necessarily calculate to the percent removal.
[b]APHA units.
[c]Nil = below quantification level.
[d]mg/L.

TABLE 6.6   Chambers Works Wastewater Treatment Plant (PACT) Operating Conditions During Priority Pollutant Sampling

	Average	Range
Flow, gpm	25,800	21,800–29,100
Feed TOC, mg/L	169	115–207
Feed color, APHA	1,120	700–1,400
Carbon dose, ppm	114	84–151
Percent virgin	63	6–97
Percent regenerated	37	3–94
Aeration temperature, °C	29	16–35
Aeration time, hr	7.7	6.6–9.0
Aerator DO, mg/L	2.9	1.5–4.3
Aerator, pH	6.6	6.4–6.7
MLSS, mg/L	24,800	21,200–29,300
Sludge age, days	54	33–75

and so on. The more oxidized a compound is, the less biodegradable and, hence, the poorer the removal. Other data show that increasing chlorination also decreases the removal of aliphatics as well. Chlorination reduces biodegradability by increasing the degree of oxidation. Aromatic compounds were very well removed. Di- and trichlorobenzenes and dinitrotoluenes are the hardest compounds to remove; yet, 70% of these were removed by PACT. Base-neutral extractable compound removal varies, depending on the compound.

Chemical Wastewater   A waste with a high VOC content and having an appreciable alcohol content was initially to be treated by air stripping alone. However, upon pilot plant testing, it became evident that stripping was unable to achieve the desired treatment without some form of iron pretreatment and biological and/or granular carbon posttreatment (Table 6.7).

The PACT system was then tested on both the raw and the air-stripped waste (Table 6.8). The PACT system was highly effective in meeting treatment objectives, even if air stripping pretreatment was not used. The PACT system, when treating raw waste directly, demonstrated:

- Removal of most toxic organics to below detection limits
- Removal efficiencies exceeding 99.9% for most organics
- Volatilization (stripping) of less than 1% of the inlet organics treated
- No iron precipitation step requirement before or after PACT treatment to control metals

TABLE 6.7   Air Stripper Performance

Analysis (mg/L)	Influent	Effluent
COD	1470	1500
BOD_5	740	850
$NPOC^a$	470	490
1,1,2,2-Tetrachloroethane	240	29
Trichloroethylene	112	5.7
cis-1,2-Dichloroethylene	107	11
Vinyl chloride	37	0.48
Acetone	22	12
Ethyl ether	5.3	0.91
Ethyl acetate	1.2	0.14
n-Butanol	145	136
Isopropanol	34	49
Methanol	40	51
Iron	50	0.65

^{a}NPOC = nonpurgable organic carbon (the fraction of TOC remaining after gas stripping).

TABLE 6.8   PACT System Treatment of Raw Waste and Air-Stripped Waste

	Raw Waste		Air-Stripped Waste	
Analysis (mg/L)	PACT Influent	Final Effluent	PACT Influent	Final Effluent
COD	1470	110	1500	155
BOD_5	740	6	850	6
NPOC	470	6	490	9
1,1,2,2-Tetrachloroethane	240	0.11	29	0.18
Trichloroethylene	112	0.085	5.7	0.019
cis-1,2-Dichloroethylene	107	0.023	11	0.016
Vinyl chloride	37	<0.01	0.48	0.004
Acetone	22	0.077	12	0.037
Ethyl ether	5.3	0.004	0.91	0.007
Ethyl acetate	1.2	0.058	0.14	0.095
n-Butanol	145	<0.5	136	<0.5
Isopropanol	34	<0.5	49	<0.5
Methanol	40	<0.5	51	<0.5
Iron	50	1.2	0.65	0.34

- Removal of 97% of the inlet iron by natural means
- Production of a single nonhazardous waste stream (sludge passed the TCLP test).

Volatile Organic Chemical Control (Odors)   Research conducted by Weber and Jones (14) on the treatment of various VOCs led to the conclusion that little stripping of VOCs occurred when powdered activated carbon was added to the activated sludge system. Without the carbon, however, stripping from the aeration basins accounted for appreciable removal of the VOCs (Table 6.9).

Previously reported work by Zimpro Environmental on a raw waste and the same raw waste pretreated by air stripping showed extremely good performance for volatile organics removal and iron removal (Table 6.8). Volatilization (stripping) accounted for less than 1% of the inlet organics applied, based on off-gas analysis. In addition, the waste PACT system sludge passed the TCLP test.

Pilot testing of a two-stage aerobic PACT system for treating refinery wastewater proved to be highly successful for removing benzene without significant air emissions. A material balance of benzene showed that an average of 0.03 wt % of the influent benzene was in the off-gas (by measurement) (Figure 6.9). The amount of benzene emitted in the off-gas did not increase when the air supplied to the aeration tanks was spiked with benzene (12).

Numerous odor control studies were conducted at the Kalamazoo wastewater treatment plant, in an effort to minimize problematic odors that plagued the activated sludge plant over the years (15). Of the systems evaluated (ozonation, granular activated carbon, and several forms of scrubbing), scrubbing with PACT system mixed liquor was the most effective (Table 6.10). Based on these results and the fact that the PACT/wet air regeneration system was the most cost-effective treatment solution for the treatment of Kalamazoo's 54-mgd flow, the plant was designed to capture all odor sources throughout the plant site, route all odors through the aeration blowers, and scrub those foul odors in the aeration tanks of the PACT system. Results have been impressive—all odors from the wastewater treatment plant site have been eliminated.

TABLE 6.9   Fate of Toxic Organics (% of Influent)

Compound	Activated Sludge		PACT at 100 ppm Carbon Dose	
	Effluent	Off-gas	Effluent	Off-gas
Toluene	<1	17	<1	0
o-Xylene	<1	25	<1	0
1,2-Dichlorobenzene	6	59	<1	6
1,2,4-Trichlorobenzene	10	90	<1	6
Lindane	>95	0	<1	0

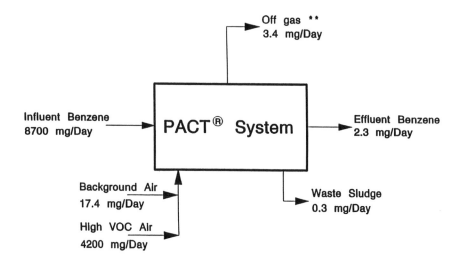

** 0.03% of the influent benzene was measured in the off gas.

FIGURE 6.9 Benzene material balance.

Color Removal An attribute of carbon is its ability to adsorb color from various types of wastewaters. Early investigations by Robertaccio and co-workers (16) showed how increased dosages of powdered activated carbon to an activated sludge system aided color removal (Figure 6.10). Experience at Chambers Works with activated sludge pilot plants showed either little removal of color or an increase in color due to biodegradation products.

TABLE 6.10 Odor Control Study Results from Two-Stage Liquid Scrubbing (Kalamazoo, Michigan)

	% Removal		
Scrubbing Liquid[a]	Odor	THC	H$_2$S
PACT–PACT	97	46	91
VC–VC	95	26	82
AS–AS	81	14	49
Water–NaOCl	68	25	96
Water–KMnO$_4$	46	55	97

[a]PACT, PACT system mixed liquor; VC, virgin powdered activated carbon; AS, activated sludge mixed liquor; THC, total hydrocarbons.

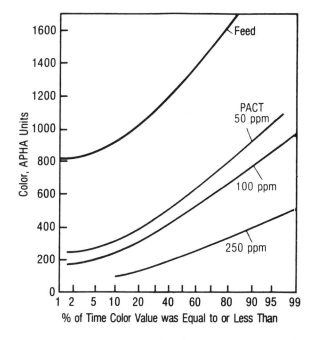

FIGURE 6.10   Effect of carbon dosage on color removal.

Metals Removal   While metals removal is not the primary purpose of the PACT system, at least one organic chemical plant achieved significant metals removal with the PACT system, as opposed to the activated sludge process (Table 6.11) (12). Similar results have been reported by Lee and co-workers (17), when treating chromium containing water and by McIntyre (18) when treating an iron-containing wastewater/contaminated groundwater.

Nitrification   As early as 1973, Burant and Vollstedt (19) showed that the addition of powdered activated carbon to a municipal activated sludge process could give a system excellent nitrification capability. Sampayo and Swets (20) reported similar results from a 0.2-mgd pilot plant study at Kalamazoo, Michigan. Attempts

TABLE 6.11   Metals Treatment Performance Comparison for Organic Chemicals Wastewater

	Influent Water	Effluents from	
		PACT	Activated Sludge
Copper, mg/L	0.41	0.07	0.36
Nickel, mg/L	0.52	0.24	0.35

to nitrify this waste using activated sludge were unsuccessful; but, the full-scale PACT system has shown 94% ammonia nitrogen removal (Table 6.12).

Ciba-Geigy Corporation personnel obtained similar results at their 1.5-mgd treatment plant in Toms River, New Jersey. When the plant was converted from an activated sludge plant to a PACT system in 1986, nitrification was established, in a short time, that reduced the effluent ammonia nitrogen concentration to less than 1 mg/L (18).

Ng and Stenstrom (21) concluded that the PACT system was able to dramatically improve nitrification rates because of its ability to adsorb nitrification inhibitors.

Process Stability   Activated sludge plants can become unstable for any number of reasons; such as changes in feed concentration, composition, or flow; temporary equipment problems; or operating mistakes. Regardless of the cause, the result of such "upsets" is loss of removal efficiency in a plant that is expected to operate 100% of the time and produce a clean effluent, irrespective of the quality or quantity of the raw waste. Whether the plant is industrial or municipal, system upsets can result in permit violations and legal action. Upsets typically occur with sudden increases in flow and/or concentration that overload system capability for removal of pollutants, because the system cannot manufacture biomass rapidly enough to respond to sudden changes.

Adding powdered activated carbon to an activated sludge plant will make the plant less susceptible to upsets (excursions from normal operation and/or performance) by adsorbing some of the excess load, and result in more consistent production of high-quality effluent. Foertsch and Hutton (22) saw this in a parallel "train" test of the PACT system at the DuPont textile fibers plant in Waynesboro, Virginia. Feed to this waste treatment plant originated from the manufacture of Orlon® acrylic fiber and Acele® cellulose acetate fiber.

Influent BODs averaged 837 mg/L. Soluble effluent BODs for the PACT system averaged 11 mg/L versus 20 mg/L for the activated sludge process. More importantly, the distribution of effluent soluble and total BOD_5 data showed significant

TABLE 6.12   Treatment Plant Performance, Kalamazoo, Michigan, Summary of Typical Results

	PACT System		Removal (%)	Permit Limits
	Influent[a]	Effluent		
BOD, mg/L	325	7	98	10
COD, mg/L	600	50	92	—
TSS, mg/L	115	4–5	96	20
P, mg/L	5–6	0.3	95	0.5–1
NH_3-N, mg/L	25	1–2	94	<2

[a]Combined waste from paper mills, organic chemical, and chemical/pharmaceutical industries and domestic after primary treatment.

improvement with PACT, making it far easier to achieve maximum permit goals (Figures 6.11 and 6.12).

Resistance to Shock Loads

Metals   Robertaccio and coworkers (16) showed, in laboratory studies, that BOD_5 removal by the PACT system is less affected by both chromium (VI) and cobalt (II) than activated sludge (Table 6.13). Both of these substances are inhibitory to an activated sludge process. Lee and coworkers (17) confirmed this point when treating wastewater having a high chromium content.

Organics   The PACT system has also been found to be far more reliable than the activated sludge process for this application. Zimpro Environmental, for instance, demonstrated that treatment of a high-strength waste with spikes (additions to the feed that were much greater than the average feed concentration) of phenol up to 150 mg/L, could be reliably treated in the PACT system at aeration times one-half that of conventional activated sludge. Figures 6.13 and 6.14 show the influence of COD and COD–BOD ratio on effluent phenol and indicate that activated sludge was far more susceptible to organic loads and ''refractories'' than the PACT system. The PACT system, in fact, was able to achieve <10 ppb phenol in its effluent, whereas the activated sludge system's effluent was greater than 100 ppb phenol during the study. Confirmation of the PACT system's ability was demonstrated near the end of the study, when 300 mg/L of phenol was added to the waste. A con-

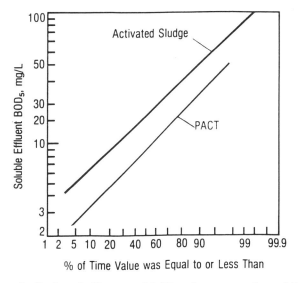

FIGURE 6.11   Distribution of effluent total BOD_5 values: comparison of PACT and activated sludge.

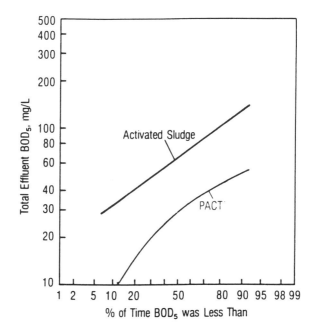

FIGURE 6.12   Distribution of effluent total BOD_5 values: comparison of PACT and activated sludge.

centration of less than 80 ppb of phenol was generated quickly in the PACT system effluent, before it again decreased to <10 ppb. At DuPont's Chambers Works facility, tank trucks containing as high as 1% biodegradable organics, such as phenol, are routinely discharged in the wastewater treatment plant without equalization and National Pollutant Discharge Elimination System (NPDES) limits are met.

Hydraulic Residence Time   Sometimes there are reasons to minimize the land area required to construct a wastewater treatment plant. This can arise from the high cost of land in the location where the treatment plant is needed or from simply

TABLE 6.13   Effects of Metals Addition on BOD_5 Removal

	BOD_5 Reduction (%)	
Metal Added (Dosage)	PACT System	Activated Sludge
None	86	77
Cr^{6+} (2 mg/L)	79	68
Co^{2+} (1.5 mg/L)	83	67

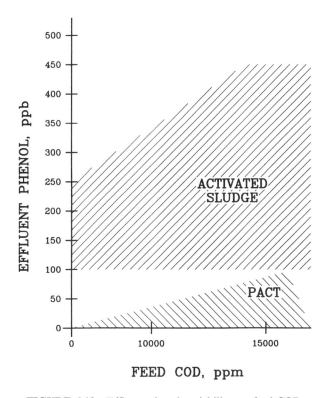

FIGURE 6.13   Effluent phenol variability vs. feed COD.

the severe lack of available land where it is needed. PACT systems can be constructed with smaller aeration tanks than comparable activated sludge plants. The principal reason for this is that the organics in the wastewater are adsorbed onto the activated carbon in a PACT system as well as sorbed onto the biomass, so that the adsorbed organics also stay in the system for a time equal to the sludge age of the mixed liquor rather than the hydraulic retention time of the water. Without activated carbon, the activated sludge plant does not tend to hold onto organics because the only mechanism for retention is the relatively ineffective sorption of organics onto the biomass. Retention of the organics on the activated carbon not only allows a PACT system to be smaller than a comparable activated sludge plant, but the added treatment time also helps improve effluent quality of a PACT system over a comparable activated sludge plant (12). Examples of improved quality were already discussed in the previous section regarding phenol treatment, and in an article (23) on the treatment of petrochemical waste where PACT and activated sludge systems were tested. Table 6.14 summarizes performance and operating conditions, showing the PACT system to be able to treat phenol-containing wastes to a higher level at less aeration time, and at the same time generate less solids for

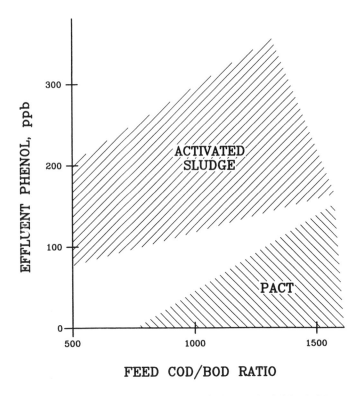

FIGURE 6.14 Effluent phenol variability vs. feed COD/BOD.

disposal. The clarifiers can also be smaller because of the enhanced settling/thickening, as discussed later.

Low-Temperature Stability   Lee and Johnson (24) reported the PACT system to be far more efficient and stable for cold-temperature nitrification than conventional biological treatment. Table 6.15 shows PACT was able to more efficiently nitrify (1.1 mg/L NH_3-N) at 5°C and 7.2 hr hydraulic detention time than activated sludge (1.9 mg/L NH_3-N) at 10°C and 22.1 hr hydraulic detention time.

Sludge Settleability and Thickening   Adding powdered activated carbon to activated sludge produces a combined carbon–biological aggregate that settles better than activated sludge alone, as described later.

Sludge Handling and Disposal   Sludge disposal from a treatment plant is often one of the most difficult aspects of wastewater treatment. The problem occurs when dewatering conventional activated sludge using polymer conditioning. Under these circumstances, it is usually impossible to reduce the water content of the sludge below 85%, unless conditioning agents such as ferric chloride and lime are

TABLE 6.14   Petrochemical Wastes Treatment

		PACT System	Activated Sludge
Operating conditions			
HDT, days		2	3
MLSS, mg/L		10,000	5,000
ML carbon, mg/L		5,000	0
Performance	Influent	Effluent	Effluent
BOD$_5$, mg/L	1517	14	32
COD, mg/L	3684	100	285
Stability		Good	Poor
Dewatered sludge concentration, % solids		40–50	10–15
Wet solids for disposal, tons/day		0.4	1.4

HDT = Hydraulic detention time.
ML = Mixed liquor.

used. Thus, incinerating sludge from most industrial wastewater treatment plants not only causes the usual problems associated with air emissions from an incinerator, but also increases fuel costs, because of the extremely high water content and/or inorganic chemical dewatering aid content of the sludge. PACT system sludges, however, will reflect far less cake moisture (40–50% water) without any

TABLE 6.15   Comparative Performance PACT System versus Activated Sludge

Operating Conditions		PACT	PACT	Activated Sludge	Activated Sludge
Aeration time, hr		3.9	7.2	22.1	17.0
SRT, days		13	38	25	36
Temperature, °C		10	5	10	10
Performance Results	Influent (mg/L)	Effluent (mg/L)			
BOD$_5$	198	10	3	—	—
COD	291	67	27	—	92
Total N	27	—	—	—	—
NH$_3$-N	—	0.5	1.1	1.9	7.7
NO$_3$-N	—	16.8	7.5	10.5	5.3

SRT = Solids retention time.

From J. S. Lee and W. K. Johnson, *JWPCF*, Tables I–III, pp. 113, 119, 120, Volume 51. Used by permission of Water Environment Federation.

dewatering aids. Figure 6.15, from work on an industrial waste by Zimpro Environmental, shows that a relationship exists between cake dryness and the amount of carbon contained in the system's mixed liquor. Activated sludge solids, on the other hand, would be extremely difficult to dewater without either polymer conditioning (to achieve 10–15% solids in the cake) or lime, or lime plus ferric chloride treatment (to achieve >20% cake solids). The low-moisture cake from a PACT system will generally contain a Btu content of 8–11,000 Btu/lb dry weight solids (12). The presence of the carbon enhances the Btu content of most sludges.

Another common sludge disposal method has been landfilling of dewatered sludge cake. Not only have landfilling costs been going up at a rapid rate, industry executives are becoming increasingly aware of the potential liabilities arising from contaminating groundwater near landfills and the continuing problems of odors that can easily be produced by wet biological sludges. Overall, there are very serious political and economic costs associated with landfilling. PACT system sludge pro-

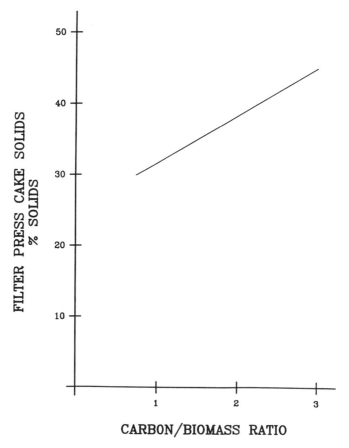

FIGURE 6.15 Influence of carbon on PACT system solids dewatering.

duces a very stable, low leachability cake that minimizes much of the landfilling problem. This will also impact sludge disposal costs. Table 6.16 illustrates how important a role low-moisture solids play in disposal costs, especially where landfill costs are high. As Table 6.16 shows, even though PACT system sludge dry-weight solids may be more than an activated sludge system's solids, the wet-weight amount from PACT will be less.

Also, wet air regeneration or other thermal regeneration methods can be used to economically recover activated carbon from the waste sludges for reuse. The amount of sludge for disposal will be reduced by over 90% and will consist, largely, of inert ash material that is very easy to dewater and dispose. Regeneration, in effect, virtually eliminates sludge disposal problems (12).

Waste sludge from the PACT system will generally pass the Environmental Protection Agency's (EPA's) TCLP test, allowing the sludge to be landfilled, or in some cases disposed of as sewage, without fear of leaching priority pollutant organics into the environment. Unlike granular activated carbon, which would not pass the TCLP test because of sorbed compounds on the carbon, PACT system sludge would pass the TCLP test because the organics have been destroyed in the biological treatment. PACT system treatment of the highly VOC-contaminated wastes (noted previously in Table 6.8) resulted in the wasted PACT system sludge passing TCLP tests.

Should spent PACT system sludge be taken to wet air regeneration, any solids generated for disposal will generally pass TCLP criteria, even when metals and/or priority pollutant organics are present in the spent sludge.

TABLE 6.16   Sludge Disposal Cost Comparison[a] PACT versus Activated Sludge[b]

	Activated Sludge	PACT System
Dry weight solids to disposal		
Biosolids, lb/day	500	500
Powdered carbon, lb/day	0	250
Total, tons/yr	91	137
% Solids to disposal	15	40
Wet weight to disposal, tons/yr	608	342
Annual disposal cost (chemicals)		
Powdered carbon at 0.40/lb	0	$36,000
Polymer conditioning at $2/lb	$2,000	0
Disposal (Class 1 landfill) at $200/ton	$122,000	$68,000
Total disposal costs	$124,000	$104,000

[a]In order to meet U.S. EPA discharge standards, activated carbon was required. The PACT system was able to meet such standards, activated sludge was not unless sand filtration plus granular activated carbon was also used. The costs for granular activated carbon disposal are not included herein.
[b]For a 0.3-mgd industrial wastewater treatment facility having an influent COD of 1000 mg/L.

Effluent Toxicity Laboratory studies on the aquatic toxicity of Chambers Works wastewater during dye manufacture showed that the PACT system greatly reduced the acute toxicity of the effluent to fish (mummichogs). The 96-hr LC_{50} (concentration that kills 50% of the fish in 96 hr) of the laboratory activated sludge effluent was 35%, but the PACT system having only 50-ppm carbon dosage removed the toxicity (Table 6.17). Both PACT treatments resulted in no fish toxicity to 100% effluent (16).

The Toms River Plant of Ciba-Geigy converted its activated sludge process for treating wastewater from the batch manufacture of basic dyestuffs, epoxy resins, and additives to a PACT system in 1986 to meet stricter permit requirements, including a mysid shrimp bioassay (18). The conversion resulted in improving an effluent that typically produced a mysid shrimp acute LC_{50} of less than 5% to one that had an effluent LC_{50} of greater than 100% (actually, 100% survival in 100% effluent). In addition, the PACT system effluent also passed the Ames mutagenicity test.

Foam Control Foam found in wastewater treatment plants is usually chemical or biological in nature. Chemical foam may be attributed to surfactants or organics, both of which can be reduced or eliminated by using the PACT system. Prior to the conversion of the Rothschild, Wisconsin, wastewater treatment plant and the Kalamazoo, Michigan, wastewater treatment plant to PACT/wet air regeneration systems, effluent foam was a significant problem. In Rothschild, severe foaming also occurred in the aeration tanks on Mondays, the normal laundry day for the residents. However, when the activated sludge plant was converted to a PACT system in 1973, water quality improved and foam was not a problem (19). A similar experience was reported from Kalamazoo (25).

Biological foam, on the other hand, can be problematic for all biological processes, including the PACT system. Proper design to minimize such occurrences, or choosing proper operating conditions, can alleviate this problem. Operation at too high of a sludge age, for instance, can contribute to biological foam problems, as the fine powdered activated carbon particles can stabilize foam under these conditions. Antifoam chemicals or mechanical devices are then needed to break the foam.

TABLE 6.17 Aquatic Toxicity of Chambers Works Wastewater at Various Stages of Treatment

Treatment	96-Hr LC_{50}, % Effluent
Raw waste	5
Primary effluent (after neutralization and settling)	16
Activated sludge effluent	35
PACT (50 ppm carbon dosage) effluent	>100
PACT (250 ppm carbon dosage) effluent	>100

Other   In addition to the advantages already discussed of using the PACT system for treating dyes-containing wastewaters, the PACT system provides more protection from toxic upsets (substances toxic to the bacteria in the treatment system that degrade performance) than is seen in other biological systems. The PACT system is compatible with the activated sludge system, so that many conventional activated sludge systems can be converted to PACT systems without undue difficulty. The addition of powdered activated carbon in the PACT system provides a new control variable, one that can provide a significant degree of flexibility to wastewater treatment plant operation. Many PACT system operators vary the powdered activated carbon dose as the waste load changes. Finally, there is the option of regenerating the carbon, depending on the economics of a given situation.

SYSTEM VARIABLES

Operational

Operation of the PACT system is controlled by adjustment of specific process parameters that include:

- Hydraulic detention time (θ_N)
- Solids retention time (θ_c)
- Mixed liquor carbon suspended solids (S_c)
- Carbon dose (C_c)
- Carbon type
- Placement (location, time) of carbon

The hydraulic detention time is a measure of the length of time that the wastewater is in contact with the powdered carbon and biological solids. The solids residence time is a measure of the average time that the solids, both powdered activated carbon and biological solids, are retained in the PACT system. The total mixed liquor suspended solids (MLSS) in the contact-aeration basin of the PACT system is comprised of the S_c (carbon in the mixed liquor), biological solids (biomass), and inert suspended solids (ash). The carbon dose is defined as the mass of powdered activated carbon that is added per unit mass of influent stream flow.

S_c is dependent on the operating parameters, as follows:

$$S_c = (C_c) \, (\theta_c)/(\theta_N)$$

The carbon type influences the operation of the system, due to differences in ash content, adsorptive capacity for a specific wastewater, specific gravity, and regenerability.

Placement of added carbon in the PACT system can influence organics removal, particularly in two-stage systems, where the carbon can be added to either the first or second stage. In single-stage systems, adding the carbon to the latter portion of a plug flow aeration tank can result in improved organics removal versus adding the carbon at the front of the aeration tank, while the PACT sludge is recycled to the front. The idea is to place the virgin carbon near the effluent, to maximize the concentration gradient for adsorption near the discharge point. The carbon is recycled so that it can be fully saturated prior to disposal. In the two-stage system illustrated in Figure 6.3, the influent BOD is low, so the suspended solids are rich in carbon. Therefore, less carbon is added to the second stage so that carbon losses in the effluent can be reduced.

Unit Processes

Aeration

Empirical Design   The heart of the PACT system is the aeration, where the biological reactions and the carbon adsorption occur. The aeration tanks are essentially reaction vessels in which organic materials in the wastewater are biologically oxidized by air or oxygen to form CO_2, H_2O, and biological sludge. Flynn (26) classified the various components of a wastewater as biodegradable, nonbiodegradable, or very slowly biodegradable, and hence, only removed if held in contact with the microorganisms for an extended period of time. A wastewater component may also be carbon adsorbable or nonadsorbable. The resulting cross categories and areas where conventional biological and the PACT system achieve treatment are found in Table 6.18.

Readily biodegradable materials are removed by both conventional biological processes and the PACT system, because they do not depend on carbon adsorption for removal. However, a nonbiodegradable material that is adsorbable on carbon will be removed by the PACT system. Slowly biodegradable and carbon adsorbable materials will be removed by being held on the carbon long enough for the bacteria to degrade it. In some cases, even nonadsorbable, nonbiodegradable components of a waste stream, such as 1,4-dioxane, can be removed by the PACT system.

Because the PACT system depends on both degradation and adsorption processes, the most important process variables are sludge age (θ_c), hydraulic detention

TABLE 6.18   Wastewater Component Removability[a]

	Adsorbable	Nonadsorbable
Biodegradable	X	X
Slow contact degradable	O	Does not exist
Nonbiodegradable	O	—

X, removed by conventional biological treatment and by the PACT system.

O, removed by the PACT system only.

—, sometimes removed by PACT.

From 30th *Annual Purdue Industrial Waste Conference Proceedings*, Table I, p. 234, 1975. Used by permission of Purdue University.

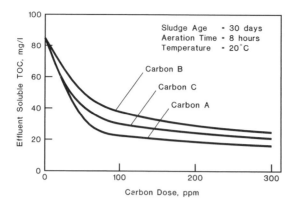

FIGURE 6.16   Effect of carbon type and dose on effluent-soluble TOC.

time (θ_N), carbon dosage (C_c), carbon type, and temperature (T). Correlations can be empirically derived between these variables, and performance from either bench-scale studies or kinetic models can be used. Both the dosage and the type of carbon used will affect the performance of a PACT system. The effect of carbon dosage on DuPont's Chamber Works wastewater containing dyes is shown in Figures 6.16 and 6.17, where zero carbon dose represents the activated sludge process. Soluble TOC (Figure 6.16) and color (Figure 6.17) are significantly affected by carbon dose. The major difference in the types of carbon used was surface area, where carbon A had the highest surface area, followed by carbon C with intermediate surface area and carbon B with the lowest surface area.

It should be noted, however, that surface area is not the sole basis for carbon selection. Other important factors are pore size distribution, surface charge, particle size, density, pH, and chemical composition.

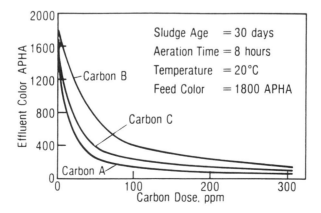

FIGURE 6.17   Effect of carbon type and dose on effluent color.

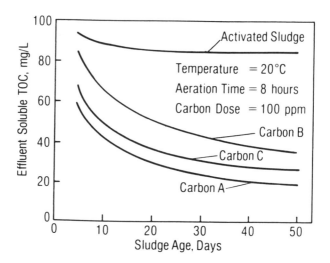

FIGURE 6.18 Effect of sludge age on effluent-soluble TOC.

The effect of sludge age on PACT system effluent TOC is similar to the effect of carbon dose. Effluent TOC will decrease as sludge age increases (Figure 6.18). Again, the type of carbon used has an effect.

Kinetic Model

MODEL DEVELOPMENT O'Brien (27) modeled the kinetics of the PACT system for priority pollutant removal by using a modified form of the equations of Stover (28) and Blackburn (29,30). The aeration tank was considered to be a completely mixed reactor, which meant that the concentration in the aeration tank was uniform and identical to the effluent concentration. The effluent and influent concentrations are related through a material balance:

Rate In by Flow	Rate Out by Flow	Rate Out by Stripping	Rate Out by Biodegradation	Rate Out by PAC Adsorption	Rate of Accumulation

$$FS_i \quad - \quad FS_e \quad - \quad K_s S_e V \quad - \quad K_b S_e XV \quad - \quad K_a S_e S_e V \quad = \quad \frac{dVS_e}{dt} \qquad (1)$$

where F = flow rate
  K = rate coefficient
  S = concentration
  t = time
  V = aeration tank volume
  X = biomass concentration in aeration tank
 PAC = powdered activated carbon
  a = adsorption

b = biodegradation
c = aeration tank PAC
e = effluent
i = influent
s = stripping

At steady state, the accumulation term equals zero, so

$$FS_i - FS_e - K_sS_eV - K_bS_eXV - K_aS_sS_eV = 0 \qquad (2)$$

And from a solids balance:

$$S_c = C_c\theta_c/\theta_N \qquad (3)$$

$$S_e = \frac{S_i}{1 + K_s\theta_N + K_bX\theta_N + K_a\theta_cC_c} \qquad (4)$$

where C_c = influent PAC concentration (carbon dose)
θ_c = solids retention time (sludge age)
θ_N = hydraulic retention time = V/F

The model combines adsorption by the bacteria into the biodegradation coefficient and assumes that biodegradation is first order in both substrate and biomass concentration. The Monod equation reduces to this form at low substrate concentrations. Hutton (31) showed this to be valid in a study on the BOD removal from Chambers Works wastewater using adenosine triphosphate (ATP) as the measures of biomass. O'Brien (27) assumed the removal of each priority pollutant was proportional to the total biomass present. The biomass concentration in the aeration tank was obtained by difference, by subtracting the PAC dose and primary clarifier overflow suspended solids from the mixed liquor suspended solids:

$$MLSS = S_c + S_p + X \qquad (5)$$

$$S_c = C_c\theta_c/\theta_N$$

$$S_p = C_p\theta_c/\theta_N \qquad (6)$$

$$X = MLSS - (C_c + C_p)\frac{\theta_c}{\theta_N} \qquad (7)$$

where MLSS = mixer liquor suspended solids concentration
C_p = primary clarifier solids carryover into the aeration tank
S_c = carbon concentration in the aeration tank
S_p = primary solids concentration in the aeration tank

Adsorption on PAC is usually assumed to follow the Freundlich isotherm. However, Sontheimer (32) reported that adsorption is linear at low substrate concentrations.

O'Brien (27) verified that adsorption for the priority pollutants was linear with respect to PAC and substrate concentrations. Synergism between the PAC and biomass was lumped into the adsorption coefficient.

Stripping was assumed to be proportional to the concentration of the pollutant in the aeration tank, as shown by Blackburn (29, 30) for coarse bubble diffusion.

The fractional removal (R) by each mechanism for a single-stage PACT system can be obtained using Equation (2) and is determined as follows:

$$\text{Stripping:} \qquad R_S = K_S \theta_N S_e / S_i \qquad (8)$$

$$\text{Biodegradation:} \qquad R_b = K_b X \theta_N S_e / S_i \qquad (9)$$

$$\text{Adsorption:} \qquad R_a = K_a \theta_c C_c S_e / S_i \qquad (10)$$

where S_e/S_i can be calculated from Equation (4).

The model was verified under steady-state conditions by O'Brien (27) and under dynamic (unsteady-state) conditions by O'Brien and Teather (33). Equation (1) is the dynamic equation for a completely mixed aeration tank. A comparison of a steady-state and a dynamic model was presented by O'Brien (34). The model is valid for compounds present at low concentration, where each removal mechanism is linear with the substrate concentration and adsorption is also linear with the PAC concentration.

FACTORS AFFECTING PRIORITY POLLUTANT REMOVAL   Each compound has associated with it a set of rate coefficients that vary with the operating conditions. Blackburn (30) showed that for coarse bubble aerators the stripping coefficient [Equation (11)] is a function of both the airflow rate per unit volume and the Henry's law coefficient, which in turn varies with the temperature:

$$K_s = 6.18 \times 10^{-5} \, (Q/V) H_c^{1.045} \qquad (11)$$

where K_s = stripping coefficient
Q = airflow, L/h
V = aeration tank volume, L
H_c = Henry's law coefficient, torr L/mol

Similarly, the biodegradation coefficient varies with the nature of the wastewater and temperature. The adsorption coefficient differs for different carbons and may vary with the composition of the wastewater. Examination of Equation (4) shows that effluent concentration will also change with hydraulic retention time, biomass concentration, sludge age, and carbon dose. The flow profile will also affect removal, whether the aeration tank is completely mixed, plug flow, and so forth. Different equations can be developed to represent pure plug flow with varying amounts of backmixing (33).

EFFECT OF PAC ON VOLATILE ORGANICS REMOVAL AND AIR EMISSIONS   A stated advantage of the PACT system was that it not only enhances total compound re-

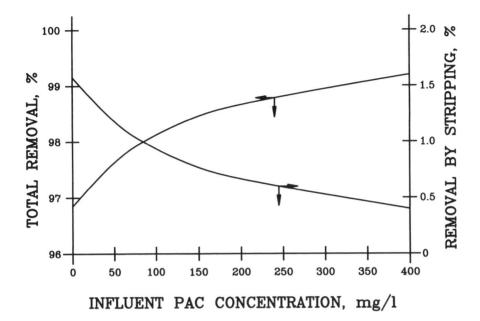

BASIS:
Θ_N = 10h
Θ_C = 480h
X = 6000 mg/L
K_S = 0.05h^{-1}
K_B = 50 x 10^{-5} L/(mg − h)
K_A = 50 x 10^{-5} L/(mg − h)

FIGURE 6.19   Effect of PAC in improving total removal and depressing stripping (K_s = 0.05 hr^{-1}).

moval but also reduces air stripping. Consider the example in Figure 6.19 where the compound is volatile, biodegradable, and adsorbable. The total removal, R_T, can be calculated from Equation (4) where

$$R_T = 1 - S_e/S_i \qquad (12)$$

Removal by stripping can be calculated from Equation (8). Figure 6.19 shows the increase in total removal and decrease in stripping as the carbon dose is increased from 0 to 400 ppm. A completely mixed aeration tank without carbon would have 96.8% removal of this compound with 1.6% released to the atmosphere. A similar aeration tank with 400 ppm carbon dosage would have 99.2% removal with only 0.4% stripped, which is a factor of 4 reduction of stripped material.

Figure 6.20 shows similar calculations where the K_s value (removal by stripping) has been increased fourfold. This could occur due to an increase in the airflow rate

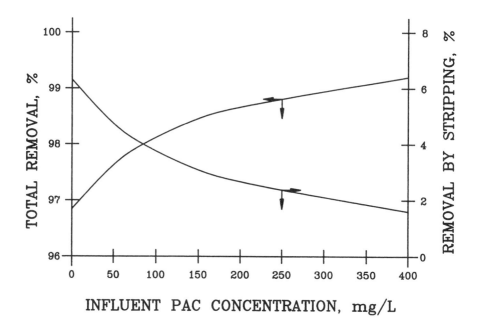

INFLUENT PAC CONCENTRATION, mg/L

BASIS:
θ_N = 10h
θ_C = 480h
X = 6000 mg/L
K_S = 0.2h^{-1}
K_B = 50 x 10^{-5} L/(mg – h)
K_A = 50 x 10^{-5} L/(mg – h)

FIGURE 6.20   Effect of PAC in improving total removal and depressing stripping (K_s = 0.2 hr^{-1}).

[see Equation (11)], a higher temperature, or represent a more volatile compound. Total removals are not altered significantly, but the activated sludge unit would have 6.1% of the compound stripped, while the PACT unit at 400 ppm carbon dose would only lose 1.6% due to stripping.

Addition of PAC to an activated sludge system is most effective in reducing air emissions when the compounds to be removed are not readily biodegradable but adsorbable or when the wastewater is inhibitory to the bacteria but PAC removes the inhibition. Figures 6.21 and 6.22 arise from K_s values of 0.05 and 0.2 hr^{-1}, respectively, but with K_b = 20 × 10^{-5} and K_a = 200 × 10^{-5} L/(mg hr). These rate coefficients were chosen to represent a readily adsorbable, biodegradable, volatile compound. Stripping is reduced from 3.7 to 0.1% for a K_s value of 0.05 hr^{-1} (Figure 6.21) and from 13.3 to 0.5% for a K_s value of 0.2 hr^{-1} (Figure 6.22) as carbon dose is increased from 0 to 400 ppm, while total removal is increased from 93 to almost 100%.

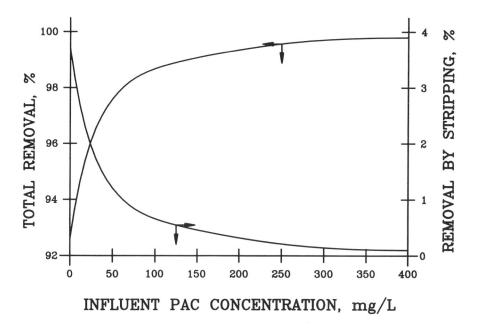

BASIS:
θ_N = 10h
θ_C = 480h
X = 6000 mg/L
K_S = 0.05h^{-1}
K_B = 20 x 10^{-5} L/(mg − h)
K_A = 200 x 10^{-5} L/(mg − h)

FIGURE 6.21   Effect of PAC in improving total removal and depressing stripping.

If the wastewater is inhibitory, the only removal mechanism remaining in the activated sludge unit is stripping. From Equation (4) it can be determined that 33 and 67% of the influent concentration would be stripped for K_s values of 0.05 and 0.2 hr^{-1}, respectively, as contrasted to the 0.1 and 0.5% stripped when the carbon dose is 400 ppm in PACT, the case where the wastewater is inhibitory without PAC. Large differences in stripping between PACT and activated sludge have been measured by O'Brien where the wastewater was partially or completely inhibitory (35).

PACT reduces stripping by providing another removal mechanism. From Equation (4), one can see that adding PAC to an activated sludge system reduces the effluent concentration if the compound is adsorbed. Equation (8) shows that the amount removed by stripping decreases with effluent concentration. Weber and Jones (14) and Dietrich (36) also observed that stripping was suppressed in the PACT system. Explanations of the mechanism, whether by simple adsorption, re-

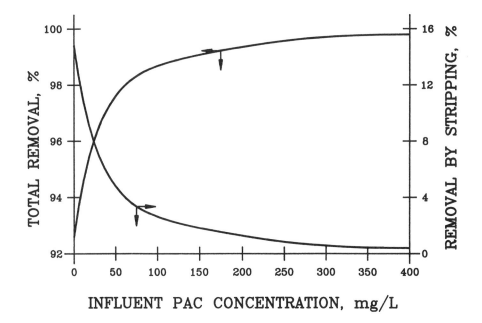

INFLUENT PAC CONCENTRATION, mg/L

BASIS:
θ_N = 10h
θ_C = 480h
X = 6000 mg/L
K_S = 0.2h^{-1}
K_B = 20 x 10^{-5} L/(mg – h)
K_A = 200 x 10^{-5} L/(mg – h)

FIGURE 6.22   Effect of PAC in improving total removal and depressing stripping.

moval of inhibition, or by synergism between the PAC and the biomass, were not offered.

EFFECT OF SLUDGE AGE   Altering the sludge age θ_c in a fixed aeration volume will change the concentration of biosolids, X, which in turn will change removal and air emissions. The effect of sludge age on biomass concentration is determined from the Monod equation and a solids balance. The rate coefficients in the Monod equation are characteristic of the particular wastewater, temperature, and so forth. Figure 6.23 shows the relationship between X and θ_c for a hypothetical industrial wastewater. Figure 6.24, which was generated from Figure 6.23 and Equations (4) and (8), shows that as sludge age is increased, the removal of a given compound (as defined by the rate coefficient in Figure 6.24) increases, and the air emissions are decreased for both an activated sludge (dashed lines) and a PACT system (solid lines). Figure 6.24 also shows that addition of PAC to activated sludge enhances

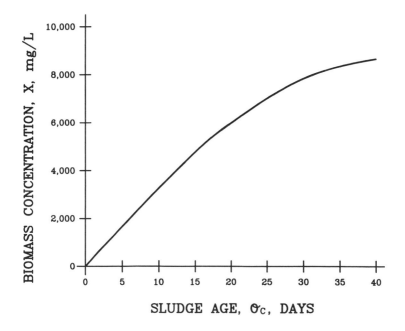

FIGURE 6.23   Biomass variability with sludge age for an industrial wastewater.

overall organics removal and suppresses air stripping over the range of sludge ages. The parameters are the same as in the first example ($\theta_N = 10$ hr, $K_s = 0.05$ hr^{-1}, $K_a = 50 \times 10^{-5}$ L/(mg hr), $K_b = 50 \times 10^{-5}$ L/(mg hr) with the carbon dose fixed at 150 ppm. Again, different results would be obtained for different values of the rate coefficient.

Clarification and Thickening   The importance of good clarification and thickening in an activated sludge process cannot be overemphasized. Bacterial particles that do not settle will be lost from the system and can lead to system failure.

Bacteria in the PACT system tend to grow on and attach themselves to the carbon particles (see normal microscopic picture in Figure 6.25 and an electron microscopic picture in Figure 6.26). Because carbon particles are heavier than sludge flocs, the mixture tends to settle more rapidly in clarifiers than bacterial flocs alone. This is seen in initial settling velocity curves generated for the DuPont Chambers Works, where PACT (at 75- and 150-ppm carbon dose) is compared with activated sludge (Figure 6.27). It is important to note that it is the carbon–biomass ratio and not just carbon dose that controls sludge settling. As a result, PACT plants can be designed at higher than normal loadings.

Polymer is often added to PACT system mixed liquor as the mixed liquor enters the clarifiers. This aids in collecting carbon fines that are not trapped in the biological floc and reduces suspended solids and carbon losses in the clarifier overflow.

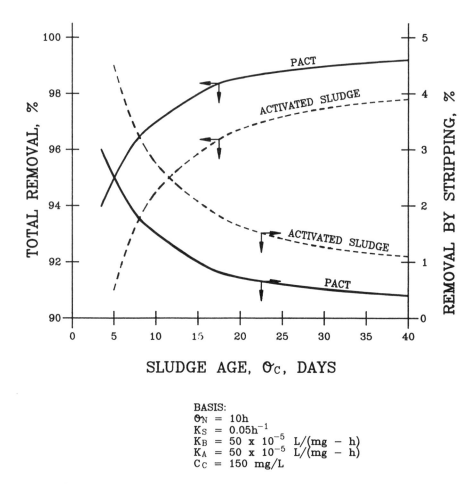

BASIS:
Θ_N = 10h
K_S = 0.05h^{-1}
K_B = 50 x 10^{-5} L/(mg – h)
K_A = 50 x 10^{-5} L/(mg – h)
C_C = 150 mg/L

FIGURE 6.24   Effect of sludge age and PAC on total removal and stripping.

The polymer has little effect on sludge settling rate; and any of several polyelectrolytes may be used for this purpose. Since effectiveness will likely be site specific, polymer screening is required. Adding polymer also helps compact the sludge, so that it does not build up in the bottom of the clarifiers and lead to higher effluent suspended solids.

Carbon Regeneration   Powdered activated carbon from the PACT system has been regenerated by many methods. DuPont's Chambers Works plant regenerated over 18 million pounds of carbon, between 1977 and 1980, in a multiple hearth furnace, having adequate quality for reuse. In addition, every year since 1980, the Vernon, Connecticut, 6.5-mgd municipal plant has regenerated more than 1.5 million pounds of high-quality carbon by the Zimpro® wet air oxidation technique.

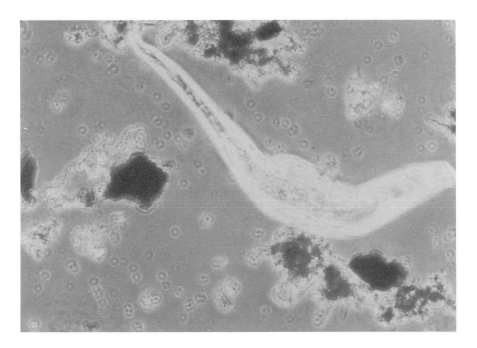

FIGURE 6.25 Microscopic picture of PACT system sludge.

Other thermal processes also have possible application, for example, the transport reactor (37), electric furnaces (38), and fluidized-bed furnaces (38).

The advantages of regenerating carbon are twofold, viz. elimination of sludge disposal problems and recovery of the economic value of the carbon. Except for the DuPont Chambers Works, all PACT systems that regenerate carbon use the wet air oxidation technique. Advantages of the wet air oxidation system over thermal methods include:

- Cost effectiveness
- Ease in permitting
- Produces no NO_x or SO_x emissions
- Autothermal operation
- Slurry regeneration without dewatering

Wet Air Regeneration   Zimpro's wet air oxidation system is a proven method for regenerating PAC from waste PACT system sludge. Experience with wet air regeneration goes back to the early 1970s, on a full-scale demonstration basis at a 1-mgd wastewater treatment plant in Rothschild, Wisconsin (19), to 1975 at the Kimitsu wastewater treatment plant in Japan (40), and to 1979, on the first full-scale U.S. facility at Vernon, Connecticut (41). A typical wet air oxidation system

FIGURE 6.26 Electron microscope picture of PACT system sludge.

is illustrated in Figure 6.28. Spent carbon from a thickener or storage tank at 6–10% solids concentration is pumped at 700–800 lb per square inch gage (psig) to a heat exchanger where the cold sludge is heated by the hot slurry exiting the reactor. Compressed air is also fed to the heat exchanger along with the sludge slurry. The air–sludge mixture then passes to the reactor where the organic material undergoes autooxidation that destroys biomass and organic compounds while regenerating carbon activity. The oxidation step produces enough heat to keep the reactor at the desired 200–260°C operating temperature. The reactor product passes back through the heat exchangers. Following heat exchange, the regenerated slurry passes through a pressure control valve, where the slurry pressure is reduced, and the regenerated carbon is discharged back into the aeration tank of the PACT system. Because the wet air oxidation system is autothermal, requiring no outside source of fuel to sustain the oxidation–regeneration reactions, the steam generator (boiler) is required only during startup periods.

The typical regeneration temperature range is 200–260°C, with the higher temperatures used to obtain a more active carbon. The effect of temperature on the chemical adsorption characteristics of carbon is shown in Table 6.19. These results indicate that the carbon's standard chemical adsorptive characteristics are easily altered by changing regeneration temperature, a design feature of all wet air regeneration systems (40). At regeneration temperatures of 200°C and above, essentially all biological solids are destroyed. Though some solubilization accompanies regeneration, it creates little problem in the PACT system, because the recycle

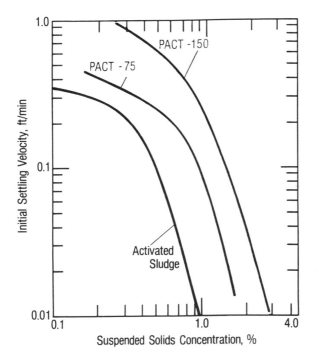

FIGURE 6.27   Comparative settling characteristics of PACT and activated sludge.

organics are composed of biodegradable, low-molecular-weight compounds, approximately 80% being volatile acids of four carbon atoms or less. Most nonbiodegradable organics are also destroyed during regeneration, with destruction being a function of temperature.

As Figure 6.28 denotes, ash elimination from the wet air regeneration (WAR) system is by controlled venting from the system's reactor. In some cases, this may be sufficient to adequately control the amount of ash accumulation in the PACT system. However, in certain cases, where the inlet ash to the PACT system is significant and where controlled venting is unable to keep up with the incoming load to the PACT system, other ash control techniques can be employed. One such method that has been used at a number of the PACT/WAR systems is the Differential Sedimentation Eutriation (DSE) system illustrated in Figure 6.29. In this method, WAR effluent is first mixed with one or more dispersants, then an anionic polymer. This mixture is fed to an elutriation column, where the carbon settles to the bottom for recycle back to the PACT system, and the ash is carried overhead. The ash stream has a cationic polymer added to it, so that the ash can be settled out in a clarifier for disposal.

Thermal Regeneration   Over a period of years, Chambers Works researchers evaluated the high-temperature thermal regeneration of several PACs from PACT

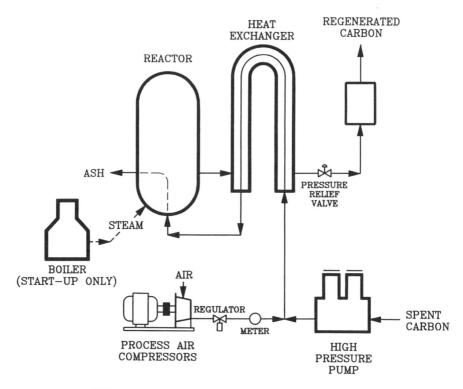

FIGURE 6.28   Typical Zimpro wet air oxidation system.

TABLE 6.19   Effect of Temperature on Chemical Adsorption of Regenerated Carbon Produced by Zimpro Wet Air Oxidation of PACT Sludges

	Relative Adsorbate Efficiency (% of parent carbon)		
	Regeneration temperature (°C)		
	200	230	250
Iodine number	20–40	45–60	60–80
Methylene blue	30–60	60–90	85–135
Erythrosin	30–60	50–130	80–110[a]
Molasses number	50–85	70–110	100–150

[a]Range smaller because of limited database.

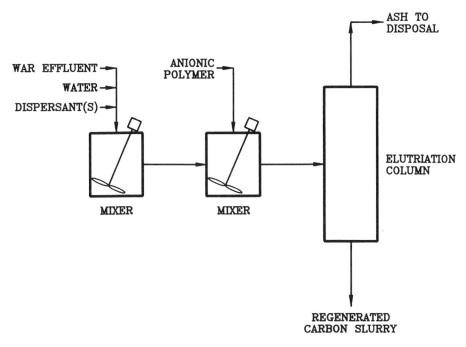

FIGURE 6.29   Zimpro® DSE ash removal process schematic.

sludge, and found that all of the carbons could be satisfactorily regenerated for reuse in the Chambers Works PACT system. Regeneration temperature and time significantly affect both yield and quality of regenerated carbon. Also, the ratio of steam purged into the furnace to the amount of carbon in the sludge sample affects carbon quality. This effect of steam on carbon regeneration is well known (42). In general, increasing temperature, time, and H_2O or CO_2 in the thermal reactor will result in a lower yield but higher quality.

DuPont's Chambers Works high-temperature PACT sludge carbon regeneration facility is the only one that utilizes a multiple hearth furnace to regenerate PAC from a PACT system. The waste sludge is simultaneously destroyed during carbon regeneration. Sludge destruction is now a more valuable aspect than carbon regeneration, since Chambers Works PACT system sludge cannot be landfilled because of RCRA (FO-39) regulations. Off-site disposal in a suitable incinerator is cost prohibitive. The overall system is illustrated in Figure 6.30, starting with cake from a belt filter press. Dewatered sludge (35% solids) from the filter press is pumped to the gatelock feeders on top of the furnace. Wet sludge drops onto the center portion of the top furnace hearth (1) and is pushed by rabble arms in furrows across the hearth, where it begins drying (Figure 6.31). When the sludge reaches the outside of the hearth it passes through drop holes onto the outside portion of hearth 2. Sludge drying is completed on hearth 2 and devolatilization of the sludge begins.

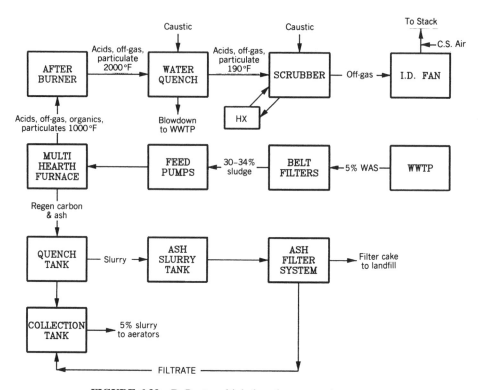

FIGURE 6.30 DuPont multiple-hearth regeneration system.

On hearth 2, sludge is rabbled from the outside toward the center drop hole. Pyrolysis and devolatilization are completed on hearths 3 to 5. This pyrolysis process will leave a carbon char deposited within the pores of the carbon. Carbon is regenerated on the lower hearths by the following endothermic reactions:

$$XCO_2 + C_{Char} \rightarrow (1 - x) C_{Char} + 2XCO$$

$$XH_2O + C_{Char} \rightarrow (1 - x) C_{Char} + XH_2 + XCO$$

Steam is injected to control the water concentration on the lower hearths, and air is injected to control the amount of combustion gases. Operation of the Chambers Works multiple hearth furnace has been sporadic, depending to a large degree on carbon dosage to the PACT system and whether the dosage warrants the cost of regeneration. Average performance data for the year 1980 are shown in Table 6.20.

Carbon recovery ranges from 40 to 75% by weight (fixed carbon basis). The iodine number of this regenerated carbon ranges from 420 to 650 on an "as is" basis. Factors that enhance iodine number reduce yield. An ash purge is also required to prevent buildup of inerts in the aeration tank, and this can be accomplished by acid washing the regenerated carbon.

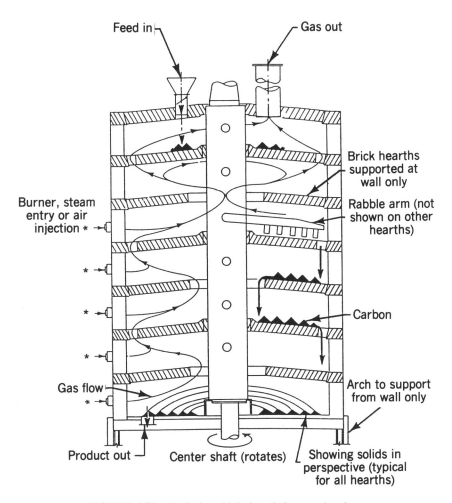

FIGURE 6.31 Typical multiple-hearth furnace drawing.

TABLE 6.20   Regeneration Furnace Performance

	1980 Performance
Regenerated carbon produced, tons/day	12.6
Carbon recovery, % fixed carbon	73
Estimated quality recovery, %	72
Regenerated carbon iodine number	420
Regenerated carbon volatiles, % volatiles	7
Regenerated carbon ash, % ash	21
Furnace in-time, %	68[a]

[a]74% excluding January when furnace was not operating.

Half of the investment in the equipment train (Figure 6.30) is in the off-gas system. Regulations require the afterburner to operate at 2000°F and afford 99.9% destruction of organics. The scrubber system consists of two reverse jets and one froth tray column followed by another reverse jet and a demister. Alternatively, a dry scrubber followed by a baghouse can be used.

OTHER CONSIDERATIONS

Materials of Construction

In general, adding carbon to the activated sludge process has little effect on materials used to construct aeration tanks or clarifiers. Aeration tank corrosion tests were made during the pilot plant program for the DuPont Chambers Works installation, where the wastewater tested averaged about 2800 mg/L total dissolved solids, including 1200 mg/L chloride. Parallel tests were conducted on an activated sludge process, with no significant difference observed between PACT and activated sludge. With the exception of 1020 steel and cast iron, which showed high corrosion rates, crevice corrosion, incipient pitting, and cratering, all materials looked satisfactory. Results of further tests on plastics and cement showed both to be satisfactory.

However, the same cannot be said for locations where there is significant flow, as in clarifier underflow lines. PACT sludges tend to become corrosive/erosive to carbon steel, as sludge concentration increases to about 5% and velocities increase to 8 ft/sec. Sludge line flows should be kept relatively low, with no restricting devices. Particular attention must also be paid to valves in PACT system service, because severe erosion has been seen in some valve designs.

Upgrading Existing Activated Sludge Units

Generally, upgrading an activated sludge unit to a PACT system is straightforward. To date, 19 conversions of existing activated sludge plants to PACT have been successfully accomplished, including 8 municipal and 11 industrial applications. However, due to questions that arise concerning proper aerator design, clarification means, and sludge transport, it is recommended that someone with experience in upgrading activated sludge to a PACT system should be contacted.

PACT System Pilot Testing

Proper pilot testing is essential in order to obtain performance descriptive of a PACT system and to obtain adequate design information. Types of evaluations used are:

- **Wastewater Characterization**

  A complete analysis of the wastewater and effluent required, to include BOD,

COD, TOC, TKN,[1] NH$_3$, P, TSS,[2] TSA,[3] TDS,[4] alkalinity, pH, priority pollutants, and metals is recommended as a first step in predicting how the PACT system will look. A database such as that maintained by Zimpro Environmental then allows estimation of PACT system size and operation and maintenance costs. Pilot plant testing may not be necessary, depending on the waste and effluent quality required.

- **Batch PACT System Testing**

  Batch PACT systems are tested where such applications warrant this type equipment. Typically, 1-liter to 1-gal size aerators are used. A well-equipped laboratory will have automated the operation sufficiently to cycle without manual inputs.

- **Continuous Testing (with and without regeneration)**

  Bench-scale, frame-mounted-scale, and pilot-plant-scale units are frequently used. Bench-scale testing typically uses approximately 7.5-liter aerators, with approximately 22 liters of flow per day. Flow volume is dependent on waste strength. Frame-mounted units typically use 25- to 100-liter aerators and treat 75–300 liters of wastewater per day. Pilot plant units have been operated with 4000- to 50,000-liter aerators and a flow volume of 8–150 liters per minute.

Suitable pilot-scale regeneration systems are also available for those cases that warrant such investigations. If regeneration is a consideration, then the selected equipment should be pilot tested in conjunction with any PACT system testing undertaken.

PACT SYSTEM TREATMENT FOR DYE/TEXTILE WASTES

Introduction

The attributes of the PACT system, as previously discussed, make it applicable to the dye and textile industries of today. Brief descriptions of pilot testing and some of the full-scale PACT system facilities treating dye/textile wastes are now described.

PACT System Applicability

DuPont Chambers Works Testing DuPont needed a process to remove color from the wastewater of the Chambers Works in Deepwater, New Jersey. This large multiproduct plant produced dyes, some of which ended up in the wastewater.

---

[1] Total Kjeldahl Nitrogen.
[2] Total Suspended Solids.
[3] Total Suspended Ash.
[4] Total Dissolved Solids.

TABLE 6.21   Chambers Works Organics and Color
Performance

Operating conditions	
Carbon dose, ppm	134
Aeration temperature, °C	23
Hydraulic residence time, hr	8.4
Feed quality	
Soluble BOD_5, mg/L	169
Soluble TOC, mg/L	195
Color, APHA	1400
Effluent quality	
Soluble BOD_5, mg/L	5.3
Soluble TOC, mg/L	35.3
Color, APHA	450

Laboratory tests showed that conventional activated sludge treatment removed essentially none of the color (16). However, adding 50, 100, and 250 ppm of PAC to the feed of laboratory activated sludge units reduced mean color levels from 1300 American Public Health Association (APHA) units to 560, 490, and 210 APHA color units, respectively (Figure 6.10). The full-scale 40-mgd plant averaged 70% color removal after startup with a continuous carbon dosage of 134 ppm. Some of the key operating conditions and results, including color removal, for the Chambers Works are shown in Table 6.21.

EPA Pilot Plant Testing   To illustrate the effectiveness of the PACT system for treating dye wastes, the U.S. EPA conducted an extensive pilot plant study on a waste stream collected downstream from the discharge of a dyes and pigments processing plant (43). Sampling showed that 95% of the COD was coming from the industrial discharge. Waste characteristics are shown in Table 6.22.

TABLE 6.22   Dyes and Pigments Plant Wastewater Characteristics from EPA Study

Parameter	Minimum	Maximum
BOD_5, mg/L	460	1900
COD, mg/L	1260	2840
TOC, mg/L	420	1080
TSS, mg/L	290	1580
Color, ADMI units	1070	9910
NH_3, as N, mg/L	22	160
Total chromium, mg/L	0.58	13.9
Total lead, mg/L	1.5	13.6
Total copper, mg/L	0.23	2.4
Total zinc, mg/L	0.27	12.2

The feed to each of the parallel activated sludge and PACT units was about 173 liters per day, with a 1.2-day hydraulic retention time. Dissolved oxygen concentrations were maintained at or above 2.0 mg/L, pH was maintained between 7.0 and 8.5, and the temperature ranged from 22 to 27°C. Five parallel activated sludge and PACT tests were conducted at the operating conditions summarized in Table 6.23.

Results of the tests with the dyes and pigments wastewater are shown in Table 6.24, where the data are summarized in order of increasing carbon dose. In every case, the PACT unit performed better than the corresponding activated sludge unit, in terms of BOD, COD, TOC, and color removal. Similarly, the PACT unit also was better at removing acute toxicity, as measured with fathead minnows, *Daphnia magna*, or the Microtox® test (Table 6.25).

EPA personnel correlated the organics, color, and toxicity removal data to removal by the PAC as follows for the 68-ppm carbon dose:

Activated sludge (AS) effluent COD = 490 mg/L

PACT system effluent COD = 440 mg/L

Assumed COD removal by PAC = 490 − 440 = 50 mg/L

$$\% \text{ COD Removed} = \frac{\text{AS} - \text{PACT}}{\text{AS}} = \frac{490 - 440}{490} = 10\%$$

TABLE 6.23   Summary of EPA Test Operating Conditions

Phase[a]	Carbon Dose (ppm)	SRT (Days)	MLSS (mg/L)	Aerator PAC (mg/L)
1				
AS	0	7	1,900	0
PACT	130	7	2,800	850
2				
AS	0	8	2,300	0
PACT	1,800	9	17,500	14,000
3				
AS	0	7	2,000	0
PACT	1,000	10	12,900	8,200
4				
AS	0	4	1,090	0
PACT	290	4	2,950	1,300
5				
AS	0	16	3,100	0
PACT	68	19	4,660	1,100

[a]AS, activated sludge; PACT, PACT system.

TABLE 6.24   Results of EPA Tests on Dyes Wastewater

Phase	Carbon Dose (ppm)	% BOD Removal	% COD Removal	% TOC Removal	% Color Removal
5					
AS	0	99	58	60	<0
PACT	68	99	63	62	<0
1					
AS	0	97	64	59	11
PACT	130	99	70	61	23
4					
AS	0	98	57	56	<0
PACT	290	99	71	70	16
3					
AS	0	98	74	64	34
PACT	1000	>99	96	87	98
2					
AS	0	98	75	61	<0
PACT	1800	>99	96	82	96

TABLE 6.25   Acute Bioassay Results from EPA Dyes Waste Tests

Phase	Carbon Dose (ppm)	Acute Bioassay, in Toxic Units[a]		
		Fathead Minnows	Daphnia magna	Microtox
5				
AS	0	1.9	2.0	<2.0
PACT	68	<1.1	1.9	<1.0
1				
AS	0	6.7	4.8	4.4
PACT	130	4.3	3.2	<3.6
4				
AS	0	1.7	2.0	<2.5
PACT	290	<1.0	1.6	<1.0
3				
AS	0	2.9	2.1	12.5
PACT	1000	<1.2	<1.1	<1.0
2				
AS	0	5.0	6.7	3.8
PACT	1800	<2.6	4.2	<1.0

[a]Toxic unit = $100/LC_{50}$; LC_{50} = concentration that kills 50% of test organisms.

The results of these calculations, for the chemical parameters and the bioassays are summarized in Table 6.26. For the chemical parameters, including color, there is a definite trend of increasing removal with increasing carbon dose. For the bioassays, a trend also exists, but with a lot of scatter. This is due in part, to the fact that the bioassays were conducted in a manner that did not give definitive results (many of the numbers generated were less than or greater than numbers).

EPA personnel then calculated the incremental amount of COD removed per weight of carbon fed to the aeration tanks, with the following results:

PAC Dose mg/L	mg SCOD Removed per mg Carbon Dosed
68	0.74
130	0.85
290	0.59
1000	0.27
1800	0.20

These data were plotted according to the Freundlich equation, as shown in Figure 6.32:

$$\frac{X}{M} = K(C_f)^{1/n}$$

where X/M = amount of SCOD removed per unit weight of carbon
C_f = amount of SCOD in the final effluent for each test
SCOD = soluble COD and n and K are constants.

Also plotted in Figure 6.32 are three lines from jar tests conducted on the effluents arising from the activated sludge unit employed for these tests. The line for the PACT unit shows significantly more removal of SCOD than was expected from the carbon adsorption of the activated sludge units alone.

TABLE 6.26   Organics, Color, and Toxicity Removed by PAC

	% Removal of					
Carbon Dose (ppm)	COD	SOC	Color	Fathead Minnow Toxicity	*Daphnia magna* Toxicity	Microtox Toxicity
68	10	10	11	>42	5	<50
130	16	7	14	36	33	>18
290	33	30	23	>41	20	>60
1000	84	65	97	>59	>48	>92
1800	85	55	96	>48	37	>74

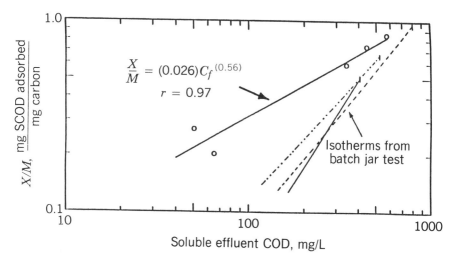

FIGURE 6.32 Carbon adsorption isotherm for SCOD.

Dyestuff Technology Evaluation Keinath (44) evaluated granular activated carbon, biological treatment, the PACT system, and ozonation for the removal of organics and priority pollutants from dyestuff manufacturing wastewaters. The PACT system evaluation, involving wastewaters from three dyestuff manufacturing sites, showed significant improvements for soluble organic carbon (SOC) and color removal for the PACT system compared to conventional biological treatment. Both processes were very effective for removing organic priority pollutants in this study. Data for metals removal was scattered, ranging from zero to 97% removal, depending on the metal and the wastewater.

Zimpro Testing Zimpro Environmental conducted an extensive testing program to develop and optimize an effective approach to removing color and COD from an industrial dye wastewater. A wide variety of conventional and advanced treatment techniques were employed and evaluated. They included chemical oxidation (Cl_2, O_3, thiourea), UV-catalyzed oxidation, wet oxidation, precipitation with organic and inorganic flocculants, and carbon treatment. Because the waste was highly variable (ranges were 390–2100 mg/L BOD_5, 760–5000 mg/L COD, 1.9–8.9 pH, 0.14–0.5 BOD/COD ratio, 0.07–0.23 TKN/COD ratio), with a maximum color intensity of 37,900 APHA units, no single technology provided a clear solution to the treatment of this particular wastewater. Most technologies were ineffective and/or were uneconomical approaches. However, a combined chlorination plus two-stage PACT system provided an effective treatment and was the most economical. The results of this evaluation indicated effluent COD and color were related as shown in Figure 6.33 and led to the design of a full-scale PACT plus chlorination system.

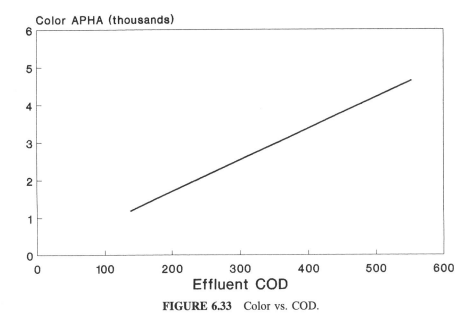

FIGURE 6.33   Color vs. COD.

Full-Scale Experience

Previous sections contained discussion of some of the investigations undertaken that ascertain the ability of the PACT system to treat dye-containing wastewater of varying organic strength and color intensities. Such capabilities have led to the design and construction of several full-scale PACT systems. A brief summary of a few of these facilities follows.

DuPont Chambers Works   This was the first and largest industrial application of the PACT system. The plant is a 40-mgd facility that was started up in 1977 and has operated continuously ever since. In 1991, this single-stage PACT system was converted to a two-stage system to improve its performance. Performance data for the single-stage system are shown in Table 6.5, and its operating conditions are given in Table 6.6.

Vernon, Connecticut   In the early to mid 1970s, Vernon's engineer investigated a number of technologies in order to upgrade an existing trickling filter biological wastewater treatment plant. At that time, the plant was badly overloaded by municipal wastes and industrial wastes from local dye/textile manufacturing operations. Inadequate treatment led to numerous permit violations, causing deterioration of the Hockanum River. Passthrough of textile dye wastes also badly colored the river, causing it to be devoid of aquatic life. Textile dyehouse operations operated 7 days a week, 24 hr per day, concentrating on the dyeing of various fabrics.

Numerous treatment alternatives were evaluated and, based on the evaluation, a PACT/WAR system was pilot plant tested at Vernon. A 6.5-mgd PACT system with wet air regeneration was selected and designed, with operation beginning in early 1979. Wastewater BOD$_5$ and COD were reduced from 200 and 840 mg/L to <5 and <80 mg/L, respectively. Wastewater color was, for the first time, essentially removed, as the PACT system reduced color levels from 5000 APHA units to less than 40 APHA units (41). Currently, although only one dyehouse remains in operation, the PACT/wet air regeneration system continues to reliably treat Vernon's wastewaters and is being expanded for the next generation of users.

Mt. Holly, New Jersey   A 5-mgd PACT/WAR system began operation in March 1982, after the previous trickling filter system was unable to meet stringent BOD$_5$, total suspended solids (TSS), ammonia, and color standards set for Rancocas Creek. A dye manufacturing operation was the major industrial contributor to the plant. Prior to installation of the PACT system, the biological filters were unable to nitrify or to remove color (45). The project engineer evaluated a number of technologies to meet the discharge goals and compared the cost of the PACT/WAR system against activated sludge plus granular activated carbon (Table 6.27) as add-ons to the existing trickling filters. Significant cost advantages existed for the PACT system in both capital and operating cost areas.

The project engineer then analyzed PACT system sludge regeneration alternatives, using both wet air oxidation and a multiple hearth furnace (Table 6.28). The analysis showed a large operating advantage for the WAR system, even without considering the superior carbon recovery ability of wet air oxidation over that of the multiple hearth furnace.

Burlington, North Carolina (East Plant)   This highly industrialized, textile manufacturing area relies on its 12-mgd PACT/WAR system to meet the BOD$_5$, TSS, color, ammonia, phosphorus, and bioassay requirements (*Daphnia magna*) of its direct discharge permit. It is accurate to say that, without the PACT system, neither color nor bioassay standards would be met.

The PACT system went on-line in 1980, after on-site pilot testing was conducted to confirm design recommendations and treatment performance. Raw waste color averages 1000 Pt-Co units at times, yet effluent remains crystal clear and colorless.

TABLE 6.27   Cost Effectiveness Evaluation at Mt. Holly, New Jersey

	PACT/Wet Air Regeneration	Activated Sludge with Granular Activated Carbon
O&M (yearly)	$ 261,751	$ 505,550
O&M (present worth[a])	$2,772,985	$ 5,355,797
Capital	$4,140,352	$ 6,983,188
Total present worth[a]	$6,913,337	$12,338,985

[a]Present worth: 7%, 20 years; O&M assumed constant.

TABLE 6.28 Comparison of Regeneration Costs at Mt. Holly, New Jersey (Cost per Ton of Dry Solids)

	Multiple Hearth Regeneration (Includes Dewatering)	Wet Air Regeneration
Power	$ 16.30	$ 18.92
Fuel	43.20	0.00
Labor	32.00	11.00
Maintenance	6.26	3.84
Total, $/ton	$ 97.76	$ 33.76
Operating cost, $/yr	$ 178,400	$ 61,600
Capital cost, $	$1,959,000	$1,500,000
Present worth, $	$3,849,000	$2,152,700

Regeneration of the spent PACT system sludge is important here, not only for the recovery of the PAC but also to minimize the amount of solids to the landfill, and to optimize performance of phosphorus control (46, 47).

Ciba-Geigy, Toms River, New Jersey The Toms River plant, a large producer of dyestuffs, epoxy resins, and additives, converted an existing activated sludge system to a PACT system in 1985, after the New Jersey Department of Environmental Protection imposed a strict aquatic toxicity limit on its ocean discharge, based on mysid shrimp and monitoring of Ames mutagenicity. The discharge permit, labeled as the most stringent in the nation, was not achievable by conventional biological treatment. The PACT system, however, was able to:

- Achieve >90% TOC removal
- Achieve >98% BOD_5 removal
- Establish and maintain nitrification, achieving <1 mg/L NH_3 in the effluent
- Achieve 100% survival of mysid shrimp in 100% effluent
- Produce an effluent that passed the Ames test and does not have a tendency to bioaccumulate

Because of the PACT system's excellent performance, the system was awarded a Certificate of Excellence by the NJWPCA in 1989, and received a National Environmental Award on Environmental Success from President George Bush in 1990 (48, 49).

Crompton-Knowles, Birdsboro, Pennsylvania In 1986, Crompton-Knowles converted an existing 0.25-mgd granular activated carbon and biological system to a PACT system. The existing system was unable to economically treat the water-soluble dyes, organics, and nitrogenous compounds. The PACT system was pilot tested at Zimpro Environmental as well as on-site.

The original system used granular activated carbon (GAC) treatment prior to biological treatment to protect the sensitive biosystem from upset. However, this arrangement caused the adsorption of so much biodegradable material onto the carbon that rapid carbon regeneration became a necessity. Reversing the order (biotreatment and GAC) was not possible, because of biotoxicity of the dye wastes themselves. Evaluation of the PACT system showed significant operating cost and capital cost benefits when compared to the original system. And with PACT it was possible to double production without expanding the original wastewater plant and to eliminate the expensive GAC columns without loss of treatment stability.

The PACT system has met National Pollutant Discharge Elimination System (NPDES) discharge requirements for BOD, TSS, ammonia, and color. Effluent color, for instance, is consistently less than 100 APHA units, well below the 600 APHA unit limit (9).

REFERENCES

1. D. G. Hutton and F. L. Robertaccio, U.S. Pat. 3,904,518, September 9, 1975.

2. J. A. Meidl and A. R. Wilhelmi, PACT®: An Economical Solution in Treating Contaminated Groundwater and Leachate, paper presented at the New England Water Pollution Control Association meeting, January 27–29, 1986.

3. G. J. O'Brien, R. A. Reich, L. M. Szabo, M. H. Feibes, C. N. McManus, and H. W. Heath, in John M. Bell, ed., *44th Annual Purdue Industrial Waste Conference Proceedings*, Lewis Publishers, Chelsea, Michigan, 1990, pp. 325–334.

4. A. D. Adams, *Water & Sewage Works*, August, 46 (1975).

5. J. A. Meidl and T. J. Vollstedt, Use of Powdered Carbon to Treat Contaminated Groundwater and Leachate, paper presented at the Haztech International Conference, Denver, Colorado, August 1986.

6. W. M. Copa, Hybrid Anaerobic Processes: The Multizone Anaerobic Reactor and the Anaerobic PACT® Process, University of Wisconsin Short Course, "Anaerobic Treatment Technologies," Milwaukee, Wisconsin, January 23–24, 1989.

7. B. P. Flynn and L. T. Barry, Finding a Home for the Carbon: Aerator (Powdered) or Column (Granular), paper presented at the 31st Annual Purdue Industrial Waste Conference, West Lafayette, Indiana, May 5, 1976.

8. D. G. Hutton, Priority Pollutant Removal—Comparison of the DuPont PACT® Process with Activated Sludge Followed by Granular Activated Carbon Columns, paper presented at the Symposium on Applications of Adsorption to Wastewater Treatment, Vanderbilt University, Nashville, Tennessee, February 16–19, 1981.

9. B. Dobinsky and C. P. Wickersham, *Chem. Proc.* **50**(13), 173 (1987).

10. F. A. DiGiano, Toward a Better Understanding of the Practice of Adsorption, paper presented at the 87th National Meeting of the AIChE, Boston, Massachusetts, August 19–22, 1979.

11. D. G. Hutton and F. L. Robertaccio, Use of Powdered Carbon for Textile Wastewater Pollution Control, paper presented at the Textile Industry Technology Symposium, Williamsburg, Virginia, December 5, 1978.

12. J. A. Meidl, An Overview of PACT® Wastewater Treatment for the Petroleum and Petrochemical Industries, paper presented at the 203rd American Chemical Society Meeting, San Francisco, California, April 6, 1992.

13. P. B. DeJohn and A. D. Adams, *Hydrocarbon Proc.*, **54**(10), 104 (1975).

14. W. J. Weber and B. E. Jones, Toxic Substance Removal in Activated Carbon and PAC Treatment Systems, EPA Report 600/2-86/045, Accession No. PB86-182425, 1986.

15. Anon., *Chem. Eng.* **84**(19), 95 (1977).

16. F. L. Robertaccio, D. G. Hutton, G. Grulich, and H. L. Glotzer, Treatment of Organic Chemicals Plant Wastewater with the DuPont PACT Process, paper presented at the AIChE National Meeting in Dallas, Texas, February 20–23, 1972.

17. S. E. Lee, H. S. Shin, and B. C. Paik, *Water Res.* **23**(1), 67 (1989).

18. D. C. McIntyre, Powdered Activated Carbon Wastewater Treatment, paper presented at the Division of Environmental Chemistry, American Chemical Society meeting in Washington, DC, August 23, 1992.

19. W. Burant, Jr., and T. J. Vollstedt, *Water & Sewage Works*, **120**(11), 42 (1973).

20. F. Sampayo and D. Swets, in *Proceedings of the 6th Annual Industrial Pollution Conference*, sponsored by WWEMA, St. Louis, Missouri, Water and Wastewater Equipment Manufacturers Association, McLean, Virginia, April 1978, pp. 401–424.

21. A. S. Ng and M. K. Stenstrom, *J. Environ. Eng.* **113**(6), 1285 (1987).

22. G. B. Foertsch and D. G. Hutton, Scale-up Tests of the Combined Powdered Carbon and Activated Sludge (PACT) Process for Wastewater Treatment, paper presented at the Virginia Water Pollution Control Association meeting, Natural Bridge, Virginia, April 30, 1974.

23. Anon., New PACT System On-Line In Israel, *Reactor*, No. 76, Zimpro Environmental, Rothchild, Wisconsin, June, 1993.

24. J. S. Lee and W. K. Johnson, *J. WPCF* **51**(1), 111 (1979).

25. R. G. Simms, Successful Removal of Toxics at Kalamazoo Using the PACT® Process, paper presented at the 59th Annual WPCF Conference, Los Angeles, California, 1986.

26. B. P. Flynn, A Model for the Powdered Activated Carbon–Activated Sludge Treatment System, paper presented at the 30th Annual Purdue Industrial Waste Conference, West Lafayette, Indiana, May 7, 1975.

27. G. J. O'Brien, *Water Environ. Res.* **64**(7), 877 (1992).

28. E. L. Stover and D. F. Kincannon, *J. WPCF* **55**(1), 97 (1983).

29. J. W. Blackburn and W. L. Troxler, *Environ. Prog.* **3**(3), 163 (1984).

30. J. W. Blackburn, *Environ. Prog.* **6**(4), 217 (1987).

31. D. G. Hutton, *Water—1973*, A.I.Ch.E., Symposium Series, 136, Vol. 70, 1974, pp. 91–100.

32. H. Sontheimer, J. C. Crittenden, and S. R. Summers, *Activated Carbon for Water Treatment*, 2nd ed., DVRW-Forschungsstelle am Engler-Bunte Institut der Universitat, Karlsruhe, West Germany, 1988, p. 119.

33. G. J. O'Brien and E. W. Teather, in Ronald F. Wukasch, ed, *Proceedings of the 47th Industrial Waste Conference*, Lewis Publishers, Chelsea, Michigan, 1993, pp. 359–367.

34. G. J. O'Brien, *Environ. Prog.* **12**(1), 76 (1993).

35. G. J. O'Brien, unpublished work.

36. M. J. Dietrich, W. M. Copa, A. K. Chowdhury, and T. L. Randall, *Environ. Prog.* **7**(2), 143 (1988).

37. C. F. Koches and S. B. Smith, *Chem. Eng.* **79**(9), 46 (1972).

38. Anon., *Environ. Sci. Tech.* **11**(9), 854 (1977).

39. Anon., 201 Facilities Plan, Kalamazoo Metropolitan Area, Jones & Henry Engineers, Ltd., June 1977.

40. J. A. Meidl, C. L. Berndt, and K. Nomoto, Experience with Full Scale Wet Air Oxidation of Spent Carbon from the PACT® Process, paper presented at the 51st Annual Conference WPCF, Anaheim, California, October 1–6, 1978.

41. C. Pitkat and C. Berndt, *Public Works*, **112**(10), 54 (1981).

42. J. W. Hassler, *Activated Carbon*, Chemical Publishing, New York, 1963.

43. G. M. Shaul, M. W. Barnett, T. W. Neiheisel, and K. A. Dostal, Activated Sludge with Powdered Activated Carbon Treatment of a Dyes and Pigments Processing Wastewater, Report No. EPA-600/D-83-049, U.S. EPA, Cincinnati, Ohio, 1983.

44. T. M. Keinath, Technology Evaluation for Priority Pollutant Removal from Dyestuff Manufacture Wastewaters, Report No. EPA-600/2-84-055, U.S. EPA, Cincinnati, Ohio, 1984.

45. Anon., Mt. Holly Meeting Nitrification, Color Removal, Sludge Disposal Challenges, *Reactor*, No. 51, Zimpro Environmental, Rothschild, Wisconsin, November 1983.

46. Anon., East Burlington—Co-systems Winner, *Reactor*, No. 67, Zimpro Environmental, Rothschild, Wisconsin, January 1990.

47. A. R. Kornegay and C. Sell, *Water Environ. Tech.* **2**(10), 26 (1990).

48. Anon., Ciba-Ceigy Meeting Tough Bio-assay Test, *Reactor*, No. 57, Zimpro Environmental, Rothschild, Wisconsin, June 1986, pp. 13–14.

49. G. Huff and S. Schexnailder, *Pollution Eng.* (7), 98 (1988).

CHAPTER 7

OPERATING EXPERIENCE WITH THE PACT® SYSTEM

DANIEL C. MCINTYRE
Ciba-Geigy Corporation, Toms River, New Jersey

INTRODUCTION

Chapter 6, written by Hutton, Meidl, and O'Brien, covered the theoretical aspects of the PACT® treatment process. Ciba-Geigy purchased the rights to install and operate the PACT system at its Toms River Plant from Zimpro Environmental, Inc., in 1985. This chapter will present (1) the reasons for selecting the PACT treatment process, (2) bench-scale and pilot plant results, and (3) our full-scale operating experience with this treatment process, with particular emphasis on effluent toxicity reduction, biological oxygen demand (BOD) and total organic carbon (TOC) removal efficiencies, and odor control benefits achieved with this process.

History

In 1977, Ciba-Geigy constructed a conventional activated sludge biological treatment plant for end-of-pipe treatment of the aqueous waste generated by the production processes at its Toms River plant. Production facilities of the plant consisted of batch-type processes producing approximately 450 different products in a given year. The main product lines were organic dyes, both finished products and intermediates, epoxy resins, and fine chemicals used by the textile, paper, and plastics industries.

Total aqueous effluent volume generated by these production facilities was typically in the range of 3–4 million gallons per day (gpd). In 1985, groundwater extraction facilities were started up, adding approximately 450,000 gpd of contaminated groundwater to the overall treatment load.

The biological treatment process constructed in 1977 is depicted in Figure 7.1. The overall treatment process included equalization basins, neutralization, metals precipitation, aeration basins, and clarification, to separate the biological sludge

Environmental Chemistry of Dyes and Pigments, Edited by Abraham Reife and Harold S. Freeman.
ISBN 0-471-58927-6   © 1996 John Wiley & Sons, Inc.

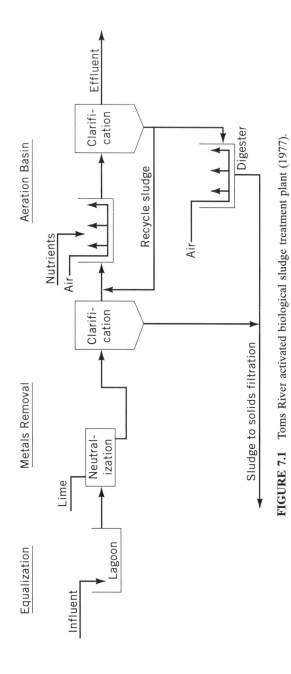

FIGURE 7.1 Toms River activated biological sludge treatment plant (1977).

from the final effluent. Typically, approximately 70–80% of the sludge was recy-cled to the aeration basins, with the remainder being pumped to an aerobic digester to reduce the sludge volume. From the digester, the sludge was pumped to a plate and frame filter to produce a filter cake for disposal in an on-site landfill.

The final effluent was pumped through a 10-mile pipeline and discharged 3500 ft offshore into the Atlantic Ocean. This pipeline was constructed and placed in operation in 1966. At that time, there was some opposition to ocean discharge in the beach communities; however, construction of the pipeline was generally her-alded as a sound environment improvement, by the Toms River community as well as by local and state officials. However, in 1984, the pipeline developed a small leak, which resurfaced the original concerns about ocean discharge, and ignited major opposition to ocean discharge by both environmental groups and beachfront residents. Ciba-Geigy's ocean pipeline, the only direct industrial ocean discharge pipeline in the state, became a natural focal point for environmentalists. In addition, odor complaints were received on a regular basis from the residential areas sur-rounding the plant.

It was in this context that the New Jersey Department of Environmental Pro-tection (NJDEP) issued a New Jersey Pollution Discharge Elimination System (NJPDES) permit renewal for continued discharge of treated effluent to the Atlantic Ocean. The NJDEP stated publicly that the permit was one of the most stringent permits ever issued in the nation. Overall, permit requirements (see Table 7.1) included the following:

- General tightening of both organic and inorganic discharge limits
- 85% removal of BOD
- Aquatic toxicity requirement of 50% survival of mysid shrimp in 50% effluent (LC_{50} 50%)[1]
- Ames test to monitor mutagenicity
- Plume-dispersing studies in the Atlantic Ocean

The initial bench-scale studies conducted relative to compliance with the new permit indicated that the Mysid shrimp toxicity requirement was the most stringent of the new requirements and that treatment requirements were a consequence of this toxicity requirement.

TOXICITY IMPROVEMENT PROGRAM

Prepermit Toxicity Test Comparisons

Inclusion of an aquatic toxicity test requirement in the permit renewal was antici-pated by the plant. Comparative toxicity tests were conducted on the treated plant

---

[1] Liquid concentration at which 50% of the test species survive.

TABLE 7.1 NJPDES Effluent Limitations for Outfall 003, Toms River Plant

Effluent Characteristics[a]	Units	Discharge Limitations				Monitoring Frequency
		Minimum	Average	Maximum		
Flow	mgd	N/A	N/A	N/A		Continuous
pH (standard units)	su	6.0	N/A	9.0		Continuous
Oil & Grease	mg/L	N/A	10	15		Monthly
5-day BOD	% removal	N/A	85	N/A		2 × week
Lead	kg/day	N/A	1.4	3.1		2 × week
Mercury	kg/day	N/A	0.23	0.27		2 × week
Zinc	kg/day	N/A	13.2	23.3		2 × week
5-day BOD						
EDP–EDP + 2 yr	kg/day	N/A	1886	3595		2 × week
EDP + 2 yr–EDP + 5 yr	kg/day	N/A	939	2412		2 × week
TOC						
EDP–EDP + 2 yr	kg/day	N/A	4291	6815		2 × week
EDP + 2 yr–EDP + 5 yr	kg/day	N/A	3467	6815		2 × week
TSS						
EDP–EDP + 2 yr	kg/day	N/A	3060	6895		2 × week
EDP + 2 yr–EDP + 5 yr	kg/day	N/A	1879	6895		2 × week
Chromium						
EDP–EDP + 2 yr	kg/day	N/A	16.0	37.0		2 × week
EDP + 2 yr–EDP + 5 yr	kg/day	N/A	7.1	11.1		2 × week
Copper						
EDP–EDP + 2 yr	kg/day	N/A	16.0	43.2		2 × week
EDP + 2 yr–EDP + 5 yr	kg/day	N/A	10.5	43.2		2 × week
Nickel						
EDP–EDP + 2 yr	kg/day	N/A	10.30	24.80		2 × week
EDP + 2 yr–EDP + 5 yr	kg/day	N/A	1.57	5.84		2 × week

			96-hr LC₅₀ ≥ 5% (by volume)		
			96-hr LC$_{50}$ ≥ 10% (by volume)		
			96-hr LC$_{50}$ ≥ 20% (by volume)		
			96-hr LC$_{50}$ ≥ 50% (by volume)		
Toxicity (bioassay)					
EDP–EDP + 6 mo					Monthly
EDP + 6 mo–EDP + 18 mo					Monthly
EDP + 18 mo–EDP + 36 mo					Monthly
EDP + 36 mo–EDP + 60 mo					2 × month
1,2,4-Trichlorobenzene	kg/day	N/A	N/A	1.46	Monthly
1,2-Dichlorobenzene	kg/day	N/A	N/A	1.46	Monthly
1,3-Dichlorobenzene	kg/day	N/A	N/A	1.46	Monthly
1,4-dichlorobenzene	kg/day	N/A	N/A	1.46	Monthly
1,2-*trans*-Dichloroethylene	kg/day	N/A	N/A	1.46	Monthly
2-Chlorophenol	kg/day	N/A	0.83	2.19	2 × week
bis(2-Chloroethoxy)methane	kg/day	N/A	1.57	6.13	2 × week
Benzene	kg/day	N/A	N/A	1.46	Monthly
Chlorobenzene	kg/day	N/A	N/A	1.46	2 × week
Ethylbenzene	kg/day	N/A	N/A	1.46	Monthly
Naphthalene	kg/day	N/A	N/A	1.46	Monthly
Nitrobenzene	kg/day	N/A	4.4	25.3	2 × week
Tetrachloroethylene	kg/day	N/A	N/A	1.46	Monthly
Toluene	kg/day	N/A	2.8	6.57	2 × week
Trichloroethylene	kg/day	N/A	N/A	1.46	Monthly
Chlorinated hydrocarbons	kg/day	N/A	1.5	4.6	2 × week
Priority pollutants (e.g. asbestos)		N/A	N/A	N/A	Monthly
Nitrogen scan		N/A	N/A	N/A	Monthly
Mutagenicity (Ames test)		N/A	N/A	N/A	Quarterly

[a]EDP, effective date of permit (July 1, 1985).

effluent over the period from October 1984 to April 1985 for three different aquatic species. Overall results were as follows (see Figure 7.2):

	$LC_{50}\%$
Grass shrimp	76–97%
Sheephead minnow	48–88%
Mysid shrimp	5.6–19%

Grass shrimp or sheephead minnow were normally the aquatic species selected by the regulatory agencies for defining toxicity discharge limits. Based on these preliminary investigations with these two species, the effluent would not be considered to be too toxic to marine life.

The treated plant effluent did, however, exhibit high toxicity to Mysid shrimp (*Mysidopsis bahia*). At this point, the use of Mysid shrimp as an aquatic test species was in the early stages of development and had not been used, for regulatory purposes, as a species for measuring the toxicity of the effluent discharges. The testing protocol of Mysid shrimp had not been officially promulgated and, for the above reasons, it was not anticipated that Mysid shrimp would be specified as the species for measuring compliance with toxicity requirements.

Surprisingly, the draft permit issued in May 1985 contained a 96-hr LC_{50} 50% (50% survival in 50% effluent for 96 hr) toxicity limit based on the Mysid shrimp species. The NJDEP was immediately informed that this requirement could not be met with the existing biological treatment plant and an extended compliance sched-

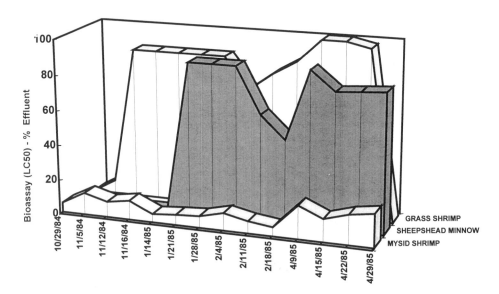

FIGURE 7.2   Toxicity test comparison, Toms River plant (prepermit).

ule was requested, to install unspecified modifications to the treatment to comply with this toxicity requirement. The request was granted, and the final permit, which became effective on July 1, 1985, contained the following compliance schedule:

Compliance Schedule	Toxicity Requirement 96-hr Mysid Shrimp Bioassay
First 6 months	$LC_{50} \geq 5\%$ by volume
6 months through 18 months	$LC_{50} \geq 10\%$ by volume
18 months through 36 months	$LC_{50} \geq 20\%$ by volume
36 months through 60 months	$LC_{50} \geq 50\%$ by volume

The NJDEP also granted compliance schedules for BOD, total suspended solids (TSS), TOC, and the metal limits in the final permit (see Table 7.1).

Literature Review

Review of the literature (1–3) did not identify any significant information regarding the factors that influence Mysid shrimp toxicity. A large body of information was available regarding factors influencing toxicity toward other aquatic species. Based on this body of literature, it was decided to concentrate the toxicity improvement efforts in the following four areas:

- Treatment and/or reduction in the volume of individual product waste streams that exhibited high toxicity
- Overall reduction of TOC concentrations in the final effluent
- Reduction in metal concentrations in the final effluent
- Reduction of the concentration of nitrogen compounds in the final effluent

Product Waste Stream Characterization

Initial review of the overall waste streams coming from each of the four main production buildings (individually) indicated that there was little difference in the toxicity characteristics of these streams, and all exhibited high toxicity to Mysid shrimp. This data initiated an extensive program designed to characterize the individual batch waste streams from the various product batches.

It is difficult to characterize the waste streams generated by individual dye or resin batches through chemical analysis. For this reason, a classification system base was developed to classify the product waste streams based on relative biodegradation rates (treatability) and their potential for inducing Mysid shrimp toxicity (4). Mysid shrimp toxicity of the individual product waste stream was evaluated by using a 1:25 dilution and using a shortened 48-hr Mysid shrimp bioassay, rather than the official 96-hr test, to reduce bioassay time requirements.

Traditionally, rapid determination of inhibition thresholds and biological kinetic removal rates have been evaluated by batch studies. Such batch studies generally consisted of batch and/or Warburg-type respirometry tests. To reduce the time requirements for measuring biodegradability, our technical consultant, Aware, Inc. (name changed to Eckenfelder, Inc.) developed a continuous-feed back reactor (FBR) procedure (5) to provide more accurate data in a short time period. The FBR method involves operating well-stirred, continuous-feed reactors without sludge recycle or effluent draw-off. This test procedure allows the measurement of relative biodegradation rates in a very short time period (2.5 hr). A 2-liter reactor was seeded with rinsed Toms River plant sludge at an initial mixed liquor suspended solids (MLSS) concentration of 2000 mg/L. The test sample containing a TOC:N:P ratio of 100:10:1 was fed to the reactor at a volumetric input rate of 0.1 liter per hour. During the 2.5-hr test period, the oxygen uptake rate, total organic carbon, and pH were measured at intervals of 15–30 min. Initial and final mixed liquor volatile solids concentration, total nonvolatile (fixed) dissolved solids, reactor volume, and influent flow rates were also monitored.

The 48-hr Mysid shrimp bioassay and FBR biodegradation results were used to classify 152 individual batch product waste streams. These waste streams represented the higher production volume products produced in the plant. Based on the results, the product waste streams were classified into the following four groups:

- *Class A (Toxic and Nonbiodegradable)* The waste streams of 17 products were placed into this classification and were considered to be the most likely to induce Mysid shrimp toxicity because they would have the tendency to pass through the biological treatment plant unchanged.

- *Class B (Toxic and Biodegradable)* The batch products of 29 streams fell into this category.

- *Class C (Less Toxic and Biodegradable/Nonbiodegradable)* The waste streams of 37 products fell into this category. The actual impact of Class C streams on effluent toxicity would require confirmation after biological treatment to determine residual toxicity.

- *Class D (Least Toxic and Biodegradable/Nonbiodegradable)* The products in 69 waste streams fell into this category and were considered to be the least probable to induce Mysid shrimp toxicity in the final plant effluent.

Due to the nonbiodegradable nature of products in Class A streams, physical/chemical processes such as carbon adsorption, wet air oxidation, and hydrogen peroxide oxidation were evaluated as methods for toxicity reduction. The effect of Class A compounds on PACT reactor performance was also evaluated. Overall conclusions reached regarding treatment of Class A streams are summarized as follows:

- The use of carbon adsorption to treat Class A streams gave TOC removals of 54–99%.

- At high-carbon dosages, toxicity reduction was achieved in 9 of the 11 waste streams tested.
- Toxicity reduction was not achieved on all 11 waste streams using carbon treatment. Other treatment technologies (e.g., peroxide treatment) investigated were also not considered effective.
- Continuous-bench-scale PACT was used to investigate treatment of combined streams from 7 of the Class A products.

Results from these initial, though limited, investigations of the PACT treatment process on Class A products were somewhat inconclusive, in that toxicity tended to increase when the PACT reactors were employed at low-carbon dosages.

Based on these results, a program designed to investigate the impact of biological degradation on the Class C and D waste streams was initiated. Ten Class C streams were combined and fed to continuous-bench-scale aerobic reactors for a period of 2 months. The initial toxicity of these combined streams was in the range of 75% (96-hr LC_{50}). The effluent bioassay results from these bench-scale reactors decreased gradually as the reactors reached steady-state conditions and stabilized in the range of 20–35%, indicating a substantial increase in Mysid shrimp toxicity. Although further work would be required to fully define the reason for this surprising increase in toxicity arising from the aerobic biological treatment process, these results conclusively demonstrated that a biological treatment process alone would not effectively treat the combined plant effluent.

Fortunately, a parallel investigation involving end-of-pipe treatment of waste streams generated by the plant production facilities using granular activated carbon (GAC) columns and PACT system was proving fruitful. Consequently, further work on the treatment of individual product waste streams or classes of waste streams was not undertaken. These classification results were used to direct and prioritize process development work on the production processes. Class A streams received the highest priority, and the insuing work was directed at minimizing TOC in the product waste streams and/or modifying the process to achieve a less toxic effluent.

Metals Removal

After reviewing the sources of metal salts in the overall influent stream to the biological treatment process, it was concluded that the most effective mechanism for reducing metal ions content in the final effluent was to include a metals precipitation step in the production process for those dyes that contained a significant quantity of unbound metal ions. The concentrated aqueous waste streams produced by the manufacturing processes of these products were neutralized to a pH that would give precipitation of the metals as inorganic salts. The resulting slurry was filtered in the production unit, to reject these salts before the wastewater was sent to the treatment plant.

End-of-Pipe Treatment   Following the initial screening studies, bench-scale and pilot plant investigations were concentrated on the full-scale plant feed streams being fed to the aeration basins in the biological treatment plant. Two treatment processes were evaluated—the PACT system and GAC columns—for treating effluent from a biologically activated sludge pilot plant. The end-of-pipe treatment process investigated did not include efforts to reduce metal content in dye wastewater.

PACT System Toxicity Reduction

Figure 7.3 depicts the statistical analysis of the continuous-bench-scale PACT reactor results, operating on full-scale plant streams being fed to the plant aeration basins (primary clarifier overflow). Effluent toxicity was maintained above an LC_{50} of 50% with a carbon dosage of 250 mg/L. Total organic carbon was reduced from an average of approximately 70 mg/L, with no powdered carbon addition, to an average of 30 mg/L at a carbon dosage of 250 mg/L. Other conclusions reached during our evaluation of the PACT system are as follows:

- Calgon RB afforded the greatest reduction in TOC per gram of carbon charged. Of the remaining two carbons evaluated, Calgon WPX was the second-most effective and Calgon HDC was least effective.
- Brief bench-scale regeneration of the spent carbon indicated that regeneration is feasible. However, regeneration reduced adsorption capacity significantly. For instance, 3 lb of regenerated Calgon RB is equivalent to 1 lb of virgin Calgon RB. Regeneration of Calgon HDC is more effective, as 1.5 lb of regenerated HDC is equivalent to 1 lb of virgin HDC.
- The ability of PACT reactors to lower wastewater toxicity was not affected by temperatures as low as 10°C or by a reduction in hydraulic retention time from 1.4 to 0.7 days.
- Effluent toxicity was not related to solids retention time (SRT), in the interval of 20–50 days. Effluent toxicity did increase slightly, however, when the SRT was decreased below 20 days.

Ammonia Removal in PACT

It is well known (2,3,6) that free ammonia can be toxic to certain aquatic life. For this reason, free-ammonia content of the reactor effluent was measured during the bench-scale studies. Data calculated from the literature (1,4) indicate that the free-ammonia content would have to be reduced to 0.4 mg/L to reach an LC_{50} of 50%. Figure 7.4 is a plot of carbon dosage and effluent ammonia content data obtained from four bench-scale reactors. At a carbon dosage of 250 mg/L, the free-ammonia effluent content in the reactor effluent reached approximately 0.4 mg/L. No further evaluations involving the impact of free-ammonia content in the bench-scale reactors were conducted because carbon addition was initiated in the full-scale plant system in July 1985, and additional data were available from the full-scale plant effluent analyses. However, it was apparent from the bench-scale data that nitrification had been achieved in the bench-scale reactors at the 250 mg/L carbon dosage.

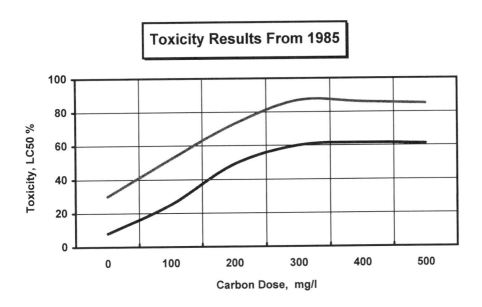

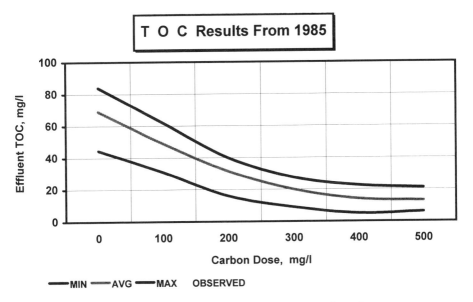

FIGURE 7.3   Bench-scale PACT studies, Toms River plant.

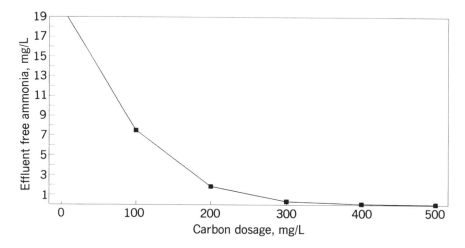

FIGURE 7.4   Effluent free ammonia, bench-scale reactors.

GAC Column Toxicity Improvement Studies

The performance of GAC columns was investigated briefly, from the standpoint of both TOC removal and toxicity reduction. A pilot-scale system was installed, consisting of a pilot-scale biological reactor and four pilot-scale carbon columns (three GAC columns in series with a fourth column used to replace the first column when the granular carbon in the first column was saturated with TOC—point of breakthrough). The columns were operated at a hydraulic load of 2 gpm/ft^2 and an empty load contact time of 30 min.

With respect to TOC removal, the GAC columns were capable of removing TOC down to 5 mg/L. At the point of TOC breakthrough, however, Mysid shrimp toxicity did not increase. The operation of the columns was continued for approximately 2.5 months to determine the point of toxicity breakthrough. During this period, no TOC reduction was occurring, but bioassay breakthrough (increase in toxicity) did not occur. Upon repeating this experiment, the same results were obtained.

At this point, the decision to proceed with the installation of the PACT treatment system was made. Only minor modifications, designed to improve agitation in the aeration basins, were required to implement the PACT process. In addition, it proved unfeasible to employ the GAC process in lieu of the PACT treatment system because of the relatively high capital cost of the GAC columns.

FULL-SCALE PACT SYSTEM

Treatment Plant Modifications

Figure 7.5 contains an outline of the modified biological treatment plant at Toms River. To address odor emissions, it was decided to close the open lagoons and

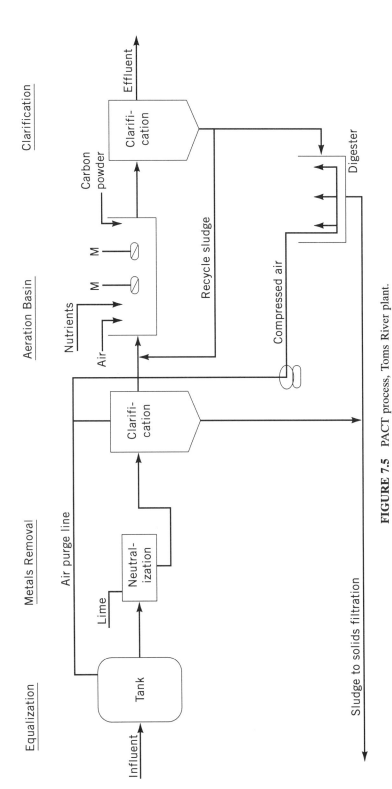

FIGURE 7.5 PACT process, Toms River plant.

177

install above-ground, covered equalization tanks. In addition, covers were installed over the neutralization equipment and primary clarifiers to eliminate volatile emissions coming from the metals removal step. Enclosing the equalization tanks and the primary clarifiers would have produced an explosive mixture in the air space between the liquid and the roof over the equipment. As an alternative, air purge lines were provided to control the volatile solvent content in these air spaces.

Several different methods were investigated as potential mechanisms for removing volatile solvents from the air purge streams, including air scrubbers, thermal oxidizers, and condensation. Scrubbing tests, conducted on spent PACT sludge mixture, indicated that the slurry mixture was an excellent scrubbing liquid and that it could achieve essentially complete removal of the volatile solvents from the airstream. The existing full-scale digester in the waste treatment plant was equipped with Kenics air sparge mixers for mixing the spent sludge in the digester. It was decided to utilize the digester for scrubbing the air purge streams. The air purge lines were routed to the suction of the digester air compressor, as shown in Figure 7.5. No modifications were made to the air purge lines in the digester.

The aeration basins in the existing treatment plant were equipped with standard Kenics air mixers, which were designed to suspend a maximum of 5000 mg/L of biological treatment system sludge. As indicated in Table 7.2, requirements for the PACT were considerably higher, 14,900 mg/L suspended solids (MLSS), than the existing air mixers could accommodate. To meet the design criteria of greater than 15,000 MLSS, the final design included four 250-hp draft tube mechanical mixers installed in each quadrant of the aeration basins. The actual amount of air required to maintain a dissolved oxygen content in the aeration basins was very low relative to the design requirement for the Kenics air mixers. Four small air compressors, located on the platform designed to support each draft tube aerator, were installed to sparge air into the aeration basin and supply oxygen to the biological process. Net energy savings achieved by replacing the Kenics air mixers with the combination mechanical mixing–air sparger was sufficient to offset the cost of the mixing equipment in approximately 3 years.

TABLE 7.2   PACT Design Criteria

Flow	4.0 mgd average
Effluent TOC	< 20 mg/L
PACT TOC removal	40 mg/L
Carbon dosage	500 mg/L
Sludge age	30 days
Carbon MLSS	10,700 mg/L
Total MLSS	14,900 mg/L
Waste carbon	16,680 lb/day
Total waste sludge	25,400 lb/day
Mixing	> 15,000 mg/L TSS

No other changes were made to the existing biological treatment plant. Sludge digestion, sludge-handling system, and the plate and frame filters were operated in the same manner employed in the conventional biological sludge system. The introduction of powdered carbon to the biological system actually improved the settling characteristics in the secondary clarifiers. Only minor increases in sludge filter cake volume were experienced because improvements in filtering characteristics increased the solids content of the filter cake by approximately 10%.

FULL-SCALE PACT PERFORMANCE

Compliance and operating data for the waste treatment plant from 1985 through 1989 are summarized in Table 7.3. Operating data for 1985 provided a basis for comparison between the conventional activated sludge biological waste treatment and PACT performance at Toms River. The addition of powdered carbon was initiated in August 1985; however, the carbon dosage was limited by the Kenics air mixers' solids suspension capabilities. Carbon addition rates were increased in 1986, based on 1985 operating experience. With the complete installation of the new draft tube aerators at the end of 1986, the operating data collected from 1987 through 1989 arise from full-scale operation of the PACT system.

Toxicity Reduction

Figure 7.6 presents the compliance data on Mysid shrimp and relative to carbon dosage rates on the full-scale waste treatment plant. At the lower dosage rates in 1985 and 1986, the Mysid shrimp bioassay data was quite variable. When the new mixers were brought on line at the end of 1986, Mysid shrimp toxicity increased immediately. Except for 1 month (March 1987), the percent LC_{50} was greater than 100%. After March 1987, the Mysid shrimp bioassay was 100% survival in 100% effluent at carbon dosages in the range of 350 mg/L. A comparison of these data with the results in Figure 7.3 shows that performance of the full-scale PACT was better than the bench-scale reactors.

TOC and BOD Removal

Figures 7.7 through 7.9 present the yearly average BOD and TOC removal data achieved using the PACT system compared to the official compliance specifications. The full-scale experience was equivalent to the bench-scale results, with the average TOC being in the range of 20 mg/L for carbon dosage in the 350 mg/L range. From 1987 through 1989, the percent TOC removal was stable at approximately 90–95% at the 350 mg/L carbon dosage (see Figure 7.8).

Total Suspended Solids

One aspect of PACT performance that was not anticipated was the substantial improvement in TSS experienced in the final effluent. In 1985, TSS concentrations

TABLE 7.3 Average PACT System Performance, Full-Scale Toms River Plant

	Unit	1985	1986	1987	1988	1989
Influent						
pH (standard units)	su	2.0–9.6	0.8–11.8	1.5–12.1	2.2–12.2	1.3–10.1
5-day BOD	mg/L	340	532	423	315	140
TOC	mg/L	187	237	272	175	63
Chromium	μg/L	416	337	87	38	23
Copper	μg/L	751	1354	1242	251	177
Lead	μg/L	55	74	48	45	37
Mercury	μg/L	2	5	2	1	<1
Nickel	μg/L	181	233	106	61	38
Zinc	μg/L	1025	1224	68	28	42
Aeration basin						
Biomass	mg/L	3160	3700	4220	2630	1350
Carbon concentration	mg/L	1350	1915	6060	7280	5190
Carbon addition per day	lb	1700	3440	10145	8770	7180
Carbon addition	mg/L	55	115	357	350	358
Final effluent						
Flow	mgd	3.7	3.6	3.4	3.0	2.4
pH (standard units)	su	6.8–7.8	7.0–7.9	7.2–8.1	7.4–8.2	7.5–8.3
Oil and grease	mg/L	0.7	1.1	0.6	0.1	0.5
5-day BOD	mg/L	18.6	13.3	9.3	4.4	2.1
5-day BOD	% removal	94.5	97.5	97.8	98.6	98.5
TOC	mg/L	82.1	49.8	24.5	20.9	6.9
TSS	mg/L	25.3	22.2	11.1	4.7	1.4
Chromium	μg/L	0.05	0.05	0.02	0.01	0.00
Copper	μg/L	173.5	180.0	58.6	17.7	12.6
Lead	μg/L	43.2	42.1	46.3	25.7	2.0
Mercury	μg/L	0.01	0.03	BMDL	BMDL	BMDL
Nickel	μg/L	92.7	69.8	50.3	25.1	8.8

		41.5	22.6	16.1	10.3	16.0
Zinc	µg/L					
1,2,4-Trichlorobenzene	µg/L	BMDL	BMDL	BMDL	BMDL	BMDL
1,2-Dichlorobenzene	µg/L	BMDL	BMDL	BMDL	BMDL	BMDL
1,3-Dichlorobenzene	µg/L	BMDL	BMDL	BMDL	BMDL	BMDL
1,4-Dichlorobenzene	µg/L	BMDL	BMDL	BMDL	BMDL	BMDL
1,2-*trans*-Dichloroethylene	µg/L	BMDL	BMDL	BMDL	BMDL	BMDL
2-Chlorophenol	µg/L	BMDL	BMDL	BMDL	BMDL	BMDL
bis(2-Chloroethoxy)methane	µg/L	BMDL	BMDL	BMDL	BMDL	BMDL
Benzene	µg/L	BMDL	BMDL	BMDL	BMDL	BMDL
Chlorobenzene	µg/L	BMDL	BMDL	BMDL	BMDL	BMDL
Ethylbenzene	µg/L	BMDL	BMDL	BMDL	BMDL	BMDL
Naphthalene	µg/L	BMDL	BMDL	BMDL	BMDL	BMDL
Nitrobenzene	µg/L	BMDL	BMDL	BMDL	BMDL	BMDL
Tetrachloroethylene	µg/L	BMDL	BMDL	BMDL	BMDL	BMDL
Toluene	µg/L	BMDL	BMDL	BMDL	BMDL	BMDL
Trichloroethylene	µg/L	BMDL	BMDL	BMDL	BMDL	BMDL
Chlorinated hydrocarbons	µg/L	BMDL	BMDL	BMDL	BMDL	BMDL
Organics MDL = 10 µg/L.						

BMDL = Below method detection limit.
MDL = Method detection limit.

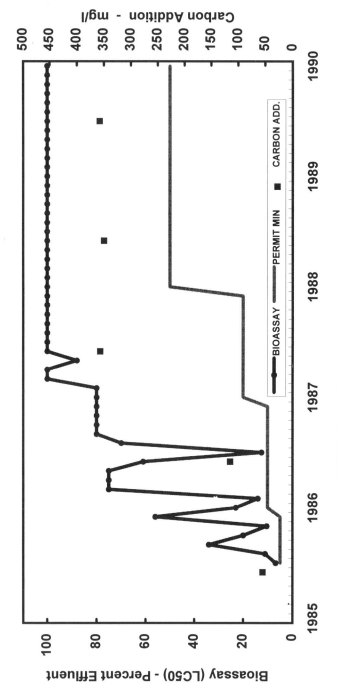

FIGURE 7.6 Full-scale PACT performance, Toms River plant.

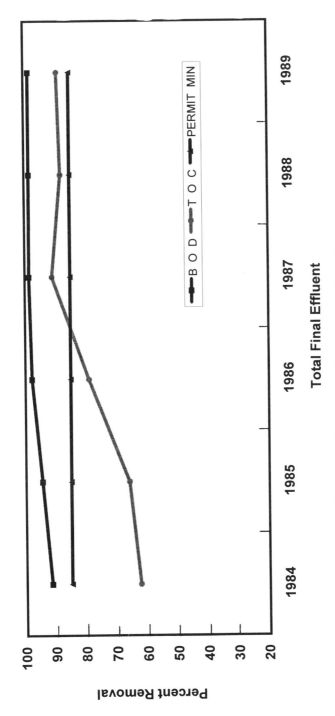

FIGURE 7.7 Full-scale PACT performance, Toms River plant.

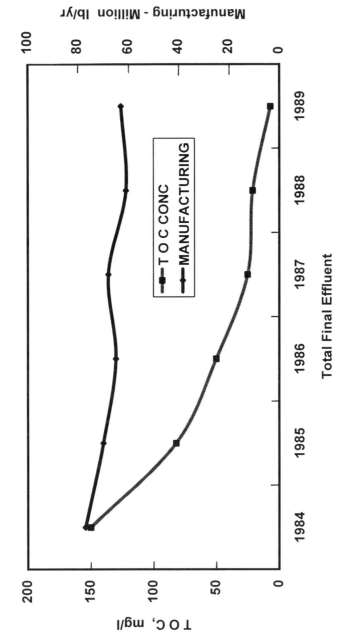

FIGURE 7.8 Full-scale PACT performance, Toms River plant.

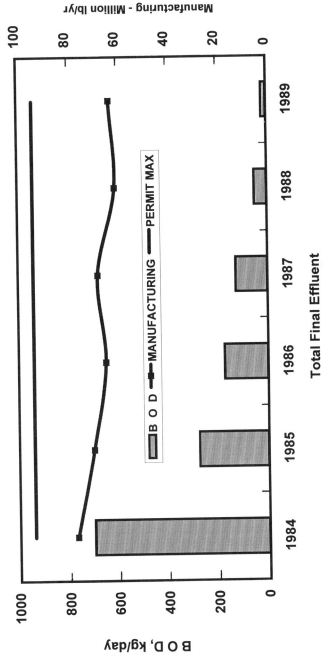

FIGURE 7.9 Full-scale PACT performance, Toms River plant.

averaged 25.3 mg/L (see Table 7.3). In 1987 when the full-scale PACT system was brought on line, TSS dropped to an average of 11.1 mg/L. Continued optimization of the system during 1988 and 1989 further reduced TSS in the final effluent to approximately 1.4 mg/L.

Nitrification

At carbon dosages in the range of 50 mg/L, nitrification was established, and effluent ammonia levels were lowered below 1 mg/L. This result was actually somewhat better than that achieved using the bench-scale reactor, where 100 mg/L was required to establish nitrification.

At the lower carbon dosages employed during 1985 and 1986, nitrification was not achieved when winter temperatures fell below approximately 27°F. When the carbon dosage was increased above 100 mg/L, nitrification was maintained during both winter and summer conditions.

Metals Removal

A metals removal step was incorporated into the production process for four metals (chromium, copper, nickel, and zinc). The effectiveness of this change can be seen in the significant reduction in the influent concentrations recorded for these four metals during 1987 and 1988. The concentration of these metals was also reduced in the final effluent. However, the total reduction in metal concentrations achieved in the final effluent was not sufficient to explain the dramatic improvements achieved in lowering Mysid shrimp toxicity. A further reduction was also achieved in 1989. This reduction was due primarily to changes in product mix.

Ames Test Results

The permit issued by the state, in 1985, required that Ames tests be conducted on the final effluent each quarter. Final effluent passed the Ames test on an as-is basis (diluted) throughout the permit period. The concentrated sample of the effluent continued to fail the Ames test until the spring of 1988. All subsequent quarterly samples passed the Ames tests—both dilute and concentrated samples. No attempt was made to correlate the effluent data with results of the Ames tests. However, it should be noted that a general improvement in overall PACT performance in terms of final effluent concentrations of free ammonia, metals, and TOC was experienced during the period from 1987 to 1989.

AIR EMISSION TEST RESULTS

As part of the odor reduction specifications, the State of New Jersey required substantial improvements in the air emission controls for both the aeration basins and digester. Based on pilot-scale test results, the state was convinced that implementation of the PACT treatment system would provide the required control of the air emissions in the two basins. The operating permit issued by the state required air emission testing be conducted on the full-scale system, to confirm the air emission control capabilities of the PACT system. The required air emission tests were performed using the test apparatus outlined in Figure 7.10 (7). This equipment employs a floating-sample platform to collect the air samples coming from the liquid surface. Sample air is withdrawn from the sample platform in sufficient

TABLE 7.4   Air Emission Test Results

	Aeration Basin	Digester Basin
Operating conditions		
Liquid depth (ft)	27	23
Basin diameter (ft)	204	140
Liquid residence time (days)	2	21
Compressed air rate (scfm)	5600	9000
Test Results		
Hourly samples		
Avg. THC emission	0.123 lb/hr	0.31 lb/hr
Removal efficiency (%)	—	98.1
24-hr samples		
Avg. THC emission	0.111 lb/hr	0.29 lb/hr
Removal efficiency (%)	—	96.9

TABLE 7.5   Aeration Basin–Digestor Component Emission Rate Calculated Emissions[a]

Component	Aeration Basin	Digestor
Methylene chloride	35.6	52.6
Trichlorofluoromethane	315.2	77.3
Chloroform	45.6	42.1
1,1,1-Trichloroethane	31.3	33.2
Trichloroethene	18.2	23.3
Tetrachloroethene	41.2	80.6
Toluene	117.0	1290.9
Ethylbenzene	2.0	34.7
Xylenes	9.4	45.1
Alkanes	79.8	149.8
Alkenes	27.5	151.5
1-Chloro-3-methylbenzene	0.5	0.0
Chloroethane	0.2	0.0

[a]Pounds per year per compound.

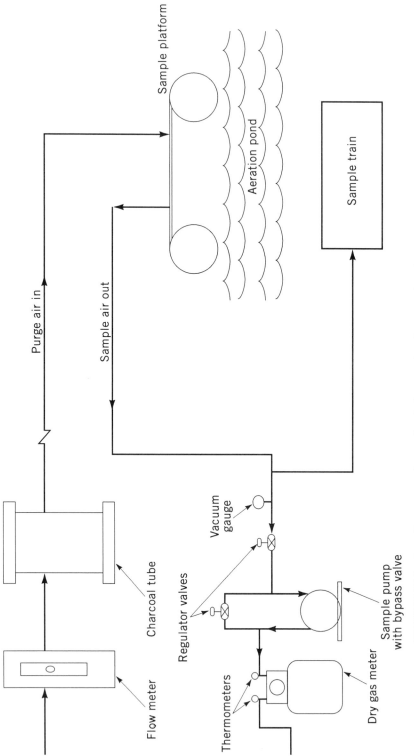

FIGURE 7.10 Sampling system for pond volatile organic substances (VOS) evaporation measurements.

quantities to establish a positive air purge flow to the sample platform. Thus, a negative pressure develops beneath the floating sample platform, collecting all the air and organics leaving the liquid surface below the sample platform. Table 7.4 provides the individual equipment operating conditions and the total hydrocarbon (THC) emission rates measured on the aeration and digestor basins. Table 7.5 provides the component emission rates for each. These results demonstrate the effectiveness of PACT for odor control. The net effects of covering the equipment upstream of the aeration basins and using the PACT system was complete elimination of all odor complaints from the neighbors.

CONCLUSION

As demonstrated, the results achieved using the PACT treatment process have been excellent. Incorporation of the PACT process into an existing activated sludge biological treatment process required only minor capital investment and reduced the overall energy costs for operating the aeration basins enough to offset these capital costs. The cost of powdered carbon is approximately one-third that of granular activated carbon. At our facility, it was not economically feasible to replace the PACT system with GAC columns as the postbiological treatment process, even though the latter process affords a lower TOC content in the final effluent.

From an operational standpoint, the PACT process proved to be very stable. Wide variation in the nature and concentration of the influent, operating conditions, and ambient temperature had only minor influences on the final effluent quality. Overall odor control in the treatment process was excellent, and the changes implemented at the end of 1986 completely eliminated odor complaints from our neighbors.

Performance of the PACT treatment process has received both state and national recognition, with the following recognition of its achievement:

May 1990   "Excellence in Industrial Wastewater Treatment Award" from the New Jersey Water Pollution Control Association.

April 1990   "National Environmental Achievement Award" from the National Environmental Awards Council's Searching-for-Success Program.

In addition, two U.S. patents (8,9) were received for the unique improvements made at Toms River's treatment plant in odor control, viz, for the use of the digestor and spent PACT liquor for scrubbing the airstream.

ACKNOWLEDGMENTS

P. Lankford and K. Torrins, Eckenfelder, Inc., and the technical consultants and technical staff at the Toms River plant.

REFERENCES

1. *Standard Methods for the Examination of Water and Wastewater*, American Public Health Association, 16th ed., Washington, D.C., 1985.

2. Mysidopsis Sp.: Life History and Culture, Workshop sponsor by the American Petroleum Institute. Gulf Breeze, Florida, October 15–16, 1986.

3. *Ambient Water Quality Criteria for Ammonia*, U.S. Environmental Protection Agency, Office of Water, Washington, D.C., January 1985.

4. *Effluent Toxicity Reduction Investigation* Vol. 1, AWARE Incorporated, Nashville, TN, 1985.

5. A. Watkin, Evaluation of Biological Rate Parameters and Inhibitory Effect in Activated Sludge, Ph.D. Thesis, Vanderbilt University, 1986.

6. *Process Design Manual for Nitrogen Control*, U.S. Environmental Protection Agency, Office of Technology Transfer, Washington, D.C., October 1975.

7. Final Summary Report of VOS Sampling and Analysis Aeration Basin and Aerobic Digestor, Engineering-Science, Inc., January 1988.

8. U.S. Pat. 4,894,162, Treatment of Volatile Organic Substances at Waste Treatment Plant, January 16, 1990.

9. U.S. Pat. 5,106,496, Treatment of Volatile Organic Substances at Waste Treatment Plant, April 21, 1992.

PART 2

WASTE MINIMIZATION/ SOURCE REDUCTION

CHAPTER 8

REDUCTION OF TEXTILE WASTEWATER USING AUTOMATIC PROCESS CONTROL, RECYCLE, AND FILTRATION

JOHN J. PORTER

School of Textiles, Fiber and Polymer Science, Clemson University, Clemson, South Carolina

Often a company's interest in wastewater treatment or recycling is initiated by the action of a state or federal regulatory agency that wants the company to reduce either the volume of wastewater or the concentration of specific pollutants being discharged in the waste stream. When this occurs, the plant manager, with the help of his or her staff, must select a plan of action that is to be taken to bring the plant into compliance with the new regulations or restrictions. Generally, the problem is solved by (1) reducing water consumption, (2) improving the waste treatment system, (3) changing the manufacturing process, or (4) ceasing the production of a specific product that caused the problem. This approach to solving wastewater treatment problems will continue as long as the regulations continue to increase and become more restrictive.

The objective of this chapter is to describe methods by which a plant may become more independent of changing governmental regulations by recycling its wastewater and, thus, reducing the volume of wastewater regulated by a state or federal agency. Besides reducing the wastewater volume and treatment cost, a plant can realize savings in process water, energy, and chemicals cost. The following section is presented to give the reader some idea of the cost of filtration methods used to recycle hot water and chemicals in textile operations.

WASTEWATER TREATMENT AND RECYCLE COST

In 1972 when the Water Pollution Control Act (1) was promulgated, the cost of treating wastewater was much less than it is today, and industry preferred at that

Environmental Chemistry of Dyes and Pigments, Edited by Abraham Reife and Harold S. Freeman.
ISBN 0-471-58927-6   © 1996 John Wiley & Sons, Inc.

time to install or upgrade conventional biological treatment systems rather than install a recycle system that was unproven and more expensive. Today, a conventional biological wastewater treatment plant is more elaborate and costly, because it must be designed to satisfy more stringent governmental regulations. This situation often makes the installation of a recycle system a less costly alternative to wastewater treatment. In fact, a plant may also save money by using the hot water and chemicals that can be obtained from the recycle system. A comparison of the cost of conventional biological wastewater treatment versus different filtration processes is given in Table 8.1, along with the estimated recycle value of the treated wastewater obtained from each process. Since the treated water from a biological treatment process is rarely suitable for recycle without additional treatment, it is shown here as having approximately zero value.

For a company to decide if recycling methods could be applied to one or more of its operations, it must know the water quality requirements for each process and the characteristics of the waste streams discharged from these processes. It is generally best to take an inventory of all process water uses and analyze all process waste streams, so that no source for potential savings is overlooked. The inventory (3) should include flow rates, temperatures, and pHs and provide other analyses that will help to characterize each waste stream. Of particular interest would be 1) a process cooling water waste stream that can either be reused directly without treatment or routed to a process water storage tank and 2) concentrated waste streams that have low flow rates and contain valuable chemicals that may be either recovered or treated more economically before they are mixed with the total plant effluent. Flow meters (4) should be installed on most processes and recording flow meters on major waste streams to determine daily water uses and variations in flow. The information on flow rates and wastewater analyses should show whether waste streams could be recycled directly or would require some type of treatment before recycling.

Because of the increasing use of filtration processes by the textile industry, several types of filtration will be described. In addition, the application of automatic process control for minimizing water flow and lowering the cost of most wastewater treatment or recycle systems will be discussed.

TABLE 8.1 Wastewater Treatment Cost and Recycle Value (2)

Method	Treatment Cost/1000 gal	Estimated Recycle Savings per 1000 Gallons[a]		
		Water[b]	Waste Treat.[b]	Energy[b]
Biological	$1.50–2.50	~0	~0	~0
Filtration	$1.00–2.00	$1.00	$2.00	0–$5.50
Ultrafiltration	$1.50–3.00	$1.00	$2.00	0–$5.50
Reverse osmosis	$2.50–3.50	$1.00	$2.00	0–$5.50

[a]Does not consider any value for recovered chemicals.
[b]Assumes water cost = $1.00/gal, waste treatment cost = $2.00/gal, energy cost = $5.00/$10^6$ Btu, Temp. range = 60–200°F.

AUTOMATIC PROCESS CONTROL FOR WASTEWATER MINIMIZATION

Many production rinsing operations use excessive quantities of water simply because it is impractical for the operator to change the water flow settings each time
the style that is being processed changes. If a plant wants to optimize its water and
energy use and each style or fabric has a different optimum water need, it would
be necessary to adjust flow settings each time the style changes. If this is not
practical, the water flow must be set at a rate that satisfies the style with the greatest
water demand, and used for all fabrics being processed. Both cases are illustrated
in Figure 8.1, where curve 1 represents the manual valve setting that is not changed
during the day and curve 2 represents the actual flow obtained as the plant water
pressure changes during periods of peak water consumption. The lower two curves
represent the optimum process water flow, curve 3, needed for each individual style
being processed, and the flow that would be supplied, curve 4, if the process were
automatically controlled. The manual flow setting represented by curve 1 must be
higher than the maximum process water flow requirements or risk going below this
level when water pressure drops. It is evident that the water and energy consumption in a plant will be greater for processes that have varying water needs during
the day but have no automatic control. The concept is illustrated in Figure 8.2,
showing how the water, steam, or chemicals needed for a process may be automatically controlled (5). When the process is automatically controlled, it can respond rapidly to style changes and change process variables such as gallons per
minute (gpm) or steam pressure (psia, pounds per square inch absolute) in a few

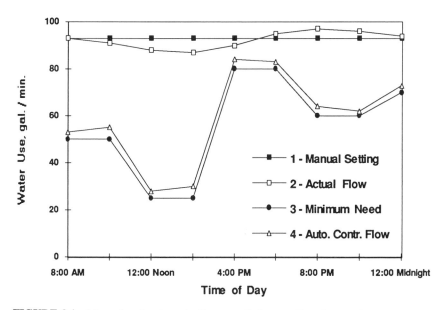

FIGURE 8.1   Manual and automatically controlled water flow for continuous finishing.

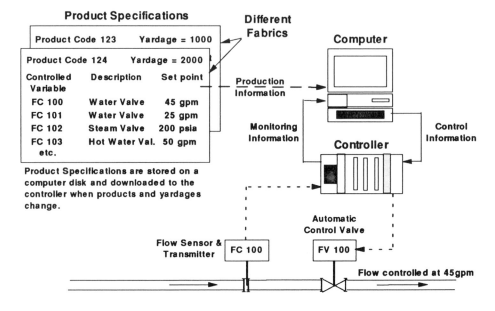

FIGURE 8.2   Typical flow control scheme (6).

seconds, allowing the process to operate at high production speeds. This is not possible when the process is controlled manually.

When a plant processes large quantities of one style of fabric, the process may be operated manually because the requirements for process water, steam, and chemicals remain constant for many hours. In this case, the need for automatic control is reduced. However, the need is not eliminated, as product quality may decrease if water and steam pressure change during the day as demand changes in a plant that has several different processes. The use of excessive quantities of water, steam, or chemicals will raise the cost of producing each yard of fabric and the cost of any wastewater treatment system that may be modified or recycle system that may be installed. Since these costs will be directly proportional to the volume of water used per day, it is imperative that water use be minimized. In many cases, this can only be accomplished if an automatic control system is an integral part of the process.

The following sections will describe 1) several factors that influence the filtration process and 2) filtration systems used to recover chemicals from wastewater and recycle them back into a plant operation.

THE FILTRATION PROCESS

Conventional Filtration

Before a waste stream can be reused or recycled, it is usually necessary to remove suspended solids or fibers from the waste stream so that they do not interfere with the operation of flow meters, valves, or pumps that may be part of a recycle system. This can usually be accomplished by using a conventional filter like the one illustrated in Figure 8.3. Before the filter is selected, the characteristics of the waste stream and the particulate or chemical to be filtered must be determined. Other useful information will be the pH, viscosity, and temperature of the wastewater. These factors can influence the rate or ease of filtration and the porosity of the filter material that must be used. While it is not the objective of this chapter to analyze in detail all of the variables that influence filtration, it is useful to look at the basic equation used to describe the conventional filtration process. For the filtration of rigid solid particles in an aqueous solution, Equation (1) can be used to describe the process (7):

$$\text{Flux} = J_F = \frac{B \times \Delta P}{\mu(L_f + L_c)} = \frac{\text{volume}}{\text{unit area} \times \text{unit time}} \tag{1}$$

In this equation, J_F is the flux or rate of filtration per unit area of filter surface, B is the Darcy permeability constant for the filter cake, and may either be obtained from the literature for known materials or be developed from laboratory experimentation (8); ΔP is the pressure drop across both the filter material and cake, in units of force per unit area; μ is the viscosity of the solution in units of mass per distance and time; L_f is the resistance of the filter material represented as equivalent cake thickness in consistent units; and L_c is the thickness of the filter cake, as shown in Figure 8.3. As the filter cake, represented by L_c in Equation (1), increases in thickness, the flow of the solution decreases if ΔP is held constant. When the filter cake is compressible, a more complex equation must be developed (9) that correlates the filtration rate to the cake thickness and filtration pressure.

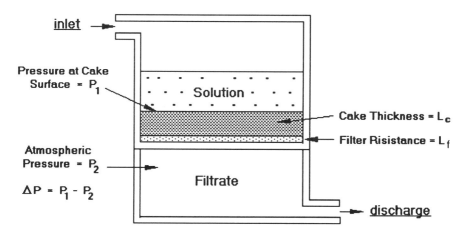

FIGURE 8.3   Conventional filtration.

Ultrafiltration

The term *ultrafiltration* (10) is generally used to describe the filtration of soluble molecules or polymers having a nominal diameter of 10–1000 Å. While these dimensions are arbitrary and have exceptions, they provide a useful range for describing the molecular sizes that fit into this type of filtration. The filtration process can be described by Equation (2), which is similar to Equation (1):

$$\text{Flux} = J_{UF} = \frac{\Delta P}{\mu(R_f + R_g)} = \frac{\text{volume}}{\text{unit area} \times \text{unit time}} \tag{2}$$

For ultrafiltration, R_f in Equation (2) represents the resistance of the membrane or ultrafilter and R_g represents the resistance of the gel or polymer layer that may form at the filter surface during filtration, in reciprocal units of thickness. Both constants must be determined for a specific membrane and solution to be filtered, in laboratory experiments. Here, the gel or polymer layer may be compressible, as opposed to the rigid filter cake required when the Darcy coefficient is used (9) in Equation (1). Because the pressures used for this type of filtration are generally low, if there is a significant difference in the osmotic pressure of the solution being filtered and that of the filtrate, filtration will occur very slowly or not at all.

The decrease in filtration rate caused by the gel layer, R_g, is commonly called *gel polarization*, and depends on the chemical or polymeric material being formed at the filter surface. The gel layer thickness will be proportional to the filtration pressure and the design of the filter system. As the material being filtered concentrates at the filter surface, the filtration rate will decrease rapidly, so that it is no longer proportional to the filtration pressure. This second effect, called *concentration polarization* (11), can be reduced if the filter is designed so that turbulent flow can be maintained at the filter surface to reduce the thickness and resistance of any gel layer at the filter surface. Ways to design the filter to prevent gel or cake formation at the filter surface are discussed later.

Reverse Osmosis

In reverse osmosis separation (10), small molecules or ions approximately 1–10 Å in diameter are filtered from a solution. The pores in the membrane or filter employed must be smaller than the actual size of the molecules or electrolyte in the solution to be filtered; and because of this the reverse osmosis process will be much slower than ultrafiltration, per unit area of filter surface. The filtration process can be described with an equation similar to (1) and (2); however, it is necessary to add a term to the numerator for the osmotic pressure of the solution being filtered. This is shown in Equation (3) (12):

$$\text{Flux} = J_{RO} = (\Delta P - \Delta \pi)A_m = \frac{\text{volume}}{\text{unit area} \times \text{unit time}} \tag{3}$$

where J_{RO} represents the flux per unit area of filter surface; ΔP is the pressure drop across the membrane surface and any gel layer formed on the membrane surface; $\Delta\pi$ is the osmotic pressure difference between the solution being filtered and that of the filtrate or permeate ($\pi = MRT$, where M is the molar concentration of the molecule or ionic salt being filtered, R is the gas constant, and T is the absolute temperature); and A_m is the membrane transport coefficient rather than the resistance term R used in Equations (1) and (2). In reverse osmosis filtration, the small molecules or ions being filtered can sometimes pass through the membrane, but at a significantly slower rate than the solvent. This enables separation to occur in a diffusion-controlled process, with the desired chemical retained and concentrated by the membrane (12). It is also possible for the molecules or ions being filtered to concentrate at the membrane surface and slow the filtration rate, as occurs in ultrafiltration. Here, as the concentration of the molecules being filtered increases at the filter surface, the osmotic pressure will increase, and the added resistance slows the filtration rate if the pressure remains constant. This phenomenon is called *concentration polarization*, just as it was in ultrafiltration, and the degree to which it occurs will depend on the flow or turbulence at the membrane surface. The filtration system can be designed to reduce concentration polarization effects. Some design techniques that are used will be discussed later.

The separation ranges for the three types of filtration: conventional, ultrafiltration, and reverse osmosis are illustrated in Figure 8.4, which shows how the three

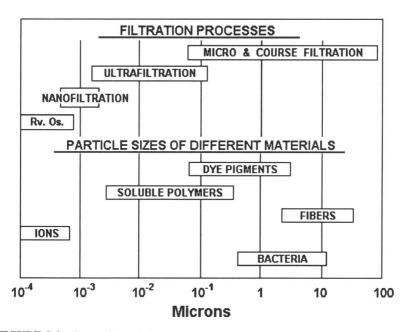

FIGURE 8.4 Comparison of the separation ranges for different filtration processes.

processes overlap each other without having discrete boundaries. Since it is possible to have a combination of particulate matter having diameters less than 0.5 μm, which compares to the dimensions of both large soluble polymers and very low molecular weight polymers that cause the solution to have a significant osmotic pressure, it is reasonable to expect the three processes to overlap as Figure 8.4 shows.

CROSSFLOW FILTRATION

In conventional filtration, shown in Figure 8.3, the rate of filtration will decrease as the filter cake is allowed to collect at the filter surface. If the filter surface is relatively small, the filter cake thickness will increase rapidly. If the filter surface is large, so that the filtration rate does not become slow, the size of the filter may be impractical. If, for example, the total quantity of solids being removed is 0.1 lb per 1000 gal (12 ppm) of wastewater, several thousand gallons of wastewater may be filtered before the filter must be backwashed or changed. When the concentration of suspended solids is significantly greater than 12 ppm, the filter must be cleaned or changed frequently. This can be avoided if the design of the filter system is modified so that the suspended solids are not separated as a filter cake. To accomplish this, the flow of the solution being filtered is directed parallel, rather than perpendicular, to the filter surface, as shown in Figure 8.5. The formation of a filter cake is minimized as the solution being filtered sweeps the membrane or filter surface to reduce the formation of a filter cake or gel layer. This type of filtration

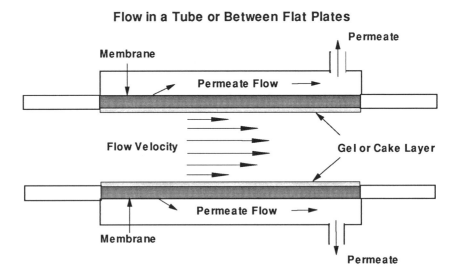

FIGURE 8.5 Velocity profile for crossflow filtration showing gel or cake layer formation.

is called *crossflow filtration* and may be illustrated by flow through a porous tube or across a flat filter surface (13). Here, the suspended solids will remain in solution, if the flow is rapid or turbulent, and insufficient cake will form at the filter surface to slow the filtration.

It is possible to reduce gel layer formation by increasing the flow velocity, as shown in Figure 8.6. The effect of increasing flow velocity on gel layer formation does reach a limit, as Figure 8.6 also shows. The optimum flow velocity that can be used to reduce the formation of a fouling layer at the filter surface will depend on the filtration pressure and the viscosity of the fluid being filtered; but it is generally between 10–18 ft/sec for aqueous systems. Higher velocities increase pumping cost with little or no increase in flux. Higher pressures increase the flux or filtration rate, but they also increase the fouling rate; therefore, it is best to get pilot plant data on a waste stream before designing a full-scale system. Obviously, all of the liquid cannot be removed from the material being filtered, as a small fraction of the original volume must be retained to carry the suspended solids past the filter surface. This is not an overwhelming consideration, usually, as the recovered chemical can also be pumped to another location where it may be reused. If it is necessary to isolate a solid material from the concentrated residue, a filter press or a spray drier may be used (14) to remove residual water.

Crossflow filters have been used to separate or recover suspended solids, soluble polymers, and inorganic salts from waste streams at room temperature and above 100°C. Commercial filters are fabricated from organic, ceramic, or metallic materials and may have a tubular or plate configuration. For textile dyeing and finishing operations, a crossflow filter may be used for filtering and recycling pigment print-

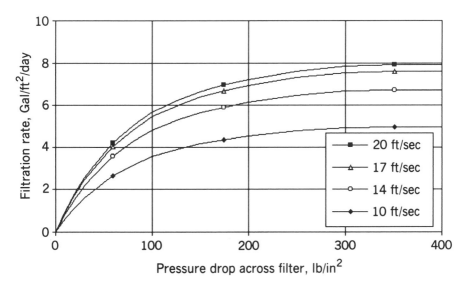

FIGURE 8.6   Effect of flow velocity and pressure on filtration rate.

ing wastewater, mercerization wastewater, or bleaching wastewater. The specific temperature and chemical composition of the waste stream will determine the type and porosity of the filter to be used. A comparison of the removal efficiency of the different methods of filtration and biological treatment for treating industrial waste streams is presented in Table 8.2.

Filter Cleaning

The filter or membrane must be cleaned if it is used continuously in a process to maintain a satisfactory rate of filtration. Information on cleaning frequency and methods should be obtained from pilot plant studies before the filtration system is designed, so that proper measures are taken to provide a cleaning system and filter surface sufficient for the plant. Usually a cleaning system is designed and installed as part of the filtration process, so that proper operation of membranes or filters is ensured. Actual data taken from a pilot plant study (16) in which dyeing wastewater from a continuous dye range was recycled using a reverse osmosis membrane is illustrated in Figure 8.7. These data show how the membrane filtration rate decreases with use, and how periodic cleaning improves the rate of filtration. The cleaning cycle can be automatically controlled to respond to pressure buildup or flow decrease or to clean sections of the filtration system at programmed intervals.

In the following sections, several plant-scale applications of filtration processes will be described in detail, along with the estimated savings obtained from each process.

PLANT APPLICATIONS OF CONVENTIONAL FILTRATION

Fabric Preparation

Fabric preparation may be the most important operation in the textile finishing plant as it can influence all of the subsequent dyeing and finishing processes. If a

TABLE 8.2   Average Removal Efficiency of Different Filtration Methods for Treating Industrial Waste Streams (15)

Parameter	Biological[a] (%)	Conventional Filtration[a] (%)	Ultrafiltration[a] (%)	Reverse Osmosis[a] (%)
BOD$_5$	50	30	50	95
COD	30	30	50	95
Solids				
Total	25	20	50	95
Dissolved	40	0	20	95
Suspended	50	30	98	98
Color	30	10	50	98

[a]These are estimates from the results obtained from several laboratory and plant evaluations. A specific case may be different.

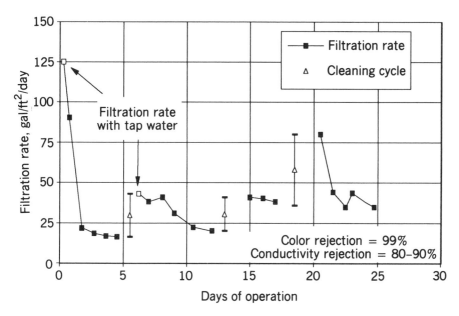

FIGURE 8.7   Operating performance of a dynamically formed reverse osmosis membrane filtering textile dyeing wastewater (16).

fabric is not consistently and uniformly desized, scoured, and bleached each time, it may be necessary for the dyeing or finishing operation to make adjustments to correct for differences in dye sorption or the wet pickup of finishing chemicals. If a process must be repeated, it adds to the cost of the fabric and decreases the profit made by the dyeing and finishing plant.

The common procedure used in the past to achieve good fabric preparation was to adjust the range controls to settings needed for the most difficult fabrics being processed and maintain these settings even though they may only be required for a fraction of the total operating time. Taking this approach will ensure that sufficient water, steam, and chemicals are always present to give good fabric preparation without having to continually change valve or flow settings for each fabric style. While this provides good fabric preparation, it also uses excessive quantities of steam and water and increases production cost. When the process can be automatically controlled, the consumption of steam and water can be minimized and fabric preparation optimized. As shown in Figure 8.1, changing to automatic control would adjust supply to correspond to the actual need rather than employing settings that may have little to do with the fabric being processed at a specific time.

A preparation range is illustrated in Figure 8.8, showing the three stages of the process: desizing, scouring, and bleaching. To reduce water usage for each rinsing operation, it is necessary for the rinse water to flow countercurrent to the fabric flow as much as possible. The peroxide wash box rinse water may be used directly

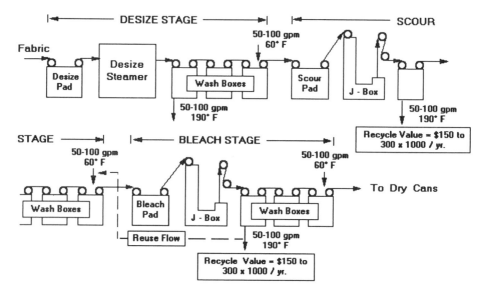

FIGURE 8.8 Three-stage preparation range.

as feed for the scouring wash boxes, as shown by the dotted line in Figure 8.8. When this is done, it is possible to save 50–100 gal/min water flow, or $125,000–$250,000 per year, depending on the flow rate used for each process. To reuse the bleach rinse water for the scour rinsing operation, it is necessary to first remove lint or suspended solids that could block valves or spray nozzles that apply the rinse water to the fabric in the scour wash boxes. This can be accomplished by conventional filtration, which was discussed earlier. The recycle process requires a pump, a suitable particulate filter, and an automatic controller that can switch from bleaching wastewater to fresh water for the scour rinsing operation when fabrics are processed that are sensitive to the bleaching wastewater. The lack of sufficient automatic process control to make rapid changes in continuous range operations has limited many textile plants from using this conservation process.

Analyses of several bleaching wastewater samples (17) show that the total solids present in the rinse wastewater is generally less than 0.5%, a level that should cause minimal problems when this is used for the scour rinsing process. The chemicals used for stabilizing the peroxide in the bleach bath should be selected carefully so that they are compatible with scouring chemicals they will contact in the scouring wash boxes. Many organic peroxide stabilizers currently in use should be compatible and suitable for the recycle process. When the fabric is to be mercerized, it must be tranferred to a separate range that contains the mercerization system, similar to the one shown later (in Figure 8.11).

A source of energy loss in preparation occurs when the fabric cools as it is skyed to straighten the fabric as it leaves the J boxes shown at two locations in Figure 8.8. If this operation could be enclosed to prevent fabric cooling and reduce energy loss, a cost savings of up to $30,000 per year could be realized. Since the required equipment is available, a plant that could apply this technique to more than one range could realize significant savings. One reason this has not been done is limited accessibility to the fabric when problems occur during processing. This limitation can be reduced if seam detection devices are used with automatic control to simplify the location of problems rather than having to open the equipment at several points along the range to identify and correct the problem.

PLANT APPLICATIONS OF ULTRAFILTRATION

PVA (Polyvinyl Alcohol) Recovery

The sizing formulations used to protect the warp yarns during the weaving process have changed a lot in the past 30 years and will continue to change as weaving speeds increase. Sizing chemicals that are resistant to degradation, such as PVA, can be recovered, when they are washed from the fabric before it is dyed, and reused. When this is done, sizing chemical cost and wastewater treatment cost are reduced. The first recycle application for the recovery of sizing chemicals used ultrafiltration for the recovery of PVA and was installed in 1971 on a continuous preparation range at the J. P. Stevens manufacturing plant in Clemson, South Carolina (18). The system is still in operation and has brought the company valuable savings in water, wastewater, chemical, and energy cost. The economics of PVA recovery for one range is shown in Figure 8.9, and can amount to over $5700 per

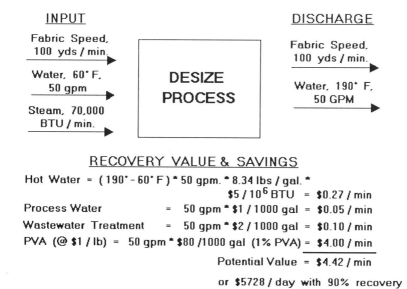

FIGURE 8.9 Value of polyvinyl alcohol recovered from the desizing operation (19).

day (19) when the range speed is 100 yards per minute. If the range speed is increased to 200 yards per minute with automatic process control, the value of the recovery process would exceed $11,000 per day. This is attainable today with available automation and processing equipment and could produce large savings in operating costs.

If the current trend continues in the textile industry toward the increased use of special synthetic sizes necessary for high-speed weaving, ultrafiltration and process automation will become an integral part of the desizing process. Because most of the synthetic sizes are difficult to treat in conventional biological wastewater treatment systems, it would be desirable to recycle and eliminate them and their corresponding wastewater volume from the waste treatment system. A flow diagram illustrating the recovery system (20) for PVA is shown in Figure 8.10. When the size is recovered and the permeate water is reused to wash out more sizing chemical, over 60,000 gallons of wastewater per day is eliminated from both the wastewater treatment system and the influence of government regulations. While some blowdown or cleanup wastes will still be discharged, the volumes should be a small fraction of the unrecycled waste stream. The costs of PVA, fresh water, and waste treatment should increase significantly in the future and make recycling more attractive. In some installations, the permeate water is not recycled but discharged directly to waste treatment because of fear of solids buildup and fabric contamination. Here, no reduction in process water needs or wastewater flow is obtained. Additional studies on permeate water reuse should provide information on the feas-

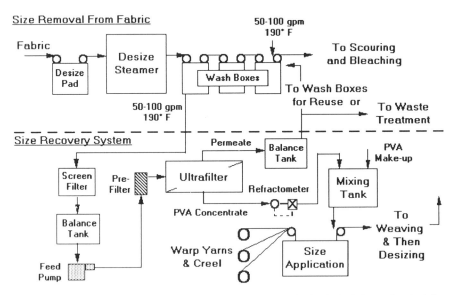

FIGURE 8.10   Flow diagram for the recovery of polyvinyl alcohol from the desizing operation.

ibility of its recycle. This has been done at some plants and several benefits are obtained when it can be achieved. It is necessary to bleed down any recycle system and replace the recycled water with fresh water when unwanted solids buildup in the system. Often there are sufficient leaks built into the system so that this is not necessary.

Caustic Recovery

The flow diagram for the recovery of sodium hydroxide from the mercerization process (21), which is used by textile plants to improve the luster and adsorption of cotton fiber, is shown in Figure 8.11. In the mercerization process, a strong sodium hydroxide solution is applied to the fabric and then rinsed from the fabric following a short dwell time. If the plant treats many yards of fabric, it would naturally require a large quantity of sodium hydroxide for the process. To avoid this expense, a multistage evaporator is generally used to reconcentrate the dilute caustic rinse solutions so that it can be reused in subsequent processing. The size and cost of the evaporator will be proportional to the volume of rinse water used to remove the strong caustic solution from the fabric. Therefore, it is important to control the water flow to the wash boxes, so that the sodium hydroxide concentration coming out of the wash boxes remains between 2 and 4%. When the concentration falls below 2%, it is usually impractical to recover the sodium hydroxide, and it must be discharged to waste treatment. When recovery is practical, the caustic rinse water is fed to an evaporator and concentrated to 30–50% NaOH. Since

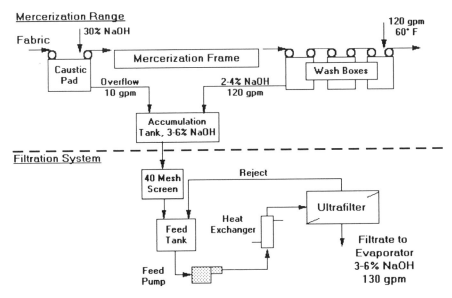

FIGURE 8.11 Filtration system for recovering mercerization rinse water (21).

textile plants use different concentrations of caustic for the mercerization process, the specific caustic concentrations will vary from plant to plant; however, the concentrations shown in Figure 8.11 are believed to be representative of what is commonly used.

The most important factor that limits the recycling and concentration of the caustic rinse water is the buildup of fabric impurities in the concentrate. After repeated evaporations, the caustic solution becomes so contaminated that it cannot be concentrated in the evaporator and must be discarded. This can be avoided if a stainless steel ultrafilter is used to remove impurities from the dilute caustic solution before it is evaporated (cf. Figure 8.11). The regenerative heat exchanger is used to maintain the dilute caustic solution at a higher temperature and reduce the surface needed for the expensive ultrafilter. One finishing plant has reported that a filtration system using a flow pattern similar to the one shown in Figure 8.11 paid for itself in 1.5 years (22) and has been operating for 8 years with an annual savings of $1.5 million in sodium hydroxide cost.

The recovery and removal of sodium hydroxide from the waste stream should reduce the concentration of salt discharged by the plant and the cost of neutralization acid that may be required if biological treatment is used to treat the plant wastewater.

Indigo Dye Recovery

Wastewater from most indigo dyeing processes contains significant quantities of indigo dye that can be recovered by ultrafiltration. The dye is present in the wastewater predominantly as a pigment, along with sodium hydroxide, lint fibers, and other chemicals used in the dyeing procedure. Because the dye has a deep blue color and is resistant to biological treatment, it may pass through conventional treatment systems and be very visible in a receiving stream. If an ultrafilter is used to recover indigo from the wastewater at the process where it is generated, coloration of a receiving stream is avoided, and considerable savings in dye cost is realized. This has been done by a textile plant in Liberty, South Carolina, which has been recovering indigo dye from the rinse water since 1981. The plant was able to pay for the recovery system before the end of its second year of operation (23).

The recovery system, illustrated in Figure 8.12, uses a vinyl-sulfone ultrafilter, which was originally manufactured by Dorr-Oliver but is now sold by the Amicon Division of W.R. Grace & Co. By modifying the dyeing process so that the indigo dyeing operation was completed before the application of any sulfur dyes, the possibility of contamination of the indigo dye by a sulfur dye was eliminated. The recovery process was designed with three stages to provide more circulation velocity and reduce membrane fouling. Each stage contains a circulation pump, with one feed pump before the first stage and a bleed valve after the last stage, to control the indigo concentration in the final discharge. Each stage automatically establishes a steady-state indigo concentration that increases from 800 up to 20,000 ppm, progressively, as the indigo dye solution passes through the ultrafilter, and the permeate water, containing the unwanted chemicals and discharge to waste treat-

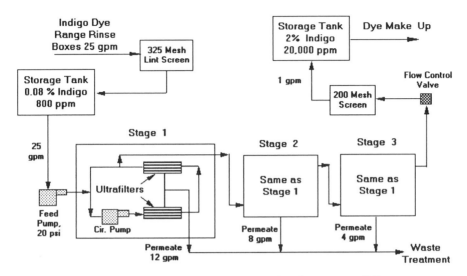

FIGURE 8.12   Filtration system for indigo dye recovery (23).

ment, is removed. The final stage reaches a concentration of 2% indigo, and the resulting solution is then filtered through a 200-mesh (74 μm) basket strainer and stored in a holding tank capable of holding 2–4 days supply of concentrate.

The recovery system is preceded by a 325-mesh (44 μm) vibrating screen to remove lint from the waste stream. Ratios of concentration flow rate to feed water flow rate is generally between 1:25 and 1:50 and depends on the particular dye concentration or strength used. The clear filtrate that comes from the ultrafilter, after the indigo dye is removed, contains a mixture of spent chemicals (sodium sulfite, sodium hydroxide, surface active agents, and dye impurities) used in the dyeing process, has no practical value, and is discharged to a conventional waste treatment plant.

For this installation, the capital cost of membranes needed for the recovery system was proportional to rinse water flow rate used on the indigo dye range. Therefore, it was necessary to reduce the rinse water flow to a minimum value that would still give satisfactory dye performance. This was accomplished by causing the rinse water to flow countercurrent to the yarn flow in the wash boxes following the dyeing step. This made it possible to reduce the flow discharged from the wash boxes to 25 gpm (gal/min), giving an indigo concentration in the rinse water near 800 ppm. By reducing the rinse water flow, the size and cost of the recovery system was reduced to almost one third of that required for conventional indigo dyeing operations that have a rinse water flow rate near 75–100 gpm and an indigo concentration near 200 ppm. This points to the importance and need for automatic process control for regulating process variables, as it is not practical in most cases to control the consumption of water and steam manually without risking product quality.

PLANT APPLICATIONS OF REVERSE OSMOSIS

Continuous Dyeing

The wastewater from dyeing processes is more difficult to recycle because most or all of the color present must be removed before the water can be reused. For indigo dyeing, discussed earlier, only the dye present in the waste stream was recovered, leaving the permeate water containing the spent chemicals from the dyeing process to be discharged to waste treatment. Because most dyeing operations use many different colors, it is not always possible to isolate sufficient quantities of the individual dyes to make their reuse practical. However, when the dyeing process uses specific mixtures of dyes to formulate their production shades, it may then be possible to reuse the recovered dyes in the formulation of a different recipe. This has been reported by a plant (16) that was able to reuse the dyes recovered from the reverse osmosis concentrate. The plant first performed laboratory dyeings using the concentrate without adding any additional color. The dyed fabric was then compared to production shades that were scheduled to be dyed. The production that was nearest to the color of the dyed fabric that employed the recovered concentrate was selected. If needed, additional dye was added to the recovered dye concentrate so that it would produce an identical shade to the one selected. Several commercial production shades were dyed this way using the recovered dye concentrate. In some cases, very little additional dye was added to the recovered concentrate to make it match a production shade. In the project, sponsored by the Environmental Protection Agency (EPA) (16), both hot rinse water and neutral, cationic, and anionic dyes for cotton, polyester, acrylic, and rayon fibers were successfully reclaimed and reused.

The reverse osmosis filtration system, shown in Figure 8.13, used membranes that are dynamically formed on porous stainless steel tubes from zirconyl hydroxide and polyacrylic acid (ZOPA) in aqueous solution. This system was able to recover over 90% of the wastewater as clear water suitable for reuse in the dyeing process. It was also possible to recover direct, acid, basic, and premetalized dyes in the concentrate and reuse them to give a wide range of commercial colors on automotive fabrics. The hot-water permeate obtained from the filtration system could be used anywhere in the dyeing process, with minimal problems occurring even when permeate from the darkest wastewater was reused.

If the system, such as the one just described was automated, dyeing formulations, known to cause problems for the filtration system, could be automatically discharged directly to waste treatment, bypassing the filtration system. Most laboratory and plant dyeings using recovered dye were successful, except for those arising from the reuse of recovered dye concentrate from very dark shades. When this dark concentrate was used to formulate dyebaths for medium to light shades, it was rarely possible to exactly match the production shade. This problem could be minimized by scheduling dyeings so that light shades do not follow the darkest production dyeings. Then, the dark concentrates would be used to formulate dyebaths for dark shades, and light concentrates would be used to formulate dyebaths

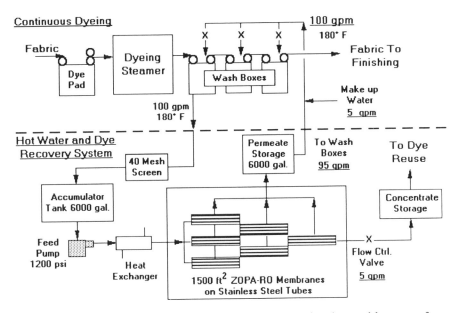

FIGURE 8.13   Reverse osmosis filtration system for recovering dyes and hot water from continuous dyeing (16).

for medium to light shades. This scheduling procedure would also minimize the need for the extensive equipment cleaning necessary when light shades follow dark shades in production.

An analysis of the savings generated by the recovery system is shown in Table 8.3. It was possible to recover over $332,000 per year from energy, water, and wastewater treatment savings. Because the actual quantities of dye that can be recovered for recycle must be demonstrated from full-scale operating experience, no value is given in Table 8.3 for dyes; however, the system was able to recover

TABLE 8.3   Annual Savings Obtained from the Filtration and Recycling of Dyeing Wastewater (16)

Assume 300 Working Days per Year, Three Shifts per Day	Savings
1. Reduction in water use from 100 to 5 gpm	
a. Process water savings ($1/1000 gal)	$41,000
b. Wastewater treatment ($2/1000 gal)	$82,000
2. Reduction in energy consumption	
a. Hot water (60–180°F) at $5/10^6 Btu	$205,400
b. Boiler chemicals	$4,000
3. Dyes and auxiliary chemicals	Depends on case
Total savings	$332,400

soluble dyes and operate at wastewater temperatures approaching 200°F. At the savings listed in Table 8.3, a recovery system could pay for itself in 1.5 years. Presently, one of the few reverse osmosis membranes capable of handling wastewater above 100°C is manufactured by Dupont Separation Systems in Seneca, South Carolina.

EFFECTS OF RECYCLE ON WASTEWATER TREATMENT

When the total wastewater flow is decreased by automation and recycle systems, not only are the water, waste treatment, and energy costs reduced, there can also be an improvement in the wastewater treatment process when a biological process is used. If the volumetric capacity of the biological treatment system is not changed when the conservation or recycle processes are installed, there will be an increase in the time for treatment, and better equalization as the water flow to the treatment plant is decreased. To take advantage of the additional treatment time, the treatment process must be carefully monitored so that chemical feeds, mixing, aeration, and settling rates are adjusted to the new flow conditions. This additional treatment time should increase the percentage of organic material removed by the biological process (24) and can also decrease the horsepower needed for aeration, in cases where more than one day's retention is used for the aeration system.

Since other factors do influence treatment efficiency, it is important that a careful analysis of the waste treatment system be made so that these benefits can be realized. This is all the more important if a chemical/physical treatment system is used for treating the plant wastewater.

SUMMARY

The potential for savings in chemicals, water, steam, and waste treatment cost by using recycle systems is very good. Recycle systems can be paid for in 1–2 years, as the examples cited show. Usually, the key to success is to design the process and recycle system so that it is fully automated. Since many textile preparation, dyeing, and finishing processes now operate at speeds of 150–300 yards per minute and treat a variety of textile fabrics, the process must respond rapidly. This necessarily includes a recycle system that knows when the wastewater can be recycled and when it cannot. When the water, steam, and chemical flow is automatically controlled, as shown in Figure 8.2, it can be changed in a matter of seconds, and the process can respond in minutes, depending on how drastic a change is made. This removes much of the human element that generally takes much longer to respond and has a higher probability for error. Many textile plants are now using process automation because of product quality control, a benefit besides those covered in this chapter.

There are several recycle systems in operation today, providing those who are considering recycling wastewater with actual data and sometimes the chance to inspect an operating system before installing one at a new location. In this way, much of the guess work is eliminated. However, each process must be carefully analyzed and future needs considered before a final selection is made. Some membranes are sensitive to specific chemicals, and others to high temperatures; therefore, a membrane or filter that has the least limitations may be the best selection in the end, even if it is more expensive.

REFERENCES

1. Public Law 92-500, 92nd Congress, 5.2770, October 18, 1972.
2. Information obtained from consulting engineers and manufacturers of filtration systems.
3. D. W. Sundstrom and H. E. Klei, *Wastewater Treatment*, Prentice-Hall, Englewood Cliffs, NJ, 1979, p. 10.
4. R. W. Miller, *Flow Measurement Engineering Handbook*, McGraw-Hill, New York, 1989.
5. N. M. Schmitt and R. F. Farwell, *Understanding Automation Systems*, Texas Instruments Information Publishing Center, Dallas, 1984.
6. Provided by K. Don Ledford, Texas Instruments, Kings Mountain, NC.
7. J. Murkes and C. G. Carlsson. *Crossflow Filtration*, Wiley, New York, 1988, p. 20.
8. S. Vigneswaran and R. Ben Aim, *Water, Wastewater and Sludge Filtration*, CRC Press, Boca Raton, FL, 1989, p. 228; J. H. Perry, *Perry's Chemical Engineers Handbook*, McGraw-Hill, New York, 1963, p. 5–53.
9. M. C. Porter, *Handbook of Industrial Membrane Technology*, Noyes Publications, Park Ridge, NJ, 1990, p. 96.
10. M. C. Porter, *Handbook of Industrial Membrane Technology*, Noyes Publications, Park Ridge, NJ, 1990, p. v.
11. S. Vigneswaran and R. Ben Aim, *Water, Wastewater and Sludge Filtration*, CRC Press, Boca Raton, FL, 1989, p. 164.
12. M. C. Porter, *Handbook of Industrial Membrane Technology*, Noyes Publications, Park Ridge, NJ, 1990, p. 264.
13. J. Murkes and C. G. Carlsson, *Crossflow Filtration*, Wiley, New York, 1988, p. 3.
14. S. Vigneswaran and R. Ben Aim, *Water, Wastewater and Sludge Filtration*, CRC Press, Boca Raton, FL, 1989, p. 204.
15. J. J. Porter and T. N. Sargent, *Textile Chem. Color.*, **9**, 38 (1977).
16. C. A. Brandon, Closed Cycle Textile Dyeing: Full-scale Hyperfiltration Demonstration, Demonstration (Design), U.S. Environmental Protection Agency, EPA-600/2-80-005, March 1980.
17. J. J. Porter, The Use of Membranes for the Recovery of Dyes and Auxiliary Chemicals and Automatically Controlling the Recycling of Bleaching Wastewater, National Textile Center Project on Source Reduction of Pollutants from Textile Processing Waste Water, Clemson University, Clemson, SC, unpublished work.
18. C. W. Aurich, "PVA Recovery by Ultrafiltration," AATCC Symposium, Textile Technology/Ecology Interface. Charlotte, NC, May 1975.
19. J. J. Porter, *Textile Chem. Color.*, **22**, 21 (1990).

20. C. R. Hoffman, *J. Coated Fabrics*, **10**(Jan.), 178 (1981).

21. J. L. Gaddis, H. G. Spenser, and P. A. Jernigan, Caustic Recovery and Recycling at a Textile Dyeing and Finishing Plant, Advances in RO and UF, Am. Chem. Soc. Meeting, Toronto, Canada. June 1988.

22. L. Johnston and T. W. Wett, *Chem. Pro.*, March, 36 (1989).

23. F. Leonard, The Design of an Indigo Dye Recovery System, Seminar on Membrane Separation Technology, Continuing Engineering Education, Clemson University, August 1981.

24. D. W. Sundstrom and H. E. Klei, *Wastewater Treatment*, Prentice-Hall, Englewood Cliffs, NJ, 1979, p. 155.

CHAPTER 9

MEMBRANE FILTRATION TECHNIQUES IN DYESTUFF RECOVERY

JOHN ELLIOTT
Ciba-Geigy Corporation, Toms River, New Jersey

INTRODUCTION

Humans have been fascinated by color since prehistoric times. This is evident from the extensive use of colorants in or on materials such as clothing, carpets, walls, plastics, food, and paper products. Prior to the mid-nineteenth century, colorants were obtained from natural sources, which were mainly of vegetable and animal origins. The discovery of the first synthetic dye, Mauveine, by W. H. Perkin in 1856, however, was the initiator of the organic chemical industry (1). Since that time, the synthetic dyestuff industry has undergone phenomenal growth, resulting in over 8000 chemically different dyes having reached commercial significance on a worldwide basis (2). In 1990, synthetic dye production in the United States alone was approximately 129,000 tons (3). A major portion of synthetic dyestuffs is consumed in the textile industry, which has an associated global fiber consumption near 30 million tons and an expected growth rate of 3% per annum (4).

The textile industry utilizes large volumes of water in its wet processing operations and, thereby, generates substantial quantities of wastewater (5,6). This is an important point, since wastewater is the principle route by which dyestuffs enter the environment. For example, of the 450,000 tons of dyestuffs produced in 1978, 41,000 tons were lost to waste streams from textile processing operations, and 9000 tons constitute the estimated loss from dye manufacturing facilities (7). During the last 20 years, the public has become increasingly aware of environmental issues pertaining to water pollution and waste disposal. This awareness has resulted in the development of more stringent environmental regulations and their enforcement. Traditional industries, such as dye manufacturing and textile processing, therefore, have been forced to change their working culture from one in which

Environmental Chemistry of Dyes and Pigments, Edited by Abraham Reife and Harold S. Freeman.
ISBN 0-471-58927-6 © 1996 John Wiley & Sons, Inc.

effluent discharge and disposal of waste were not issues to one involving a high priority on waste minimization through: 1) reduction of the volume and toxicity of discharges, 2) adoption of alternative processing methods and chemicals, and 3) recycling and reuse of water, chemicals, and colorants (8). To meet the afore-mentioned challenges, much work has been done by chemical companies to pro-duce environmentally friendly biodegradable auxiliaries in the fabric preparation field (9). In the coloration sector, dyers began using lower liquor ratios in dyeing processes (10), recycling dyebath liquors (11–14), while dye manufacturers began to actively promote bifunctional reactive dyes, which address both environmental and economic concerns (15).

It is well known that effluents from dyehouses often contain a wide range of dyestuffs combined with a variety of other products such as dispersing agents, dyebath carriers, salts, emulsifiers, levelling agents, and heavy metals. The presence of these chemicals in the wastewater often has a pronounced effect on one or more water quality parameters such as pH, biological oxygen demand (BOD), chemical oxygen demand (COD), total organic carbon (TOC), total dissolved solids (TDS), and color. Although most dyestuffs have a low BOD value, they add TOC and noticeable color to the water, which is objectionable to the public for aesthetic reasons (16). In addition, the possible accumulation of dyestuffs and pigments or their metabolites in aquatic organisms must also be borne in mind (17).

The methods for the removal of color from wastewater may be classified into two main categories: 1) decomposition of the color component and 2) removal of the color component (15). Colorant decomposition involves methods such as oxi-dative degradation (chlorine, ozone, and Fenton's reagent), reductive degradation (sodium hydrosulfite, formamidine sulfinic acid), and biological degradation. Re-moval of color involves methods such as adsorption (activated carbon, ion-pair exchange), and filtration (precipitation, flocculation, and membrane technology).

During the past decade, the utilization of membrane-based separation processes has been under active development in both the textile and dyestuff industries. This particular technology has progressed to the point of producing high-quality waste-water, which either satisfactorily meets the stringent requirements for discharge or is acceptable for reuse in fabric preparation, dyeing, and finishing processes.

MEMBRANE SEPARATION TECHNOLOGY

The Process

Osmosis is a natural process involving the diffusion of a fluid through a semiper-meable membrane separating two solutions of different concentrations and tending to equalize the concentration on both sides of the membrane. Semipermeable mem-branes have the ability to prevent the passage of dissolved molecules while allow-ing passage of a solvent (usually water) through the membrane under a pressure gradient (18). If pure water is in contact with both sides of a semipermeable mem-brane, at equal pressure and temperature, then there is no net flow across the membrane, because the chemical potential is equal on both sides. If a soluble salt is added to one side of the membrane, the chemical potential of the water on that side is reduced. Osmosis occurs when the pure water passes through the membrane to the salt-solution side (Figure 9.1a), until equilibration of the solvent chemical potential is restored. Equilibrium occurs when the pressure differential from the volume change on the two sides is equal to the osmotic pressure; a solution property that is independent of the membrane (Figure 9.1b). If a pressure higher than the osmotic pressure is applied to the salt-solution side, it will cause a reversal of the osmotic flow, and drive the water back through the membrane to the pure-water side (Figure 9.1c). This phenomenon is called *reverse osmosis* (19).

There are four commercially significant membrane separation processes, namely ultrafiltration, reverse osmosis (hyperfiltration), gas separation, and electrodialysis. Reverse osmosis (RO) and ultrafiltration (UF) are pressure-driven membrane processes that remove solutes from solution based on particle or molecular size differentials. In practice, however, the degree of separation achieved and the pressures associated with the two are different. The separation mechanism characterizing ultrafiltration is based on membrane pore size and the difference in size between the solute and solvent. The membrane pores used in reverse osmosis, however, are so small that they may not actually exist in a discrete physical sense. The reverse osmosis mechanism is believed to be a solution diffusion type mechanism where the solvent ''dissolves'' in the membrane material and then diffuses through more easily than the solute.

Generally, *ultrafiltration* involves the separation of macromolecules from solvents of lower molecular mass, whereas reverse osmosis involves the separation of solute and solvent molecules of comparable molecular mass (7,20,21). There is no clear distinction between ultrafiltration and reverse osmosis, because preferential sorption can occur simultaneously with sieving action and diffusion. As a result, when the ultrafiltration membrane pore size is similar to the pore sizes used in

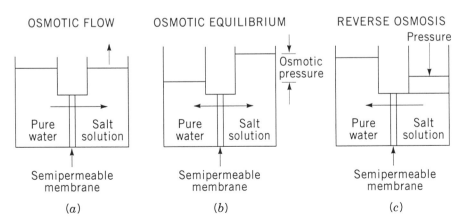

FIGURE 9.1   Process of reverse osmosis.

reverse osmosis, the process is sometimes referred to as *nanofiltration*. The range of applications for these membrane systems is illustrated in Figure 9.2 (18,21).

Ultrafiltration can be used to separate all solutes and particles larger than 0.001 μm; however, the low flux rate through the membrane limits its practical use to particles less than 0.1 μm in diameter. A related method, called *microfiltration*, is used for faster filtration of larger suspended and colloidal particulates greater than 0.1 μm. The undisputed realm of separations covered by ultrafiltration involves solutes of molecular weight (MW) between 1000–100,000 dalton (0.001–0.1 μm). Below MW 1000, osmotic pressure begins to increase significantly and reverse osmosis or nanofiltration comes into play. Nanofiltration retains small organic molecules having a MW >300 and even polyvalent ions, while allowing smaller ions to pass through the membrane. "True" reverse osmosis membranes, however, reject even the smallest ions and allow the passage of pure water only. Hence, the operational pressures for reverse osmosis systems are high, in the range of 100–1000 psi, whereas ultrafiltration systems operate at pressures usually in the range of 10–100 psi (19,22,23).

In all pressure-driven membrane processes, the liquid to be treated flows in a turbulent manner over the surface of the membrane by means of a *cross-flow* technique. This process distinguishes membrane filtration from conventional filtration, where the flow is perpendicular to the separation medium as illustrated in (Figure 9.3). In the latter case, an accumulation of particles or the formation of a gel occurs at the surface of the filter reducing the filtration rate and producing an effect known as *concentration polarization*. In cross-flow filtration, the momentum of the flow parallel to the membrane reduces the likelihood of particle or gel accumulation on the membrane surface thereby reducing the tendency for *membrane fouling* to occur. If such fouling occurs, it may be reversed, however, by backflushing of the membrane (22).

Membrane Materials

Early in their development, membranes were composed of a single substance having a relatively uniform microporous structure throughout the entire thickness of the membrane. The first commercial reverse osmosis membranes were made from cellulose acetate and, thus, were susceptible to extremes of pH and temperature, and also relatively sensitive to chlorine. Reverse osmosis and ultrafiltration membranes are now produced from a variety of other polymeric materials such as polyamides, polyacrylonitriles, polysulfones, polycarbonates, and fluorocarbon-based polymers (24). These new materials allow systems to be operated at higher flux rates and are resistant to wider ranges of pH, temperature, and solvents. The development of asymmetric membranes by Loeb and Sourirajan (25) in the 1960s advanced the use of membranes in membrane separation processes. Homogeneous asymmetric membranes have a nonuniform pore size throughout their structure. They are composed of a very thin dense skin layer that functions as the barrier to the solute transport and a thicker more porous layer that acts as the support (Figure 9.4*a*). In the early 1970s, development of heterogeneous thin-film composite membranes significantly advanced the use of membrane technology in many fields (Figure 9.4*b*).

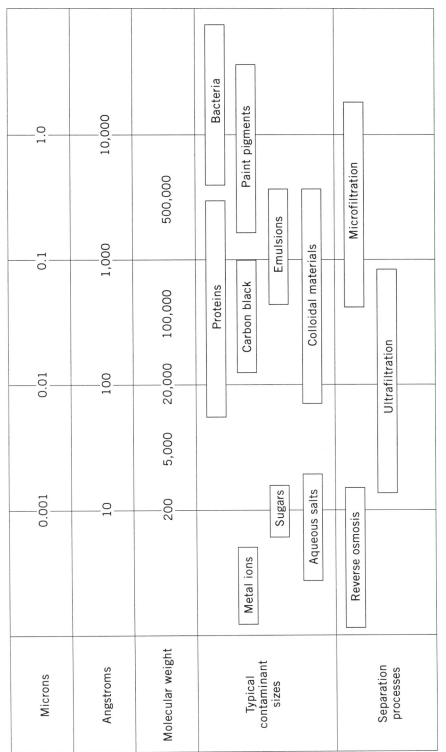

FIGURE 9.2 Range of applications of membrane technology. (Reprinted with permission from the American Association of Textile Chemists and Colorists.) (18)

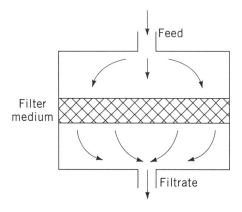

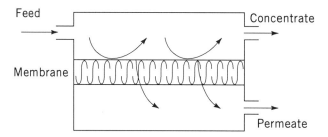

FIGURE 9.3 Difference between conventional and membrane filtration.

Another type of membrane deserves mention because it differs considerably from the conventional polymeric barriers. This membrane is known as a *dynamic membrane* and is formed by coating a selective membrane layer on a porous support such as porous carbon or sintered glass. Typical materials used to form this membrane layer include several hydrous oxides of Fe(III), Zr(IV), Th(IV), and a variety of synthetic polymers or polyelectrolytes such as poly(styrenesulfonic acid), poly(vinylpyrrolidone), poly(acrylic acid), and poly(vinylbenzyltrimethylammonium chloride) (26).

Membrane Configurations

A membrane separation process requires not only a membrane having the desired characteristics but also supporting equipment that is reliable, compact, and economically attractive. Since membrane separation processes differ significantly in their mode of operation, it is not surprising that each membrane module has its

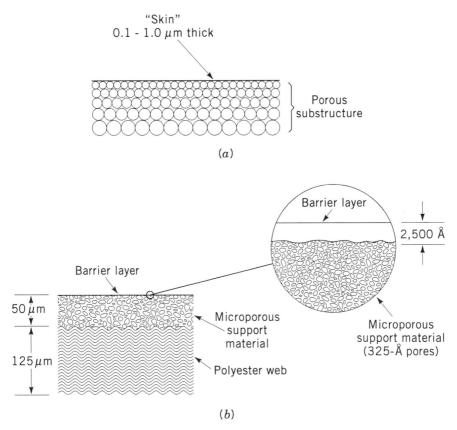

FIGURE 9.4 Composition of (a) asymmetric and (b) thin-film composite membranes. (Reprinted by special permission from CHEMICAL ENGINEERING, May 30, 1983, June 11, 1984. Copyright © 1983, 1984 by McGraw-Hill, Inc., New York, NY 10020.)

own specific features. In addition, the commercial viability of these processes was enhanced with the advent of membrane modules that could be combined together readily and were easily replaced at the end of their useful life. Membranes are packaged in modules that control the pressure, feed stream velocity, and turbulence in order to reduce concentration polarization effects. There are four basic kinds of modules, namely plate and frame, tubular, spiral-wound, and hollow-fiber modules. Description of the different modules and their operational characteristics are beyond the scope of this chapter but is well documented in the literature (19,27–29).

Modes of Operation

For efficient membrane system operation, the fluid must be passed over the membrane surface at a high velocity. Based on this requirement, there are three different modes of operation: single pass, batch, and continuous. In the single-pass mode, the feed circulates through the membrane and leaves as two streams (concentrate and permeate). In the batch process, the concentrate is circulated back to the feed tank, increasing the concentration of the feed stream. Permeate is either recycled or discharged. In a continuous operation, otherwise known as *feed and bleed*, the concentration of feed takes place entirely within the recirculation loop until the desired level is achieved. At this point, a portion of the concentrate is bled from the system and fresh feed is added to the holding tank (18,30). In this manner, the quantity of feed added must be equal to the amount of permeate flowing out of the loop plus the amount of concentrate being bled from the system. The basic

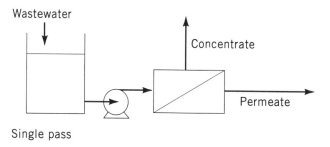

Single pass

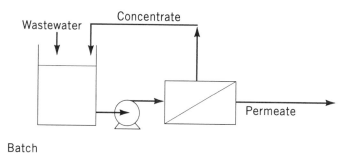

Batch

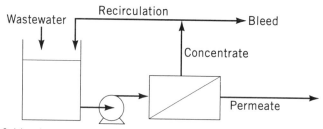

Feed & bleed

FIGURE 9.5   Basic operational modes for membrane systems. (Reprinted with permission from the American Association of Textile Chemists and Colorists.) (18)

modes of operation are illustrated in Figure 9.5. Multistage operation is achieved by combining two or more feed-and-bleed stages in series. Each stage operates at constant concentration, which increases from the first to last stage, with the latter unit operating at the desired product concentration (30).

TREATMENT OF WASTEWATER FROM TEXTILE DYEING OPERATIONS USING MEMBRANE FILTRATION TECHNIQUES

Many textile mills discharge their wastes to local sewage treatment plants, usually following some form of pretreatment, while others operate their own effluent treatment facilities and provide complete on-site treatment prior to discharge. Although biological treatment remains the most widely used treatment method, increasingly stringent environmental regulations, particularly for the removal of toxic organics and color, have induced many mills to also install advanced physical/chemical methods such as carbon adsorption, ozonation, and reverse osmosis (20). Combinations of processes are practically necessary to achieve the most efficient decolorization and lowest environmental load. All are, however, end-of-pipe treatments based on the philosophy that waste streams cannot be avoided, hence the need for additional processing to deal with them (7). The textile industry has given high priority to the reduction of pollution at source, recycling or reusing waste streams that cannot be avoided, treating waste streams after recycling possibilities have been exhausted, and deposition of the residual waste. These waste minimization efforts result in a better use and conservation of resources, lower environmental hazards risks, and reduced costs associated with waste treatment. The application of these principles has been demonstrated, using membrane technology for the recycling of water in dyehouses and the concentration and reuse of dyestuffs.

Recovery of Indigo

The economical recovery of the vat dye, indigo, has been achieved through the use of ultrafiltration membranes in several textile mills. In 1980, pilot-plant ultrafiltration trials were conducted at a textile-dyeing plant to recover indigo, using cartridges containing bundles of hollow-fiber polysulfone membranes (31). In the conventional indigo-dyeing process, significant quantities of excess indigo are present in the wash water from the rinsing operations, which become part of the plant's effluent. The first step in the recovery of this dyestuff is prefiltration, which is required to remove the cotton lint from the waste stream prior to entering the membrane system. After prefiltration, the water is transferred to a surge vessel and then fed directly to the ultrafiltration unit. The concentrate stream exiting the ultrafiltration unit represents only a small fraction of the rinse water volume fed to the system. However, this stream contains virtually all of the indigo dyestuff, which is then collected in a storage tank prior to delivery to the dyebath mixing station. The recovered indigo is mixed at this station with fresh indigo and additional dyebath chemicals prior to feeding to the dye machines. This completes the loop for indigo recovery. The permeate liquors contain caustic soda, salts, low-molecular-weight contaminants, and any sulfur dyestuffs, if they were used in the dyeing process. The permeate is treated using the plant's normal waste treatment practices and subsequently discharged. The results of several pilot-plant runs, with and without sulfur dyestuffs present, provided the data required for the design of a full-scale system to process the wastewater from five indigo-dyeing machines. A continuous three-stage hollow-fiber ultrafiltration system for the recovery of indigo was placed in operation in 1982 (31). A similar indigo-recovery process has been described by Porter (5). This recovery system has been in operation since 1981, using a multistage feed-and-bleed ultrafiltration system fitted with vinylsulfone membranes. Each stage automatically establishes a steady-state concentration that becomes progressively higher as dye concentration increases. The final stage achieves the maximum concentration of indigo needed for reuse. The use of counter-current-flow wash boxes ensures a feed stream concentration of about 800 ppm indigo prior to ultrafiltration (Figure 9.6). This particular indigo-recovery system is reported to have paid for itself in less than 2 years of operation (32). Numerous membrane filtration systems for the recovery of indigo have been reported in the literature (33–40).

Treatment of Wastewater Containing Reactive Dyestuffs

In all dyeing processes, dyestuff losses occur that are influenced by the type of dyestuff, the depth of shade, the application method, and the volume of water used. It is well known that during the application of reactive dyestuffs to cellulosic fibers a competing hydrolysis reaction also occurs, that inactivates some of the reactive dyestuff. This undesirable competitive reaction results in unfixed hydrolyzed dyestuff in the water and produces unacceptable levels of color in mill effluents. For example, the color losses of reactive dyestuffs in exhaust liquors and the associated wash liquors from dyeing cellulose have been estimated to be about 20–50% (7,41). A recent study also showed, by calculation, that the average concentration of reactive dyestuffs in the effluent from a jet-dyeing machine was about 60 ppm, while that from a pad-batch process can be on the order of 250 ppm (10).

Reverse osmosis and ultrafiltration are very effective for the removal of color from dyehouse effluents regardless of the type of dyestuff used. Decolorization for these procedures is in the range of 95–100% (20,42,43). The application of membrane technology specifically for the treatment of wastewater from reactive dyeing processes has been studied by Tegtmeyer (44) and Sewekow (45). Nanofiltration membranes were used in a pilot-plant unit to concentrate the combined wastewater from the dyebath of a deep shade reactive dyeing and four rinsing baths. The results of the trial showed that the dyestuff could be concentrated to about 0.5–1.0% of its initial volume. The resulting permeate stream was almost colorless, containing salt and low-molecular-weight impurities, and could be discharged or reused for the rinsing bath operations. Extrapolation of the pilot-plant results to the daily dyeing of about 5 tons of fabric would result in a wastewater volume of about 200 m^3 per day. Treating this wastewater by nanofiltration would separate it into a

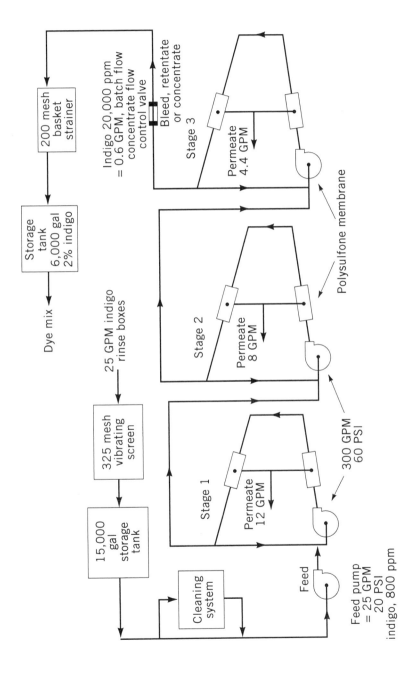

FIGURE 9.6 Indigo range recovery system. (Reprinted with permission from the American Association of Textile Chemists and Colorists.) (5)

200 mesh basket strainer

Storage tank 6,000 gal 2% indigo

Dye mix

Indigo 20,000 ppm = 0.6 GPM, batch flow concentrate flow control valve

Bleed, retentate or concentrate

Stage 3

Permeate 4.4 GPM

Stage 2

Permeate 8 GPM

Polysulfone membrane

Stage 1

Permeate 12 GPM

300 GPM 60 PSI

25 GPM indigo rinse boxes

325 mesh vibrating screen

15,000 gal storage tank

Cleaning system

Feed

Feed pump = 25 GPM 20 PSI indigo, 800 ppm

permeate and a concentrate containing virtually the entire load of dyestuff in a volume of about $1-2$ m^3 (Figure 9.7) (44,45). Additional calculations indicated that a 200 m^3 wastewater load could be treated using about 130 m^2 membrane area. The hydrolyzed dyestuff concentrate isolated in this manner could be disposed of by other chemical techniques. The authors also proposed the separation of the total reactive dyestuff wastewater into two streams, one having a low-salt content and the other having a high-salt content, since 96% of the salt load is contained in the dyebath liquor and the first rinsing bath. The two streams could then be processed independently by different methods. The liquors from high-salt dyebath and the first rinsing bath could be concentrated by nanofiltration, while the remaining wastewater containing the low-salt load could be treated by reverse osmosis. This proposal was also tested in the pilot-plant unit. Concentration of the latter three rinsing baths by reverse osmosis produced a totally colorless permeate having a salt content of less than 100 ppm. This permeate liquor could be recycled as process water, leading to about a 60% reduction in wastewater. The concentrate from the reverse osmosis process is added to the high-salt waste stream and retreated by the nanofiltration process. In this manner, the majority of the salt is found in the nearly decolorized permeate from the nanofiltration stage. Since this permeate now contains about 3% salt, the possibility of recycling some of this salt solution into the dyeing process is an attractive possibility (44,45).

Reactive dye liquors from exhaust-dyeing processes have been decolorized effectively in a closed-loop recycle system, using an ultrafiltration membrane carrying ionic charges (42). The study showed the system was technically feasible for producing decolorized permeates, with water recoveries >90% and of a quality suitable for reuse in the dyehouse. It was estimated that a full-scale ultrafiltration operation that could treat 200 tons/month of reactive dyestuff effluent should give full recovery of the initial capital outlay in under 3 years.

Treatment of Dyehouse Wastewater Using Dynamic Membranes

Reverse osmosis (hyperfiltration) utilizing dynamically formed dual-layer hydrous zirconium oxide–polyacrylate membranes on porous ceramic and carbon support tubes removed >85% TOC, with >99% color rejection, from the composite wastewater of a dyehouse processing synthetic upholstery fabrics (43). In addition, 87% water recovery was achieved without any deterioration in membrane performance. Similar dual-layered zirconium-based dynamic membranes have been used successfully to treat dye-containing effluents from both alkaline viscose and acidic polyester dyeing operations (46). Color rejection >99% has been achieved from highly colored liquor, at pH 4.3, from polyester dyeing. In the same process, water recovery on the order of 70% was achieved and was found suitable as process water. Treatment of highly colored alkaline (pH $7.8-9.3$) wastewater from viscose dyeing showed membrane rejection of 99% for color and 97% for TOC, with an overall water recovery of 80%. Brandon (47), using similar zirconium oxide–polyacrylic acid dynamic membranes on stainless steel support tubes, conducted a pilot-scale study on the renovation of wastewater from a continuous textile dye

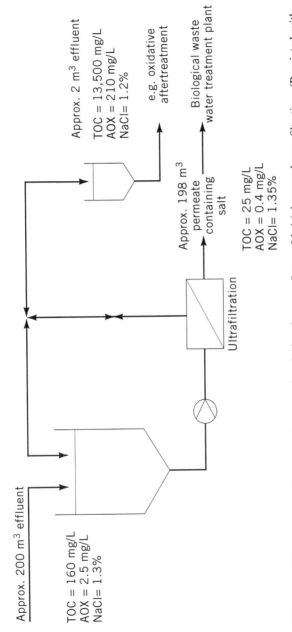

Approx. 200 m³ effluent

TOC = 160 mg/L
AOX = 2.5 mg/L
NaCl= 1.3%

Ultrafiltration

Approx. 198 m³
permeate
containing
salt

TOC = 25 mg/L
AOX = 0.4 mg/L
NaCl= 1.35%

Approx. 2 m³ effluent

TOC = 13,500 mg/L
AOX = 210 mg/L
NaCl= 1.2%

e.g. oxidative
aftertreatment

Biological waste
water treatment plant

FIGURE 9.7 Treatment of effluent from reactive dyeing (approx. 5 tons fabric) by membrane filtration. (Reprinted with permission from the American Association of Textile Chemists and Colorists.) (45)

range. The results showed high color rejection, and the treated water was suitable for recycling as process water. Reuse of dyestuff concentrates in the dyeing process appeared to be feasible by careful planning of the sequence of the dyeing processes. This pilot-plant study led to full-scale plant trials, where recovery and recycle of 90–95% of the treated water was achieved (48).

Treatment of Dyehouse Wastewater Using Polysulfone Membranes

The use of ultrafiltration polysulfone membranes has gained importance in the separation of dyestuffs and other chemicals from textile mill wastewaters, due to their high resistance to strong acids and bases, various solvents, and elevated temperatures. The efficiency of dyestuff removal from dye-containing wastewater has been studied, using an ultrafiltration system fitted with a tubular membrane module consisting of a polysulfone membrane cast on a porous poly(methylmethacrylate) support bar. The results indicated that dyestuffs having a MW >780 are retained at least 97% under an optimized feed flow of 0.76 m/sec (49). In a 200-hr ultra-filtration trial on textile wastewater, removal efficiency was maintained at a level of 75–85% and 50–60% for COD and TOC, respectively. Additional studies, using polysulfone membranes formed on a glass support, showed that the membranes were capable of up to 100% dye retention for dyestuffs of MW >800, irrespective of the casting parameters and the applied pressure (50,51). Long-term studies on different types of dyehouse wastewaters also showed that ultrafiltration involving tubular polysulfone-based membranes could be utilized effectively, provided that turbulent fluid flow was maintained (52). Ultrafiltration feasibility studies also have been conducted on mixed-dyestuff effluent from a textile facility, using a polysulfone hollow-fiber membrane. Rejection coefficients of 30–90% were attained; however, membrane fouling was encountered in some of the trial runs due to the nature of the contaminants in the combined waste stream (53).

Membrane Treatment of Direct and Acid Dyebath Liquors

The use of other membrane types in various configurations for the separation of dyestuffs from dyehouse liquors has been studied on both laboratory and commercial scales. For example, ultrafiltration studies on dyestuffs such as C.I. Direct Orange 26, C.I. Direct Brown 2, and C.I. Acid Green 25 have been conducted by Zuk and co-workers, using polyacrylonitrile hollow-fiber (54) and cellulose acetate membranes (55). Their results indicated that the high water recovery and concentration of the dyestuffs were due to aggregation of the dyestuff particles and the good separation characteristics of the ultrafiltration membranes. Similarly, other direct dyestuffs were removed from dyebath liquors effectively using cellulosic membranes, with a simultaneous reduction in the COD level (56). Laboratory studies on several anionic and cationic dyestuffs have been carried out using both cellulose acetate and aromatic polyamidohydrazide membranes of various average pore sizes (57). This study showed that the interaction forces between the dyestuff molecule and membrane material were governed by the net electric charge, the polar and nonpolar functional groups of the dyestuff molecules, and also by the acidity–basicity of the membrane materials. As a result, differences in performance were observed between membrane materials even when the average pore size remained constant.

Treatment of Wastewater from Textile Printing Operations

The removal of dyestuffs and chemicals from textile printing operations has been successful, through the design of a two-stage ultrafiltration/nanomembrane filtration system (40). This system, designed from pilot studies, has a capability of treating 288,000 gal of wastewater per day. Reuse of the treated water is expected through 10 recycles before requiring discharge. Dye recovery >95% and 60–90% COD reduction from dye-printing wastewaters using a polysulfone ultrafiltration system have been reported also in China (58).

TREATMENT OF WASTEWATER FROM DYESTUFF MANUFACTURING OPERATIONS USING MEMBRANE FILTRATION TECHNIQUES

Membrane separation technology has been utilized for several years in dyestuff manufacturing operations, producing high-quality liquid dyestuffs having excellent stability characteristics. This is particularly the case with reactive dyestuffs, since the health issues related to their usage in dry powder forms has contributed to their marketing as liquid brands. The production of stable liquid reactive dyestuffs, through the removal of dissolved electrolytes, is easily achieved by membrane filtration techniques (59–62). The use of reverse osmosis/ultrafiltration is also well documented for the preparation of highly concentrated dispersions of vat and disperse dyestuffs (63,64), in the production of liquid basic dyestuffs (65), leather dyestuffs (66), and food colorants (67).

It is a well-known fact, therefore, that the manufacture of dyestuffs creates substantial quantities of wastewater that must be treated to various regulatory standards prior to discharge into municipal waste treatment plants or directly into aquatic systems. Such was the case at a dyestuff standardization facility in New Jersey. Prior to 1992, over 350,000 gal/day of groundwater and river water were utilized in the cleaning and cooling of vessels, spray dryers, and process piping. The resulting wastewater, containing about 400–600 lb dyestuff, was processed in the plant's tertiary waste treatment facilities prior to discharge into the Atlantic Ocean, via a pipeline. In 1989, the New Jersey State Senate passed legislation requiring the plant's ocean discharge pipeline to be closed down by December 31, 1991. This decision necessitated an aggressive wastewater minimization program be implemented that focused on reducing the flow of wastewater from the dyestuff-processing operations. This program provided the basis for a concept of *standardization without effluent* (68–70). The original concept was designed to use a combined waste stream strategy and separate the water and the dyestuff using membrane technology. The water would then be recycled back to the process, and the dyestuffs would be concentrated, dried, and ultimately disposed. Laboratory and pilot-plant screening studies, using various types of membranes and equipment configurations, encountered problems with membrane fouling phenomena. However, the results of the original studies using a combined stream approach led to a new direction of point source treatment of individual product streams and minimization of the hydraulic load. The objective of this twofold approach was to reduce the wastewater from 300–500,000 gal/day to 0–10,000 gal/day. The hydraulic load studies were concerned with defining the source, quantity, and nature of the effluent, while the membrane filtration studies concentrated on the recovery of the individual dyestuffs from classes such as acid, direct, reactive, disperse, and vat dyestuffs. Pilot studies indicated that 55% of the wastewater, which resulted from the insoluble disperse and vat dyestuffs, required processing through both ultrafiltration and reverse osmosis membranes. The remaining 45% from the water-soluble acid, direct, and reactive dyestuffs required only a reverse osmosis (nanofiltration) treatment.

Various hydraulic load reduction programs were installed by the January 1, 1992, deadline, resulting in a 94% reduction in wastewater to about 20,000 gal/day. This quantity was more appropriate for treatment by membrane filtration equipment. Wastewater from the insoluble dyestuff operations is treated in an ultrafiltration unit using multichannel ceramic membranes to produce a concentrate containing about 15% solids and a permeate containing <1% solids. The concentrate is stored and, subsequently, blended with the next batch of the same color. The permeate is directed through the reverse osmosis filtration system using polysulfone-based membranes, to remove any small molecular weight dispersants and any residual color. The final concentrate from this reverse osmosis operation is unusable and is discharged. The resultant permeate, containing <1% salt, is used for cleaning equipment and the wet-air scrubbing system. Any dyestuff-contaminated water from these operations is recycled back through the complete process. Reverse osmosis treatment of the remaining water-soluble dyestuffs process streams results in a concentrate of reclaimed dyestuff, which is reused. The permeate, containing about 1–2% salt, is discharged to the waste treatment plant (68,69). Membrane technology also has been used successfully for more than 10 years at a organic chemical manufacturing plant in Germany, in the areas of dyestuff production, dyestuff standardization, and wastewater disposal. Due to more stringent TOC discharge requirements, the plant was challenged to eliminate or drastically reduce the level of nonbiodegradable refractory organics. This was accomplished through the utilization of a membrane filtration (nanofiltration) system that concentrated the large nonbiodegradable organic molecules prior to their destruction by wet-air oxidation (WAO) (71). This volume reduction process was necessary since WAO operates more efficiently at higher organic concentrations than those found in the waste stream prior to membrane concentration. The resultant treated permeate from the nanofiltration contained salts and lower-molecular-weight molecules that are biodegradable by biological treatments.

Nanofiltration is only one of the unit operations of a process that involves solvent extraction in addition to WAO and biological waste treatment. The industrial wastewater at the plant was divided into five segregated streams. The first, containing biologically degradable TOC, flows directly to the biological treatment plant. A second stream is treated via the WAO system and then by biological degradation. Two streams undergo solvent extraction whereby the resulting concentrate is reprocessed by WAO, while the extracted aqueous waste liquor (raffinate) is treated in the biological treatment plant. The fifth stream is processed by the nanofiltration membrane system. The permeate from this operation is sent for biological treatment. The concentrate, which now contains up to 79% of the original quantity of organics in the influent feed, is pumped directly into the WAO system. The total waste treatment system is illustrated in Figure 9.8. The nanofiltration

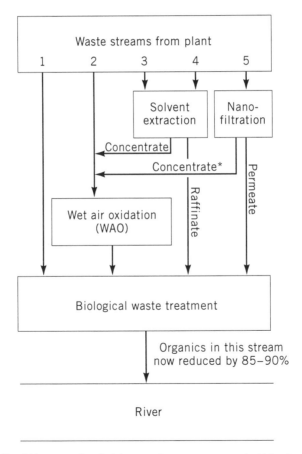

*79% of the organics fed to membranes are now in this stream.
Stream is $^1/_{15}$ the size of the stream fed to membrane system.

FIGURE 9.8 Treatment of segregated waste streams from a dyestuff/chemical manufacturing plant. (Reprinted with permission from *Chemical Processing*.) (71)

process is carried out in two stages, the first at 370 psi and the second at 440 psi. A high permeate flux rate, up to 39 gal/ft^2/day, is realized with an average feed rate of 70 gal/min producing a concentrate volume that is one-fifteenth of the influent-feed volume (71). Aqueous dye-manufacturing waste streams, containing both dyestuffs and inorganic salts of polyvalent inorganic acids (e.g., sulfates) have been treated at pH values <4.0 using membrane filtration technology (72). Under these acidic conditions, the relatively low-molecular-weight ionic organic compounds can be concentrated effectively without the disturbing increase in osmotic pressure caused by the presence of salts of polyvalent inorganic acids. This technique has been used to separate and concentrate dyestuffs, particularly reactive dyestuffs, and optical brightener intermediates such as dinitrostilbenedisulfonic acid from wastewater containing sulfates, through utilization of acid-resistant polysulfone UF–RO membranes.

A two-stage UF–RO system also has been developed to treat about 500 m^3/day of wastewater from dyestuff-manufacturing operations (73). The system utilizes polymeric membranes, preferably polysulfone or polyacrylonitrile types modified with cationic and/or anionic groups. The wastewater, containing about 10 g/L TOC and 160 g/L salts, is adjusted into the pH 4–5 range at 45°C, and fed from a pretreatment vessel through a seven-stage ultrafiltration unit. Each stage consists of membrane modules connected in parallel, and contains 120 m^2 of membrane surface area. After passage through the first seven-stage unit, the concentrate is retreated through a second four-stage unit to further reduce the volume. The permeate from this second membrane filtration stage is recycled back to the pretreatment vessel. The color rejection was about 99%, and the retention of heavy metals such as copper and chromium was in the range of 75–99% and 85–99%, respectively (73). A study of the removal of azo dyestuffs from wastewater via ultrafiltration using cellulose acetate membranes indicated that the structural features of the dyestuff, such as the arrangement of sulfonic acid groups and the presence of heterocycles in the conjugated system, have a greater influence on their selective removal by the membrane than molecular weight (74). Membranes based on vinylidene fluoride–tetrafluoroethylene copolymer and treated in a high-frequency discharge plasma have been used in the ultrafiltration treatment of dyestuffs. The plasma modification of the membranes induced changes in the chemical nature of their surfaces, which improved the ultrafiltration rate of ionic dyestuffs (75). Recently, experimental studies involving polyacrylate membranes have demonstrated their potential for ultrafiltration applications, particularly in the area of treatment of wastewaters from the dyestuff industry (76).

MEMBRANE FILTRATION OF DYE WASTEWATERS CONTAINING HEAVY METALS

In addition to residual color in wastewater from dyestuff manufacturing and application operations, other pollutants such as heavy metals may be present. This is particularly true in the application of metal-complex reactive dyestuffs where, due to fixation yields on the order of 50–90%, high concentrations of unfixed hydrolyzates may be present in the waste liquor. Depending on the type of chromophore used, metals such as copper and nickel could also be present. In these cases, strict regulations must be satisfied prior to the discharge of the wastewater produced.

Several methods have been utilized to reduce the impact of metal-complex reactive dyestuffs in wastewater. The methods include the development of high fixation bifunctional dyestuffs (15,77), use of short liquor-ratio dyeing machines, and replacement of exhaustion dyeing methods with continuous processes (45). An alternative treatment for these highly polluted waste liquors that are generated is membrane filtration, since the recovery of heavy metals by this technique is well known in the electroplating industries (78,79). The reduction of heavy metals in textile process wastewaters has been demonstrated by several workers. Gaddis and Spencer (80) showed that reverse osmosis membranes exhibited high rejection rates for color and toxic metals present in low concentrations in textile process water. Brandon and Gaddis (81) also evaluated reverse osmosis applications in textile dyeing plants and showed that rejection of salts, organic species, and toxic metals increased significantly during the first 100 hr of operation. The effective removal of copper and chromium from effluents arising from the manufacture of copper-complex direct and reactive dyestuffs, and from $1:1$ and $1:2$ chromium-complex acid dyestuffs was extremely effective using modified polysulfone membranes in a multistage UF–RO membrane unit (73). While the removal of copper from wastewater containing reactive dyestuffs can be achieved using membrane filtration, the overall economics for this particular operation must be compared with that arising from the use of other effective metal removal techniques such as coagulation (82), adsorption on carbon (83) or polymeric materials (84), chelation (85), ion-pair extraction (86), and electrochemical processes (87).

ECONOMIC CONSIDERATIONS

The use of membrane technology to treat colored wastewater from either dye manufacturing or textile dyeing operations is a viable technique that significantly reduces the volume of wastewater and provides possibilities for the reuse of dyestuffs, chemicals, and water. However, the capital costs for the membrane filtration equipment can be high, depending on the size of the plant, the operating conditions employed, and the associated wastewater treatment costs. Currently, in most cases, the economics of treating composite wastewaters from textile processes are not favorable. Therefore, full-scale commitment to membrane processes generally has been made to treat smaller, fully segregated streams. For example, a dye standardization facility successfully achieved at least a 94% reduction in the daily water usage and utilized a membrane filtration system to recover an average of 500 lbs/day dyestuff, which was lost previously to the waste treatment plant (68,69). This has resulted in substantial annual savings from recovered dyestuff and in wastewater treatment costs. In addition, over 97% of the water used in the dyestuff processing operations is recycled using this technology (69).

The use of indigoid dyestuffs poses a major waste treatment problem, especially for textile "denim" mills throughout the world. Ultrafiltration has been used successfully to recover the expensive indigo for reuse and at the same time reduce the pollution burden on wastewater treatment systems (5,40). Ultrafiltration of the indigo-containing effluent results in a dyestuff concentrate that can be recycled directly into the plant. Significant annual processing savings are realized since >98% of the indigo is recovered (40). Additional savings also are realized since the hot-water permeate can be reused. The value of the recovered indigo offsets the installation costs, and reuse of the hot water reduces operating costs. Excellent payback times can be realized from this product recovery process. Similarly, significant savings were estimated by Brandon and co-workers (48) after a full-scale reverse osmosis study was conducted on the wastewater from a continuous dye range at a textile dyeing plant. The savings emanated from recycling about 90–95% of the treated water, reduced energy usage from the hot-water recycle and lower chemical requirements (47).

CONCLUSIONS

Membrane filtration has been demonstrated to be a commercially viable technique for the effective removal of residual color from wastewater arising during the manufacture and application of dyestuffs. The practicality of membrane filtration is evident from its ability to separate various types of wastewater into two streams, viz, a concentrate and permeate. The concentrate, containing the organic dyestuff, can either be reused directly or its volume is such that it can be disposed efficiently by WAO or high-temperature incineration. In the majority of cases, the permeate, consisting essentially of water, can be recycled back into the system, to afford economic benefits from reduced energy usage and waste treatment costs.

This technology will continue to be considered one of the few treatment methods that can achieve satisfactory color removal from dye-containing wastewater at source and produce water of reusable quality. Improvements in membrane composition, characteristics, and performance will no doubt enhance the effectiveness of this technology for use in waste minimization processes. In addition, the use of integrated systems such as a combined ultrafiltration–reverse osmosis membrane system coupled with an activated sludge bioreactor (18) may have significant technical and economic benefits for the textile industry.

All the treatment methods for reducing pollution at source have advantages and disadvantages; however, they will become increasingly important, as methods for minimizing waste streams, reducing effluent treatment costs, and reducing resource consumption. No doubt, membrane filtration techniques will continue to play an important role in these areas within the dyestuff manufacturing and textile dyeing industries.

REFERENCES

1. J. G. Evans, *Perkin Centenary*, Pergamon Press, London, 1958, p. 60.

2. D. W. Bannister, A. D. Olin, and H. A. Stingl, *Kirk-Othmer Encyclopedia of Chemical Technology*, Vol. 8, Wiley, New York, 1978, p. 159.

3. Synthetic Organic Chemicals US Production & Sales–1990, U.S. International Trade Commission Publication 2470, Washington, D.C., 4-1, Dec. 1991.

4. B. Glover and J. H. Pierce, *J. Soc. Dyers Colourists* **109**, 5 (1993).

5. J. J. Porter, *Text. Chem. Color.* **22**, 21 (1990).

6. W. C. Tincher, *Text. Chem. Color.* **21**, 33 (1989).

7. I. G. Laing, *Rev. Prog. Coloration* **21**, 56 (1991).

8. M. Lomas, *J. Soc. Dyers Colourists* **109**, 10 (1993).

9. P. I. Norman and R. Seddon, *J. Soc. Dyers Colourists* **107**, 150 (1991).

10. B. Glover and L. Hill, *Text. Chem. Color.* **25**, 15 (1993).

11. F. L. Cook, R. M. Moore, and G. S. Green, *Text. Chem. Color.* **21**, 11 (1989).

12. F. L. Cook, W. C. Tincher, W. W. Carr, L. H. Olson, and M. Averette, *Text. Chem. Color.* **12**, 15 (1980).

13. F. L. Cook and W. C. Tincher, *Text. Chem. Color.* **10**, 21 (1978).

14. W. C. Tincher, F. L. Cook, and L. A. Barch, *Text. Chem. Color.* **13**, 266 (1981).

15. K. Imada and S. Hashizume, Fiber Reactive Dyes for Cotton: Reducing the Level of Color in Wastewater Effluent, AATCC Book of Papers, 1993 International Conference and Exhibition, Montreal, Quebec, Oct. 3–6, 1993, p. 220.

16. N. L. Nemerow and T. A. Doby, *Sewage Ind. Wastes* **30**, 1160 (1958).

17. J. Park and J. Shore, *J. Soc. Dyers Colourists* **100**, 383 (1984).

18. J. Carriere, D. Mourato, and J. P. Jones, Answers to Textile Wastewater Problems: Membrane Filtration and Membrane Bioreactor Systems, AATCC Book of Papers, 1993 International Conference and Exhibition, Montreal, Quebec, Oct. 3–6, 1993, p. 224.

19. L. E. Applegate, *Chem. Eng.* June 11, 64 (1984); ibid., May 30, 27 (1983).

20. P. J. Halliday and S. Beszedits, *Can. Text. J.* **103**(4), 78 (1986).

21. J. H. Steven, *Chem. and Ind.* May 2, 346 (1983).

22. H. Moes, *Chem. Process.* Feb., 62 (1986).

23. P. R. Klinowski, *Chem. Eng.* May 8, 165 (1978).

24. A. S. Michaels, *CHEMTECH* **11**(1), 36 (1981).

25. S. Loeb and S. Sourirajan, *Advan. Chem. Ser.* **38**, 117 (1962).

26. A. R. Cooper, *Chem. Britain*, Sept., 814 (1984).

27. C. H. Gooding, *CHEMTECH*, June, 348 (1985).

28. C. H. Gooding, *Chem. Eng.* Jan. 7, 56 (1985).

29. B. S. Parekh, *Chem. Eng.* Jan., 70 (1991).

30. P. H. Fergusson, *Inst. Chem. Eng. Symp. Series No. 96* (Effluent Treat. Disposal), April 15–17, 403 (1986).

31. A. Fuchs, B. R. Breslau, and A. J. Toompas, *Text. Ind.* **147**(12), 44 (1983).

32. F. Leonard, The Design of an Indigo Dye Recovery System, Seminars on Membrane Separation Technology, Clemson University, Aug. 1981.

33. W. Rummele, *Textilveredlung* **24**, 96 (1989).
34. P. Rivet, *Eau. Indust. Nuisances* **121**, 48 (1988).
35. F. Scuderi, *AES.* **4**(6–7), 63 (1982).
36. M. Frigerio, *Tinctoria* **90**(5), 81 (1993).
37. M. Crespi, J. Valldeperas, M. C. Gutierrez, C. Gibello, and B. Escribano, *Simposo. Europeo. sobre Efluentes Textiles–Espana* **9**, 1 (1988).
38. R. Rajagopal, *Colourage* **38**, 53 (1991).
39. M. T. Pessoa de Amorim, *Proc. Int. Conf. Environ. Poll·t.*, Vol 2, Nath Bhaskar, ed., Inderscience Enterprises, Geneva Switzerland, 1991, p. 686.
40. W. N. Rozelle, *Text. World* **144**(1), 67 (1994).
41. S. J. Hobbs, *J. Soc. Dyers Colourists* **105**, 355 (1989).
42. A. Erswell, C. J. Brouckaert, and C. A. Buckley, *Desalination* **70**, 157 (1988).
43. C. A. Brandon, J. S. Johnson, R. E. Minturn, and J. J. Porter, *Text. Chem. Color.* **5**, 134 (1973).
44. D. Tegtmeyer, *Melliand Textilberichte* **74**(2), 148 (1993).
45. U. Sewekow, Minimizing Effluent Load in Reactive Dyeing, AATCC Book of Papers, 1993 International Conference and Exhibition, Montreal, Quebec, Oct. 3–6, 1993, p. 235.
46. R. Groves, C. A. Buckley, J. M. Cox, A. Kirk, C. D. Macmillan, and M. J. Simpson, *Desalination* **47**, 305 (1983).
47. C. A. Brandon, *Nat. Water Supply Improvement J.* **7**, 34 (1980).
48. C. A. Brandon, D. A. Gernigan, J. L. Gaddis, and H. G. Spencer, *Desalination* **39**, 301 (1981).
49. K. Majewska-Nowak, T. Winnicki, and J. Wisniewski, *Desalination* **71**(2), 127 (1989).
50. K. Majewska-Nowak and T. Winnicki, *Stud. Environ. Sci.* (Chem. Prot. Environ.) **23**, 387 (1984).
51. K. Majewska-Nowak, *J. Membrane Sci.* **68**(3), 307 (1992).
52. K. Majewska-Nowak, *Desalination* **71**(2), 83 (1989).
53. J. C. Watters, E. Biagtan, and O. Senler, *Sep. Sci. Technol.* **26**(10–11), 1295 (1991).
54. J. Zuk, M. Rucka, and J. Rak, *Chem. Eng. Commun.* **19**, 67 (1982).
55. J. Zuk and M. Rucka, *Environ. Prot. Eng.* **13**(1), 93 (1987).
56. G. Song, Z. Jiang, and Y. Li, *Shuichuli Jishu* **14**(3), 189 (1988).
57. L. Tinghui, T. Matsuura, and S. Sourirajan, *Ind. Eng. Chem. Prod. Dev.* **22**, 77 (1983).
58. Z. Pei and S. Han, *Huanjing Kexue* **4**(2), 1 (1983).
59. U.S. Pat. 4,523,924 (June 18, 1985), R. Lacroix (to Ciba-Geigy Corp.).
60. U.S. Pat. 4,500,321 (Feb. 19, 1985), P. Hugelshofer, B. Bruttel, H. Pfenninger, and R. Lacroix (to Ciba-Geigy Corp.).
61. Ger. Offen. DE 3,301,870 (July 26, 1984), J. Koll, H-H. Mols, R. Ditzer, G. Martiny, and K-H. Steiner (to Bayer A-G.).
62. Ger. Offen. DE 3,148,878 (June 23, 1983), H. Pohlmann, W. Gruenbein, A. Walch, J. Wildhard, F. Meininger, K. Opitz, and E. Junghanns (to Hoechst A-G.).
63. U.S. Pat. 4,436,523 (May 13, 1984), P. Hugelshofer, P. Zbinden, and Z. Koci (to Ciba-Geigy Corp.).

64. U.S. Pat. 4,329,145 (May 11, 1982), J. Koll, H-H. Mols, R. Hornle, E. Schuffenhauer, H. Brandt, F. Bremer, K. Wolf, and W. Schiwy (to Bayer A-G).

65. Ger. Offen. DE 3,311,294 (Oct. 4, 1984), H. Moser, W. Samhaber, and H. Schmid (to Sandoz).

66. U.S. Pat. 4,402,701 (Sept. 6, 1983), H. Gleinig, H-H. Mols, J. Koll, G. Dick, R. Hornle, and F. Kunert (to Bayer A-G).

67. U.S. Pat. 4,560,746 (Dec. 24, 1985), R. W. J. Rebhahn, and W. L. Cook (to Hilton-Davis Chemical Co.).

68. R. Ballina and P. Douvres, Reducing Wastewater Effluent—The S.W.E.A.T. Project, ACS Symposium, Div. of Environ. Chem., Chicago, Illinois, Aug. 22–27, 1993, p. 270.

69. Effluent Reduction in Dye Processing, *TechApplication*, Electric Power Research Institute (EPRI), Process Industry Publication, **4**(5), 1 (1992).

70. D. Rotman and A. Wood, *CHEMICAL WEEK* Dec. 9, 66 (1992).

71. A. E. Hodel, *Chem. Process.* Nov., 99 (1993).

72. U.K. Pat. Appln. GB 2,207,618A (Feb. 8, 1989), H. Pfenninger, (to Ciba-Geigy A-G).

73. Jpn. Pat. 62186986 (Aug. 15, 1987), A. Henz and H. Pfenninger, (to Ciba-Geigy A-G).

74. A. E. Kashcheev, D. V. Mozharovskii, and G. A. Voloshin, *Khim. Tekhnol. Vody.* **12**(6), 520 (1990).

75. A. F. Burban, E. A. Tsapyuk, and M. T. Bryk, *Khim. Tekhnol. Vody.* **11**(8), 754 (1989).

76. A. S. Bal, C. G. Malewar, and A. N. Vaidya, *Desalination* **83**, 325 (1991).

77. S. Fujioka and S. Abeta, *Dyes Pigm.* **3**, 281 (1982).

78. L. J. Kosarek, Removal of Various Toxic Metals and Cyanide from Water by Membrane Processes, in W. J. Cooper, ed., *Chem. for Water Reuse, Vol 1*, Ann Arbor Science, Ann Arbor, MI, 1981, p. 261.

79. S. S. Kremen, C. Hayes, and M. Dubos, *Desalination* **20**, 71 (1977).

80. J. L. Gaddis and H. G. Spencer, *Proc. Symp. Text. Ind. Tech.*, EPA-600/2-79-104, U.S. Environmental Protection Agency, Cincinnati, 1979, p. 107.

81. C. A. Brandon and J. L. Gaddis, *Desalination* **24**, 19 (1978).

82. H. A. Fiegenbaum, *Am. Dyest. Rep.* **67**(3), 43,46 (1978); *Ind. Wastes* **23**(2), 32 (1977).

83. U.S. Pat. 4,005,011 (Jan. 25, 1977), C. D. Sweeney, (to American Color and Chemical Corporation).

84. *Am. Dyest. Rep.* **61**(8), 57 (1972).

85. U.S. Pat. 3,778,368 (Dec. 11, 1973), Y. Nakamura, A. Umehara, and I. Yamada, (to Sankyo Co Ltd.).

86. I. Steenken-Richter and W. D. Kerner, *J. Soc. Dyers Colourists* **108**, 182 (1992).

87. T. R. Demmin and K. D. Uhrich, *Am. Dyest. Rep.* June, 13 (1988).

CHAPTER 10

CHEMICAL REMOVAL OF PHOSPHATE IONS FROM DISPERSE DYE FILTRATES

JOLANTA SOKOLOWSKA-GAJDA, HAROLD S. FREEMAN, and
ABRAHAM REIFE
North Carolina State University, Raleigh, North Carolina

INTRODUCTION

During the past two decades, red, violet, and blue azo disperse dyes (cf. Figure 10.1) based on heteroaromatic amines (cf. Figure 10.2) have been developed as alternatives to red and blue anthraquinone dyes, which are more expensive and of lower tintorial strength (cf. Figure 10.3). The diazotization of the required heteroaromatic amines (weak bases), however, requires special conditions. Since the corresponding diazonium salts are unstable in the acidic media normally used to diazotize simple anilines, the diazotization of heteroaromatic amines is usually conducted in the presence of H_2SO_4, H_2SO_4/H_3PO_4 (1–3), $H_2SO_4/HOAc$ (4,5), or HNO_3 (6,7). In most cases, nitrosyl sulfuric acid (8) is employed as the diazotizing agent, even when the diazotization process is conducted in an organic acid medium. Among the useful organic acids is HOAc, and it is sometimes used in combination with propionic acid (9,10) because HOAc freezes at typical diazotization temperatures.

Recently, Sokolowska-Gajda and Freeman reported (11) an effective method for the diazotization of 2-amino-6-nitrobenzothiazole in a mixture of acetic acid and dichloroacetic acid, followed by coupling with N-β-cyanoethyl-N-β-acetoxyethylaniline to form Disperse Red 177. Although the yield of this two-step synthesis was good (87%), it was somewhat lower than anticipated from the work of others, in which the diazotization was conducted in the phosphoric acid (12). In addition, there is now reason to be concerned about the toxicity of an $HOAc/CHCl_2CO_2H$ effluent. Similarly, the use of H_3PO_4 causes an environmental problem known as eutrophication, owing to the release of significant amounts of phosphate ions into

Environmental Chemistry of Dyes and Pigments, Edited by Abraham Reife and Harold S. Freeman.
ISBN 0-471-58927-6 © 1996 John Wiley & Sons, Inc.

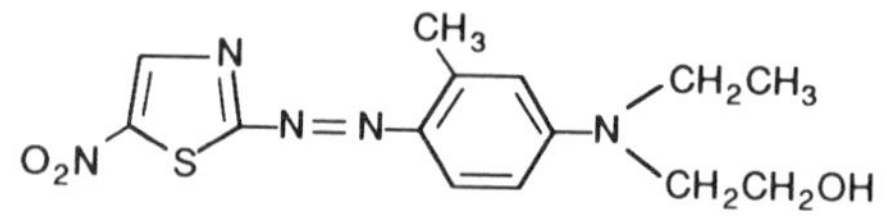

Disperse Blue 106

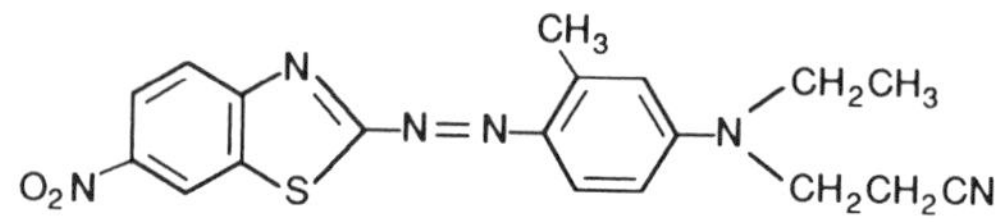

Disperse Violet 52

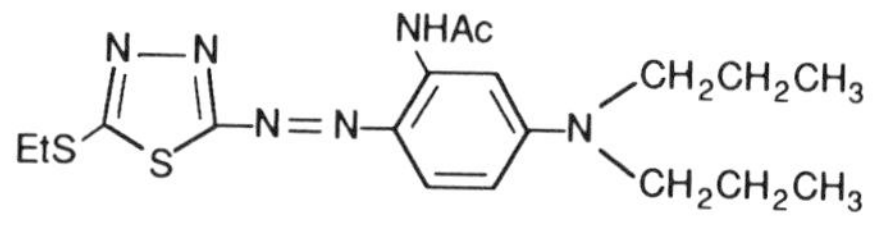

Disperse Red 338

FIGURE 10.1 Examples of commercial disperse dyes based on heteroaromatic amines.

public waters. The phosphate-rich waters are depleted of oxygen, in response to an increase in algae, and aquatic life is injured. One reasonable approach to addressing the latter problem is the development of a method for the environmentally friendly use of H_3PO_4 as the medium for diazotizing heteroaromatic amines. It is apparent that in cases involving a high concentration of phosphate, the chemical treatment of the ensuing effluent will play an important role, as conventional biological treatment will be insufficient.

Although the chemical literature contains data from studies indicating the most practical ratios of coagulants to phosphate and the optimized pH conditions to employ for efficient phosphate removal, such conditions should be regarded as empirical, and serve only as a guide for developing a protocol for treating a specific waste stream. Therefore, the goal of this study was to develop an efficient, general procedure for the precipitation of phosphate ions from disperse dye wastewater. The method chosen for this investigation involved the precipitation of orthophosphates, using metal coagulants such as aluminum and iron(III) salts, from filtrates

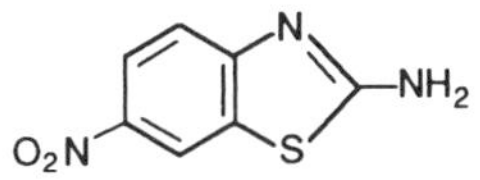

2-Amino-6-nitrobenzothiazole

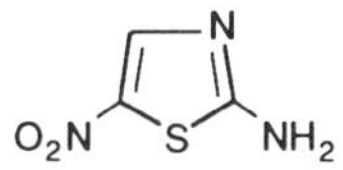

2-Amino-5-nitrothiazole

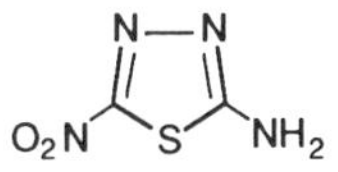

2-Amino-5-nitrothiadiazole

FIGURE 10.2 Examples of heteroaromatic amines used in disperse dye synthesis.

generated in the synthesis of commercial azo disperse dyes. Disperse Red 177 was chosen as the prototype for this study.

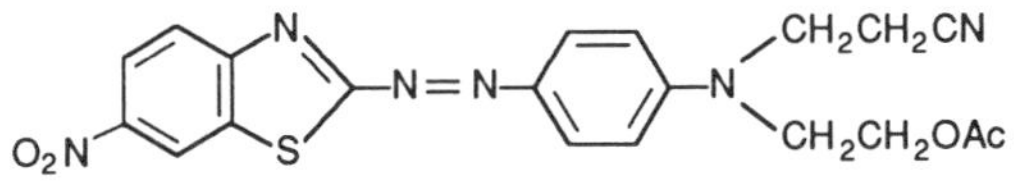

Disperse Red 177

Although aluminum and iron (III) salts have been considered as coagulants for the precipitation of phosphate from solution, the optimum acidity, efficiency, and mechanism of phosphate removal reported in literature have varied considerably. It is clear, however, that both aluminum and iron (III) salts react with phosphate ions with $1:1$ stoichiometry; but, the coagulant doses required, in actual practice, to achieve efficient removal are usually higher. Recht and Ghassemi (13) found that 1.4 mol of aluminum was required to effect satisfactory precipitation of 1 mol

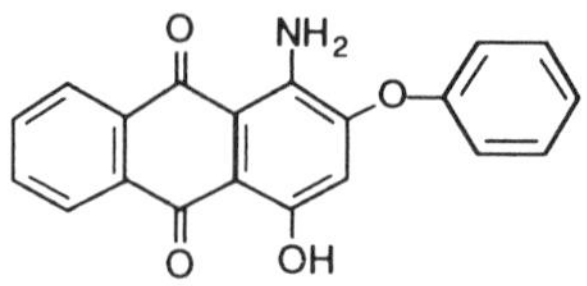

Disperse Red 60

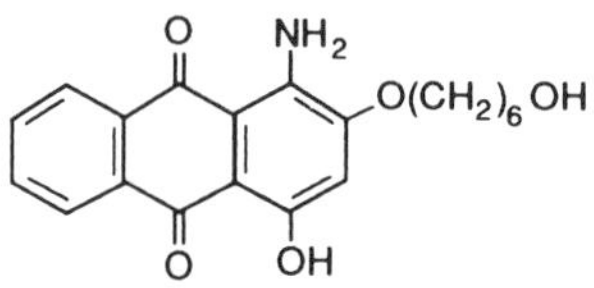

Disperse Red 91

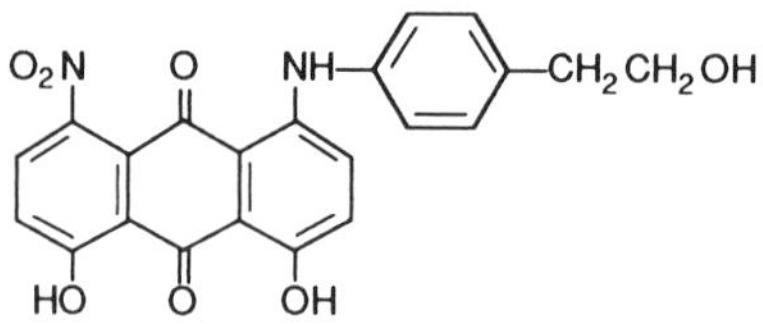

Disperse Blue 27

FIGURE 10.3   Examples of commercial disperse dyes based on anthraquinone chemistry.

of phosphate at pH 6, but 100% removal required an aluminum to initial phosphorus ratio higher than 1.8. In another study (14) involving phosphate precipitation, the required ratio was determined to fall in the range of 1.4:1.9. Iron (III) doses higher than a stoichiometric amount are also required for satisfactory phosphate removal. For example, Hsu (15) reported efficient phosphate precipitation when the iron (III)–phosphorus molar ratio was 2:1.

Concerning the relative efficiencies of different coagulants, results from three research groups (13,16–18) suggest that the iron (III) ion is less effective than the aluminum ion. On the other hand, Hsu's (15) results suggest the reverse and are based on the higher affinity of iron (III) ions for PO_4^{-3} than the corresponding affinity exhibited by aluminum ions.

It is also clear that pH is an extremely important parameter in the efficiency of phosphate removal. In the case of aluminum salts, the most favorable pH is in the

5.5–6.5 range, whereas the optimum pH when using iron (III) salts is about one unit lower. It was found, however, that in some cases, for instance at metal–P ratios (15,19), other than optimum or in the presence of other anions (19) (e.g., sulfate), the highest removal efficiency was achieved sometimes at pH values different from those indicated above.

The general conclusion to be made from these studies is that the amount of phosphate remaining after precipitation strongly depends on the initial wastewater composition. In addition, the lack of accurate formation constant data for metal complexation of phosphate in different types of waste streams makes it necessary to modify reported conditions for phosphate precipitation to reflect the specific contents of a given wastewater.

The goal of this study was to establish the most efficient conditions for phosphate precipitation from the filtrate obtained following the synthesis of Disperse Red 177 in a medium containing H_3PO_4. For this purpose, two commonly used coagulants were employed. Specifically, $Al_2(SO_4)_3 \cdot 14 H_2O$ (alum) and $FeCl_3$ were employed at different molar ratios of coagulating metal (Met) to the initial moles of phosphorus (i.e., Met–P ratio).

RESULTS AND DISCUSSION

In all cases, the samples of the filtrate were treated in the same way, that is, a predetermined amount of NaOH solution was added prior to and along with co-agulants to achieve the final pH value, and the time of mixing was the same (15 min). This is an important point because the adjustment of pH *after* coagulant addition resulted in a drastically lower efficiency of phosphate precipitation. Also, a further increase in mixing time did not improve the efficiency of precipitation. The results of this investigation are summarized in Table 10.1 and Figures 10.4 and 10.5. It is clear from these data that phosphate removal depends heavily on pH and that both coagulants afforded maximum phosphate removal at specific pH values. In the case of alum (Figure 10.4), optimum phosphate precipitation occurred at pH 6 and corresponds to the pH of minimum $AlPO_4$ solubility (pH 5.5–6.5). For $FeCl_3$ (Figure 10.5), however, it was found that the precipitation of phosphate was best at pH 3, a value considerably lower than the pH of minimum $FePO_4$ solubility (pH 4.5–5.5). It is also clear that the efficiency of both coagulants relates directly to the metal-to-phosphorus molar ratio employed for the precipitation. The results obtained indicate (cf. Table 10.1, Figures 10.4 and 10.5) that $FeCl_3$ is not as efficient as alum at metal (Fe, Al):phosphorus (P) ratios of 1:1 (cf. 250 vs. 160 ppm of residual phosphate) and 1.4:1 (cf. 3.96 vs. 2.97 ppm of residual phosphate), but works as well as alum when the metal:P molar ratio is 1.6:1 and the optimized pH (pH 3 for $FeCl_3$ and pH 6 for alum) is used. When higher metal:P ratios were employed, the following observations were made: (1) at a 1.8:1 metal:P ratio, $FeCl_3$ is more efficient than alum (cf. 0.37 vs. 0.56 ppm); (2) the use of $FeCl_3$ at a 2:1 Fe:P ratio afforded maximum phosphate removal (99.9994%), leaving a residual phosphate concentration of only 0.075 ppm; and (3) employing alum at a

TABLE 10.1 Comparison of Phosphate Removal Using $Al_2(SO_4)_3 \cdot 14 H_2O$ and $FeCl_3$

Met:P Molar Ratio	Type of Coagulant	pH	Percent Removal of Phosphate[a]	Residual Phosphate Concentration (mol/L)	Residual Phosphate Concentration (ppm)
1.0	Al	3	94.6000	2.16×10^{-2}	670
	Fe		98.0000	8.00×10^{-3}	250
1.0	Al	4	97.3000	1.08×10^{-2}	330
	Fe		97.8700	8.49×10^{-3}	260
1.0	Al	5	98.0500	7.80×10^{-3}	240
	Fe		97.8100	8.76×10^{-3}	270
1.0	Al	6	98.6600	5.36×10^{-3}	160
	Fe		97.1400	1.14×10^{-2}	350
1.0	Al	7	96.0500	1.58×10^{-2}	490
	Fe		95.7500	1.70×10^{-2}	520
1.4	Al	3	93.7500	2.50×10^{-2}	770
	Fe		99.9700	1.28×10^{-4}	3.96
1.4	Al	4	99.8500	6.00×10^{-4}	18.50
	Fe		99.9600	1.60×10^{-4}	4.95
1.4	Al	5	99.9750	1.00×10^{-4}	3.10
	Fe		99.9200	3.20×10^{-4}	9.91
1.4	Al	6	99.9760	9.60×10^{-5}	2.97
	Fe		99.8700	5.20×10^{-4}	16.10

1.4	Al	7	99.9400	2.40×10^{-4}	7.43
	Fe		99.8550	5.80×10^{-4}	17.90
1.6	Al	3	80.5800	7.77×10^{-2}	2405
	Fe		99.9900	4.40×10^{-5}	1.36
1.6	Al	4	99.8810	4.76×10^{-4}	14.70
	Fe		99.9860	5.60×10^{-5}	1.73
1.6	Al	5	99.9800	8.00×10^{-5}	2.47
	Fe		99.9780	8.80×10^{-5}	2.72
1.6	Al	6	99.9880	4.80×10^{-5}	1.48
	Fe		99.9600	1.60×10^{-4}	4.95
1.6	Al	7	99.9600	1.60×10^{-4}	4.95
	Fe		99.9500	2.00×10^{-4}	6.19
1.8	Al	3	73.9000	1.04×10^{-1}	3230
	Fe		99.9970	1.20×10^{-5}	0.37
1.8	Al	4	99.7400	1.04×10^{-3}	32.30
	Al^b		99.8700	5.20×10^{-4}	16.1
	Fe		99.9960	1.60×10^{-5}	0.5
1.8	Al	5	99.9900	2.44×10^{-5}	0.75
	Al^b		99.9940	2.44×10^{-5}	0.75
	Fe		99.9930	2.67×10^{-5}	1.36
1.8	Al	6	99.9900	1.80×10^{-5}	0.56
	Al^b		99.9950	2.00×10^{-5}	0.62
	Fe		99.9890	4.40×10^{-5}	1.36
1.8	Al	7	99.9930	2.80×10^{-5}	0.87
	Al^b		99.9900	4.00×10^{-5}	1.23
	Fe		99.9770	9.22×10^{-5}	2.85

TABLE 10.1 Continued

Met:P Molar Ratio	Type of Coagulant	pH	Percent Removal of Phosphate[a]	Residual Phosphate Concentration (mol/L)	Residual Phosphate Concentration (ppm)
2:1	Al	3	37.5000	2.50×10^{-1}	7742
	Fe		99.9994	2.428×10^{-6}	0.075
	Fe[b]		99.9996	1.60×10^{-6}	0.050
2:1	Al	4	98.1250	7.50×10^{-3}	232.0
	Fe		99.9994	2.428×10^{-6}	0.075
	Fe[b]		99.9995	2.00×10^{-6}	0.062
2:1	Al	5	99.9500	2.00×10^{-4}	6.19
	Fe		99.9975	1.00×10^{-5}	0.31
	Fe[b]		99.9970	1.20×10^{-5}	0.37
2:1	Al	6	99.9951	1.96×10^{-5}	0.61
	Fe		99.9921	3.156×10^{-5}	0.97
	Fe[b]		99.9940	2.40×10^{-5}	0.74
2:1	Al	7	99.9887	4.52×10^{-5}	1.40
	Fe		99.9830	6.796×10^{-5}	2.00
	Fe[b]		99.99805	7.80×10^{-5}	2.41

[a] Average values from at least two determinations.
[b] "Real" filtrate.

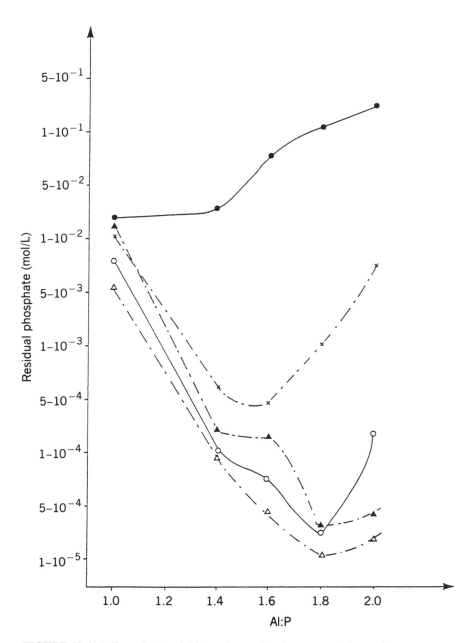

FIGURE 10.4   Effect of pH and Al:P ratio on phosphate removal from Disperse Red 177 filtrates. (●) pH = 3, (×) pH = 4, (○) pH = 5, (△) pH = 6, (▲) pH = 7.

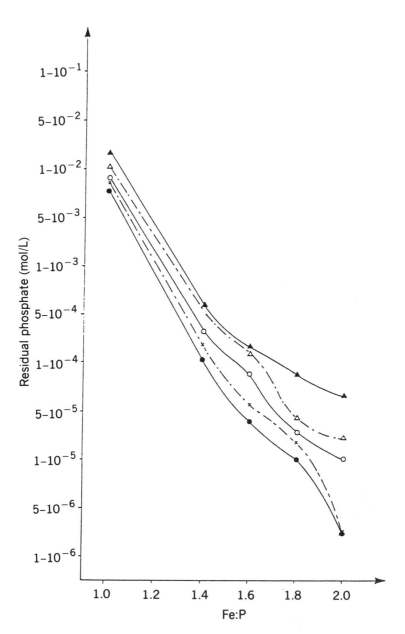

FIGURE 10.5   Effect of pH and Fe : P ratio on phosphate removal from Disperse Red 177 filtrates. (●) pH = 3, (×) pH = 4, (○) pH = 5, (△) pH = 6, (▲) pH = 7.

2:1 Al:P ratio and optimum pH did not improve phosphate removal versus what was achieved at a ratio of 1.8:1. In addition, the use of this higher ratio at pH 3–5 caused a significant decrease in the phosphate removal. These results are consistent with literature data (14). In that study, it was shown that when the pH drops below 5.5, the amount of residual phosphate increases, no matter how much aluminum is used. The high residual phosphate is caused by the formation of metastable aluminum phosphate complexes such as $AlH_2PO_4^{2+}$ and by the consumption of aluminum during formation of the polymeric aluminum complex (14) $Al_3(OH)_{20}^{4+}$ at moderately acidic pH values.

EXPERIMENTAL

General

2-Amino-6-nitrobenzothiazole and ammonium sulfamate were purchased from Aldrich Chemical Company (Milwaukee, WI) and N-β-cyanoethyl-N-β-acetoxyethylaniline was obtained from Emery Industries, Inc. (Cincinnati, OH). Triton X-100 and Surfynol 104 (surfactants) were obtained from Union Carbide Chemicals and Plastic Company, Inc. (Danbury, CT) and Air Products and Chemicals, Inc. (Allentown, PA), respectively. All other chemicals were obtained from Fisher Scientific Company.

Synthesis of Disperse Red 177

2-Amino-6-nitrobenzothiazole (3.59 g, 0.0184 mol) was dissolved in 49.87 g 85% H_3PO_4 (0.4 mol) containing 0.048 g Triton X-100, at 35–40°C. The reaction mixture was cooled to −5°C, whereupon 0.008 g of Surfynol 104 and 1.37 g (0.02 mol) $NaNO_2$ were added over 1 hr. Diazotization was completed by stirring the solution for an additional 1 hr. The diazonium salt solution was slowly added to 4.548 g (0.0196 mol) N-β-cyanoethyl-N-β-acetoxyethylaniline dissolved in 29 mL H_2O containing 0.6 g ammonium sulfamate and 5.75 g HCl (20°Bé). Ice (29 g) was added to keep the internal temperature at −5–0°C. The coupling step was continued for 2 hr, then the reaction mixture was diluted to 360 mL, with H_2O and stirred for 2 hr at 25°C. The dye was collected by vacuum filtration, washed with 640 mL H_2O, and dried, to give 8.024 g (99.56%) of crude Disperse Red 177, melting point 162–165°C; Rf = 0.30 on silica gel using n-butyl acetate:PhMe/2:3.

^{1}H-NMR, δ2.06 (s, 3 H), δ2.76 (t, J = 7.5 Hz, 2 H), δ3.70–4.08 (m, 4 H), δ4.35 (t, J = 7.5 Hz, 2 H) δ6.67–7.05 (m, 2 H), δ7.92–8.22 (m, 3 H), δ8.44 (dd, J = 9.0/2.0 Hz, 1 H) δ8.82 (d, J = 2.5 Hz, 1 H).

The above filtrate (1000 mL) contained 49.9 g 85% H_3PO_4 (0.4 mol), 0.048 g Triton X-100, 0.008 g Sulfynol 104, 5.75 g 20°Bé HCl, and 0.296 g N-β-cyanoethyl-N-β-acetoxyethylaniline. The filtrate was light red and its pH was 1.8.

Precipitation of Phosphate

Aluminum sulfate [$Al_2(SO_4)_3 \cdot 14 \, H_2O$; *alum*] and iron (III) chloride ($FeCl_3$) were used as the coagulants at metal:phosphorus (Met:P) ratios of 1.0–2.0 and at pH values of 3–7. A 20-mL sample of the filtrate was used for the precipitations and a sufficient amount of NaOH solution was added to achieve the desired pH. The coagulant was then added, and the mixture was stirred for 15 min at 15 rpm, filtered through a 0.45-μm membrane filter, and the amount of phosphate remaining in the filtrate was determined.

Analytical Procedure

To develop a procedure for the spectroscopic analysis of residual soluble phosphate, a "synthetic" filtrate was used. It contained H_3PO_4, HCl, surfactants (Surfynol 104, Triton X-100), and coupling component in amounts identical to that arising from the synthesis of a commercial disperse dye in H_3PO_4. To avoid interference arising from the presence of color, the model experiments did not contain dye. Ammonium molybdate was used to convert phosphate to phosphomolybdic acid, followed by reduction with stannous chloride in glycerol (20) to molybdenum blue (21–23) ($\lambda_{max} = 690$ nm) and comparison of the resulting absorbance value with those produced by standards prepared using a known amount of phosphate.

To avoid interference caused by Fe(III) ions (24) during the spectroscopic analyses, when $FeCl_3$ was used as the coagulant, the phosphomolybdic acid formed was first extracted into benzene:isobutanol (1:1) (25) and then reduced by $SnCl_2$ to produce molybdenum blue, which in this case gave an absorption at $\lambda_{max} = 625$ nm. Also, in this case, the amount of phosphate was determined by comparing the absorbance value recorded on the unknown with values produced using standards prepared by extracting solutions containing molybdenum blue produced from the reduction of a known amount of phosphomolybdic acid. The method used to prepare ammonium molybdate, $SnCl_2$ solution, and standard phosphate solutions is provided in the analytical literature (26).

Having developed an effective method for phosphate precipitation from "synthetic" filtrates, filtrates ("real") derived from commercial synthesis of Disperse Red 177 were engaged. Since Red 177 is also blue when dissolved in an acidic medium, the absorption of the analyzed filtrate was compensated by using, as the comparative solution, the corresponding sample of the filtrate treated with ammonium molybdate but not reduced with $SnCl_2$.

CONCLUSIONS

It is clear from this study that both alum and iron (III) chloride are suitable coagulants for the efficient precipitation of phosphate ions from filtrates derived from the synthesis of disperse dyes in an H_3PO_4 medium. The experimental work conducted shows that the use of the optimum pH for each coagulant and a metal to initial phosphorus ratio of 1.8:1 for alum and 2:1 for $FeCl_3$ provides very efficient phosphate removal, and that in the latter case phosphate ions are practically eliminated from the disperse dye filtrate. Further, the authors believe that the results are applicable to pilot-scale operations and that it will be possible to demonstrate the utility of the ensuring precipitates as a component in fertilizer. With regard to the latter point, it is anticipated that the precipitate derived from the use of $FeCl_3$ would be more appropriate.

REFERENCES

1. O. Annen, R. Egli, R. Hasler, B. Henzi, H. Jakob, and P. Matzinger, *Rev. Prog. Coloration* **17**, 72 (1987).

2. M. A. Weaver and L. Shuttlerworth, *Dyes Pigments* **3**, 81 (1982).

3. P. W. Leadbetter and A. T. Leaver, *Rev. Prog. Coloration* **19**, 33 (1989).

4. H. H. Hodgson and J. J. Walker, *Chem. Soc.* 1620 (1933).

5. H. H. Hodgson and J. Walker, *Chem. Soc.* 530 (1935).

6. W. Staedel, *Lieb. Ann. Chem.* **217**, 197 (1883).

7. O. N. Witt, *Berichte* **42**, 2953 (1909).

8. H. E. Fierz-David and L. Blangey, *Grundlegende Operationen der Farbenchemie*, 8th ed., Springer, Wien, 1952, p. 244.

9. M. A. Weaver and D. J. Wallace, U.S. Pat. 3,423,394 (1969).

10. M. F. Sartori, U.S. Pat. 3,405,118 (1968).

11. J. Sokolowska-Gajda and H. S. Freeman, *Dyes Pigments* **20**, 137 (1992).

12. A. Reife, private communication.

13. H. L. Recht and M. Ghassemi, Rept. No. 17010 EKI, Federal Water Quality Administration, Washington, D.C., 1970.

14. J. F. Ferguson and T. King, *J. Wat. Pollut. Control Fed.* **49**, 646 (1977).

15. P. H. Hsu, *Wat. Res.* **10**, 903 (1976).

16. A. Henriksen, *Schweiz Zeits. Hydrol.* **24**, 253 (1961).

17. A. Henriksen, *Schweiz. Zeits. Hydrol.* **25**, 380 (1962).

18. M. C. Mulbarger and D. G. Shifflet, *Chem. Eng. Prog. Symp. Ser.* **107**, 107 (1970).

19. E. Diamadopoulos and A. Benedek, *J. Wat. Pollut. Control Fed.* **56**, 1165 (1984).

20. O. Sletton and C. M. Bach, *J. Am. Wat. Wks. Assoc.* **53**, 1031 (1961).

21. D. F. Boltz and M. G. Mellon, *Ind. Eng. Chem. Anal. Ed.* **19**, 837 (1947).

22. H. Levine, J. J. Rowe, and F. S. Grimaldi, *Anal. Chem.* **27**, 258 (1955).

23. L. Duval, *Chim. Anal.* **45**, 237 (1963).

24. J. D. Burton and J. P. Riley, *Microchim. Acta* **9**, 13 (1956).

25. Association of the Amer. Soap and Glycerine Products Inc., Subcommittee on Phosphates, *J. Am. Wat. Wks. Assoc.* **50**, 1563 (1958).

26. M. Halmann, ed., *Analytical Chemistry of Phosphorus Compounds*, Wiley-Interscience, New York, 1972, p. 49–59.

PART 3

WASTEWATER ANALYSIS

CHAPTER 11

MASS SPECTROMETRY IN THE ANALYSIS OF DYES IN WASTEWATER

L. DON BETOWSKI[1], JEHUDA YINON[2], and ROBERT D. VOYKSNER[3]
[1]U.S. Environmental Protection Agency, 944 East Harmon Avenue, Las Vegas, Nevada
[2]Department of Environmental Sciences and Energy Research, Weizmann Institute of Science, Rehovot, Israel
[3]Analytical and Chemical Sciences, Research Triangle Institute, Research Triangle Park, North Carolina

INTRODUCTION

Dyes have long been a large and important group of industrial chemicals. More than a decade ago, dye production in the United States (excluding pigments) had reached 245 million pounds, with an additional 29 million pounds being imported (1). Although most of this quantity was used as colorants in the textile industry or as fluorescent brighteners (colorless dyes) in detergents, dyes are also used in products such as paper, leather, and foodstuffs. Interestingly, the dyes in use include many examples of colorants whose environmental fate is unknown (2). The development of effective analytical tools is essential to addressing issues such as those that follow.

Epidemiology studies of workers in the dye industry have historically shown an increased incidence of bladder tumors above that of the general population. For instance, an increase in bladder cancer has been shown to occur in kimono painters who had a practice of licking their brushes (3). In these cases, azo dyes and pigments were implicated as the causative agents, either directly or indirectly, of the observed carcinogenesis. Specifically, benzidine and benzidine-congener-derived azo dyes were implicated in both cases. As part of a long-standing effort, the U.S. Environmental Protection Agency (EPA) is currently undertaking a study involving benzidine-based dyes and pigments. This study is designed to outline the hazard case for benzidine, its congeners, and associated dyes and pigments by reviewing all the known cancer studies as well as metabolism and mechanism of action, epidemiological data, occupational and drinking water exposure, and environmental fate. The first step in the

Environmental Chemistry of Dyes and Pigments, Edited by Abraham Reife and Harold S. Freeman.
ISBN 0-471-58927-6 © 1996 John Wiley & Sons, Inc.

regulation of the benzidine and benzidine congener dyes is to issue a "dead chemical" Significant New Use Rule (SNUR) under the Toxic Substances Control Act (TSCA) for the benzidine dyes. This rule will essentially state that to the agency's knowledge, there are no benzidine dyes currently in commerce (imported or manufactured) and that any future activity with them would constitute a "new use" and require a premanufacture notice (PMN) submission. The PMN process allows the EPA to review information available on new substances prior to the substance being manufactured for commercial purposes or distributed in commerce.

Other azo dyes have also been the subject of federal investigations. Because of its chronic genotoxicity to both dogs and rats, Yellow OB [1-(o-tolylazo)-2-naphthylamine], has been effectively banned from use in foods, drugs, and cosmetics; and Acid Red 27 (FD&C Red 2) has been delicensed by the Food and Drug Administration for this application. As an added issue, it is known that some of the manufacturing precursors to dyes are genotoxic and are not always completely removed from the final dye product. This can result in a complex mixture comprising not only the dye itself but also by potentially harmful aromatic amines.

Upon chemical reduction, azo dyes can be transformed into the corresponding aromatic amines. Some of these aromatic amines, particularly lipophilic anilines, are cancer-suspect agents. The metabolism of some of these azo dyes has been studied (4–6), and certain of the metabolites formed are also either carcinogenic or cancer-suspect agents.

The Environmental Defense Fund (EDF) has engaged the EPA in litigation pertaining to wastes generated in various industries. Consequently, the EPA is under a court-ordered consent decree to determine whether to list, or not to list as hazardous, those wastes generated during the manufacture of three classes of dyes and pigments: azo (including benzidine-based types), anthraquinone, and triarylmethane.

If the water from waste treatment processes is to be used for domestic purposes, there is a need to characterize effectively that wastewater before it is discharged into the aquatic environment.

Chromatography

Clearly, analytical methods must be developed and used to help identify and quantify dyes and their by-products, to assure compliance with regulations pertaining to color and priority pollutants. Present-day analytical procedures for the analysis of dyes in complex mixtures at parts-per-million levels or lower are limited. In examining the available analytical tools, it is worthwhile to point out that the correlation between color and the structure of dyes began to develop in the late 1850s. The work of Witt (7) and others was based on visually perceived color. It was only in the 1940s and 1950s that ultraviolet (UV) and visible spectroscopy became a routine technique. Subsequently, much analytical dye work has utilized information based on the wavelengths of the absorption maxima, molar absorptivities, and bandwidths derived from UV and visible spectroscopy.

The characterization of complex mixtures, however, only became feasible with the development and use of chromatography. Although thin-layer chromatography

(TLC) is commonly used in the analysis of dyes and dye mixtures, this method does not define structural differences between different dyes. Additionally, TLC requires a relatively large amount of sample, and the reproducibility of separations may not be satisfactory (8). High-performance liquid chromatography (HPLC) affords much better sensitivity and reproducibility than TLC in the analysis of dyes. Indeed, HPLC has been used with diode array UV–visible detectors, with good precision and accuracy, in the quantification of dyes in wastewater and other media (9,10). High-performance liquid chromatography is often a suitable way to screen complicated mixtures of synthetic dyes, as ion-pair reversed-phase HPLC has been found convenient for the analysis of food dyes (11,12) and cosmetic dyes (13). Also, HPLC usually characterizes a dye based on retention time only. With the use of diode array detectors, a relationship between absorbance and wavelength can also be generated. Since dyes can differ very little in their absorption spectra, these HPLC-associated techniques are not always sufficient to unambiguously confirm the presence or absence of a particular component, especially in a complex mixture. Although modern capillary column gas chromatography (GC) has even more resolving power than HPLC, GC not only lacks specificity, it is also inappropriate for the many nonvolatile polar and ionic dyes.

Mass Spectrometry

Mass spectrometry has both the sensitivity and the specificity to be the method of choice for the analysis of dye wastewater. Mass spectrometry has the additional benefit of having a multitude of interfaces available for sample introduction. This feature differentiates the mass spectrometer from the infrared (IR) spectrophotometer and the nuclear magnetic resonance (NMR) spectrometer, which often serve as complementary techniques to mass spectrometry. The key advantage of mass spectrometry (MS) is its effectiveness in analyzing volatile and nonvolatile compounds, whether pure or in mixtures, because of the various interfaces that can be used. A particularly important combination is HPLC/MS, a technique that has revolutionized the field of mass spectrometry over the last decade. Thermospray, electrospray or ion spray, particle beam, and dynamic fast atom bombardment (FAB) are all important types of HPLC/MS interfaces. Each has contributed to the development of trace dye analysis in the last 10 years. The mass spectrometer is able to analyze for impurities, nonvolatiles, and, with a judicious choice of interface, thermally labile substances that would be subject to thermal degradation. Although pigments still present a problem, due to lack of solubility, they are not expected to be present in wastewaters; special techniques should be applied to these colorants, and they should be physically separated and analyzed off-line.

TECHNIQUES IN MASS SPECTROMETRY

Mass spectrometry is a technique requiring only picogram-to-nanogram quantities of material to obtain information regarding the molecular weight and structure of a compound. The compound of interest can be introduced directly or via a chromatographic inlet into the mass spectrometer for analysis. The mass analysis of the compound, in all cases, requires some form of energy transfer to ionize the compound. There are numerous ionization processes that enable the generation of molecular ions or structurally informative fragment ions from compounds ranging from thermally labile to extremely stable and nonvolatile. The molecular ion and resulting fragment ions are mass analyzed to generate a conventional mass spectrum consisting of mass and intensity information. The following section provides a brief synopsis of the various inlets, ionization choices, and mass analyzers used in MS. Emphasis will be directed toward techniques that are relevant for analyzing dyes, particularly nonvolatile dyes in complex mixtures. Newcomers to the field of MS may want to read the general text on this subject edited by Watson (14).

Inlets and Ionization Techniques for Volatile and Thermally Stable Compounds

The choice of inlet should always take into consideration the ionization process to be used. For instance, ionization processes that require the sample to be in the gas phase are not always compatible with liquid separation techniques. Likewise, gas-phase separation techniques are not compatible with liquid desorption processes. The most straightforward MS approach has involved direct sample introduction, where a sample (1 μg to 10 ng) is placed on a probe and introduced into the mass spectrometer. Typically, the probe is heated to vaporize the sample in preparation for gas-phase ionization. Since the mass spectrometer will detect most components present in the sample, high purity would ordinarily be required to obtain a "clean" and meaningful mass spectrum. In practice, samples are relatively impure, limiting the usefulness of this approach.

The use of gas chromatography coupled to MS (i.e., GC/MS) is one proven method for characterizing a complex sample containing numerous volatile and thermally stable components. Capillary GC is routinely used with MS to separate the components of a mixture prior to introduction into the MS for identification. The coupling of GC with MS is straightforward, involving the placement of the end of a fused silica capillary column into the MS ion source.

Direct probe MS and GC/MS approaches are typically used with electron impact (EI) ionization or chemical ionization (CI) processes. Electron impact and CI processes require the sample to be in the gas phase, a requirement shared by GC. The choice of ionization process depends on the lability of the analyte and the amount of structural information desired on the compound of interest. The most energetic ionization process is EI, where 70 eV electrons bombard the sample and impart a large excess of internal energy into the molecule, resulting in fragmentation:

$$e^- + M \rightarrow M^{-\bullet} + 2e^-$$

$$M^{+\bullet} \rightarrow m_1^+ + m_2^+ + m_3^+ + \ldots$$

where M = analyte of interest
     m = fragments of the analyte

Because of its ability to generate information, ease of use, and reproducibility, EI–MS is one of the most commonly used techniques. There are MS libraries containing over 100,000 EI spectra for computer searching, and computer programs, to aid in the interpretation of EI spectra. The other ionization processes that will be discussed do not afford this luxury.

Chemical ionization processes are used for compounds that are relatively labile or when molecular weight information is important but the molecular ion does not survive EI. Chemical ionization is a softer technique than EI, basing ionization on proton transfer from a reagent gas:

$$G + e^- \rightarrow G^{+\bullet} + 2e^- \tag{1}$$

$$G^{+\bullet} + G \rightarrow GH^+ + G - H^\bullet \tag{2}$$

$$GH^+ + M \rightarrow MH^+ + G \tag{3}$$

where G = reagent gas, typically at 1 torr pressure in the ion source. The gas can
      be methane, isobutane, ammonia.
    M = analyte of interest

The ionization specificity and extent of fragmentation of the analyte depends on the exothermic proton transfer reaction (step 3), which is based on the proton affinity difference between the reagent gas and sample. Since the operator is free to select the reagent gas, the desired specificity and fragmentation can often be obtained. Typically, methane, which has a proton affinity of 127 kcal/mol is used and affects the protonation and fragmentation of many samples. Reagent gases have increasing proton affinity (e.g., isobutane 195 kcal/mol, ammonia 207 kcal/mol) and are more specific in that they ionize analytes of higher proton affinities (e.g., amines) and give rise to minimal fragmentation. Chemical ionization techniques are appropriate for determining molecular weight of hydrophobic, azo dyes.

Ionization Techniques Compatible with Direct Probe Introduction for the Analysis of Thermally Unstable and Nonvolatile Compounds

Most dyes, in particular sulfonated azo dyes, fall under the category of nonvolatile or thermally unstable, and, therefore, are not amenable to GC or gas-phase ionization processes. Fortunately, there are several desorption ionization techniques that can be employed to desorb ions from solid or solution matrices and avoid the need for the sample to be in the gas phase for mass analysis. Techniques including (i) field desorption (FD) (15), (ii) fast atom bombardment (FAB) (16), (iii) secondary ion mass spectrometry (SIMS) (17), (iv) laser desorption (LD) (18), and (v) ^{252}Cf particle bombardment (19) are the more common desorption techniques used on samples introduced by a direct probe. A brief description of these ionization processes is covered here. Field desorption involves placement of the sample in a high electric field (10^7–10^8 V/cm^2) followed by desorption of the sample from the surface of the electrode as an ion. Fast atom bombardment and SIMS are similar processes in which high energy (6–10 keV) neutrals (Xenon for FAB or Cs^+ ions for SIMS) are used to bombard a liquid matrix (e.g., glycerol, thioglycerol, or nitrobenzyl alcohol) that contains the compound. This bombardment disrupts the surface, ejecting ions and neutrals from the liquid matrix. The liquid matrix is important for forming precharged ions in solution and provides a mechanism by which the surface of the droplet can be refreshed with sample to achieve a longer signal duration. Laser desorption, or matrix-assisted laser desorption ionization (MALDI), relies on rapid energy transfer from photons (typically 532 or 337 nm) to the sample via a sample matrix. The matrix plays a vital role in the chemistry that leads to sample ionization and insulates the sample from thermal degradation by the laser (20). The rapid heating of the sample using a laser pulse can result in sample ionization without thermal decomposition or fragmentation. The MALDI procedure has been used to ionize compounds having a molecular weight above 200,000. The fifth desorption process, ^{252}Cf ionization, makes use of the ^{252}Cf fission fragments that are ejected at a kinetic energy of 10^6 eV toward a sample matrix. As in the case of FAB or SIMS, the resulting energetic projectiles rapidly deposit energy into the matrix and sample, resulting in the formation of molecular ions. Typically this process is better suited than FAB or SIMS for higher molecular weight, nonvolatile compounds.

Liquid Chromatography/Mass Spectrometry

The fact that commercial samples of modern complex dyes can be of low purity often precludes their analysis by the probe techniques discussed above. Also, GC/MS techniques are often unsuitable for the analysis of dyes due to thermal lability, low volatility, or high polarity. However, the combination of liquid chromatography (LC) and MS (i.e., LC/MS) enables the purification of nonvolatile, thermally unstable and polar dyes prior to introduction into the MS to obtain structural information.

The coupling of LC to MS systems has provided many challenges, the nature of which has been extensively reviewed (21,22). Most of the initial work in LC/MS focused on the incompatibility of the LC mobile phase flow with the vacuum requirements of the MS system. An aqueous reversed-phase LC mobile phase, at a flow rate of 1 mL/min, can generate 1–4 liters of gas when introduced into a 10^{-6} torr vacuum of a mass spectrometer. This exceeds the operational requirements of most MS systems. Additionally, the thermal lability or low volatility of the compounds of interest may impede their transformation into the vapor state and subsequent ionization. Direct vaporization for EI or CI may be impossible, and consequently alternate routes of ionization are required. Research in the field of LC/MS has been responsible for the evolution of four types of interfaces currently used for LC/MS: (i) thermospray, (ii) particle beam, (iii) continuous-flow FAB, and (iv) electrospray.

Thermospray Interface   Thermospray LC/MS provided a major breakthrough in the application of LC/MS to analytical problems, enabling compound analysis at normal LC flow rates with sensitivity comparable to GC/MS. Thermospray is a widely accepted technique because it can handle most conventional LC solvents and flow rates, as well as provide a way to gently ionize most nonvolatile or thermally unstable compounds. The interface is simple to use (tolerant of user mistakes) and commercially available for most MS instruments.

The thermospray interface (Figure 11.1) evolved from its initial use of lasers (23) to electrical heating (24), the latter of which is currently used to vaporize and ionize the LC effluent. The LC effluent enters a vaporizer where the mobile phase is superheated to form a high-velocity spray. The statistical distribution of ions in this spray results in some of the micrometer-size droplets being electrically charged (25). When the field surrounding these droplets is high enough, ion formation ensues either by desorption ion evaporation or CI processes (26). The droplets enter the source, where ions in the spray are extracted through the ion exit cone, while the neutral molecules go to a cold trap connected to a rough vacuum pump. This extraction process allows the introduction of a total of $1-2$ mL/min of LC effluent, while maintaining a MS pressure of 10^{-5} torr.

The thermospray interface can accept flow rates of 0.1 mL/min to near 2 mL/min of LC effluent. Operation of the thermospray interface is possible at lower flow rates if either the vaporizer temperature is changed and the capillary diameter reduced or a makeup flow is introduced. The thermospray interface can accommodate most normal or reversed-phase solvent systems and any volatile buffer.

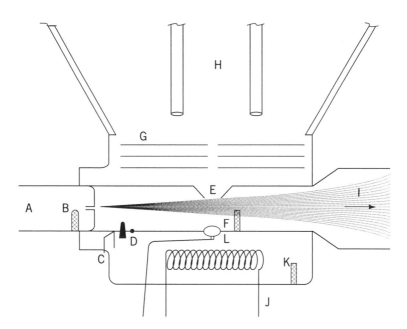

FIGURE 11.1   Schematic diagram of a thermospray interface.

This is the only interface that operates optimally under highly aqueous conditions, usually exhibiting the best compound ion currents at 100% water (27). The interface can be operated smoothly through a solvent gradient LC analysis, if the vaporizer temperature is adjusted to compensate for changes in the heat of vaporization for the solvent in sequence with the LC gradient. Buffers necessary for thermospray ion formation need not interfere with a separation, because they can be added postcolumn (28), resulting in optimal LC and MS performance.

Thermospray is an ionization technique as well as an enrichment technique. Ions may be produced by conventional CI methods, initiated by a filament or discharge, or through ion evaporation. Thermospray, however, cannot provide EI-type ionization. Ion evaporation (26) is of great interest because it is a solution-phase ionization technique applicable to most nonvolatile compounds. In ion evaporation, the addition of a volatile buffer (typically 0.1 M NH$_4$OAc) to the LC mobile phase can result in statistical charging of the micrometer-size droplets arising from vaporization (25). These droplets can have a mean surface field strength of 10^8 V/cm^2 that increases as the droplet continues to desolvate. At sufficiently high fields, evaporation of ions occurs directly from solution. Essentially, ion evaporation is analogous to FD and is well suited for LC because there is no need for the compound to be in the gas phase. Typically, ion evaporation gives rise to $[M + H]^+$ ions for compounds of high proton affinity. Otherwise, $[M + NH_4]^+$ ions (NH$_4^+$ from NH$_4$OAc) are detected. In negative ion detection, $[M - H]^-$ or $[M + \text{buffer or solvent anion}]^-$ ions are observed. The appearance of the fragment ion is compound dependent and difficult to predict. In the absence of buffer ions, polar compounds can generate a field to ionize themselves. Non-polar molecules may require that an auxiliary ionization method (such as CI) be employed.

Particle Beam Interface   Particle beam LC/MS has provided a mechanism for obtaining both EI and CI class spectra of dyes that can be brought into the gas phase but may not be amenable to GC, due to involatility or thermal lability. Particle beam EI mass spectra are more easily referenced, using mass spectral databases that assist in the identification of unknown components in sample extracts. Additionally, the fragmentation produced in EI offers valuable information for structure elucidation. The added ability to generate CI spectra from the same interface offers the opportunity to obtain molecular weight information that complements the structural information gained through EI analysis.

Mass spectrometric analyses using the PB interface (29,30) (Figure 11.2) involve four discrete steps: (1) aerosol formation, (2) desolvation, (3) separation of solvent from analyte, and (4) ionization of the analyte for the acquisition of mass spectra. The initial step, aerosol formation, is accomplished by a coaxial helium nebulization of the LC effluent. The resulting aerosol is passed through a heated desolvation chamber held near atmospheric pressure. Solute molecules in the desolvating droplets precipitate to form particles. The particle's momentum enables it to transverse the separator and transfer line to the MS source. The gaseous solvent molecules, which have a momentum many orders of magnitude lower than the solute, are

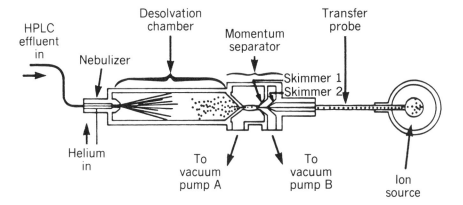

FIGURE 11.2   Schematic diagram of a particle beam interface.

removed in the two-stage separator. An enrichment of 10^4-10^5 particles relative to solvent can be achieved in the two-stage momentum separator. Finally, the particles migrate through the transfer line to the MS source. The heat in the source vaporizes for ionization (EI or CI).

The particle beam interface can handle solvents commonly used for normal or reverse-phase separations, volatile mobile phase additives, and flow rates up to 1 mL/min. However, the performance of the particle beam interface is known to vary with LC conditions (31), and spectra can be enhanced through the coelution of compounds such as ammonium acetate (32) and through the use of additional nebulization (ultrasonic and heat of desolvation) (33). The current sensitivity of 10–20 ng in the full-scan mode suggests that parts per billion (ppb) detection limits in water can be achieved with proper sample concentration steps.

The use of EI or CI in the particle beam offers advantages in interpretation and quality control but limits the applicability of the techniques to compounds that can be vaporized. This is a major limitation, inasmuch as nonvolatile and thermally unstable dyes are often are found in LC separations. For compounds that can be vaporized, the EI and CI mass spectra are interpreted in a manner analogous to GC/MS or probe MS techniques employing those ionization processes.

Continuous-Flow FAB   The coupling of a probe that allows a continuous flow of LC effluent to be introduced into a FAB ion source (34,35) provides the ability to obtain LC/MS information on nonvolatile samples. Samples that are chromatographed on 2.1- or 4.6-mm columns must use a solvent split to insure that a flow of 2–7 µL/min enters the FAB probe (26). However, a capillary (0.32 mm i.d.) column can be directly connected to the probe without the need for a solvent split (37,38). The solvent is carried to the end of the probe for desorption by the FAB beam. The matrices essential for FAB analysis are added (typically 1–5%) to the

LC mobile phase (39) or added postcolumn (40). The solvent and matrix are evaporated immediately, upon reaching the tip of the probe, allowing for minimal band broadening. The use of a frit tip (35) and heated FAB probes stabilizes the total ion current, making the signals more reproducible and giving a slight increase in the upper flow rate of the system.

Continuous-flow FAB exhibits significantly lower background compared to FAB from a probe (static FAB). The lower background greatly enhances the signal–noise ratio, permitting lower detection limits. The FAB spectra usually consist of $[M + H]^+$ or $[M - H]^-$ ions for positive and negative ion detection, respectively. Often adduct ions with salts or matrix are formed, but there is a significant reduction in their formation in continuous-flow FAB compared to static FAB.

Electrospray Interface   Electrospray, a more recently developed LC/MS technique, offers extraordinary potential for high sensitivity and specificity. The electrospray technique, pioneered by Dole and co-workers (41) and combined with mass spectrometry by Fenn and co-workers (42,43), has recently become attractive to the chemist for its ability to ionize a wide variety of nonvolatile and polar compounds and to detect program quantities of compound.

The electrospray interface (Figure 11.3) accepts a flow rate of $1-10$ μL/min into a chamber through a stainless steel hypodermic needle employed at ground potential. A tube placed coaxially around the needle permits the addition of solvents (sheath flow) and maximizes the ionization efficiency of the interface. A high electrical field (2–4 kV) on the cylindrical electrode charges the surface of the emerging liquid to form a fine spray of charged droplets. Driven by electric fields, the droplets migrate to a capillary tube through a stream of nitrogen gas (known as "bath gas"). The bath gas helps evaporate the solvent from the droplets as well as carry uncharged sample away, to prevent fouling of the source. Free expansion of the bath gas causes ions to be carried through the capillary to the inlet of the first vacuum stage. Ions in the free jet flow are carried through the two vacuum stages to the lens system and, finally, to the mass analyzer.

The key feature of electrospray is the formation of an ion through the ion evaporation process. The charged droplets, formed by the high-voltage electrode at the hypodermic needle, desolvate in the bath gas to a point where repulsive Coulombic forces exceed the droplet cohesive forces. The two mechanisms proposed for ion formation from these charged droplets are direct field evaporation of ions and droplet fission at the Rayleigh limit. It has been postulated that micro-sized droplets generated in electrospray undergo a cascade of fission processes yielding smaller and smaller droplets until the electric field at the droplet surface is sufficient for direct ion evaporation (44). It appears that almost any analyte carrying a net charge in solution can form ions in the gas phase in a variety of solvents, indicating broad applicability of the technique (45,46).

There are operational limits on electrospray posed by physical parameters that result in nebulization and desolvation. Limitations in the droplet formation and desolvation processes required to achieve ion evaporation reduce electrospray performance in highly aqueous solutions, in the presence of highly conductive mod-

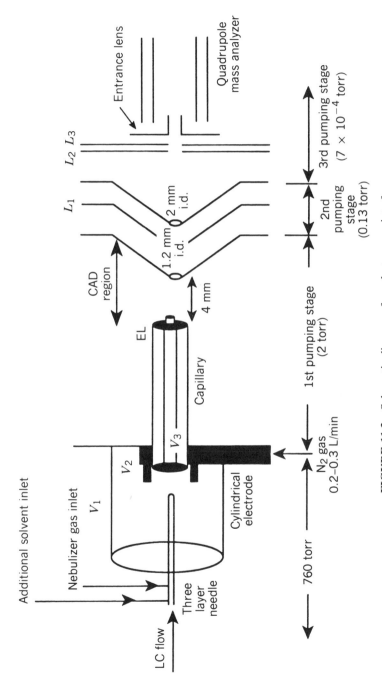

FIGURE 11.3 Schematic diagram of an electrospray interface.

ifiers, and at a flow rate much above 10 μL/min. Some of these limitations can be overcome using capillary LC, by splitting the effluent from 2.1- or 4.6–mm i.d. columns, or through postcolumn solvent modification. However, more versatility for combining MS with the variety of LC conditions can be achieved through assisting nebulization or desolvation. Techniques such as ion spray (47) or ultraspray (48) decouple the nebulization and charging process that occurs simultaneously in electrospray. Either technique expands the operation window, allowing higher flow rates and higher percentages of water to be used.

There are features of the ion evaporation ionization process that are common to electrospray: (1) the ionization process is very soft, generating $[M + H]^+$ or $[M - H]^-$ ions for even the most thermally unstable and nonvolatile compounds. (2) The transport region in the electrospray interface can be used to generate collisionally induced decomposition (CID) spectra of ions useful in obtaining structural information (49–51). This is of vital importance in structural determinations, since such a soft ion evaporation ionization process does not yield fragment ions. Although more costly tandem MS techniques can be used to provide this confirmation, it is possible to obtain analogous information through the generation of CID spectra in the electrospray transport region. (3) Finally, the ionization process has the ability to form multiply charged ions, depending on the acid–base chemistry and hydration energy of the molecules. The ability to increase charge (z) permits the analysis of larger molecular masses (m)—up to 100,000 on a conventional quadrupole analyzer, which is ordinarily limited to m/z of 2000 da'.. as for singly charged ions (52).

Mass Analyzers

There are numerous mass analyzers, including quadrupole, sector, traps, and time of flight, all of which are described in a general book by Watson (14). There are two areas pertaining to mass analyzers that deserve mentioning here because of their importance in the structural elucidation of dyes, viz. high-resolution MS and tandem MS.

High-resolution MS (HRMS) enables accurate determination of the m/z value of the ion, typically to 1 part in 10^6. This ability allows the assignment of the m/z value of an ion such that only a select few (sometimes only one) empirical formulas correspond to that mass measurement.

A good book on tandem MS (53) gives the reader pertinent details concerning the associated instrumentation, operation, and capability. Tandem MS employs a mass analyzer to select a particular m/z value (usually the molecular ion) for CID in a collision cell. The fragment ions (called product ions) formed in the collision cell are mass analyzed, to record a product ion spectrum. This spectrum contains structural information about the ion selected in the tandem MS experiment. One can appreciate the major advantages tandem MS offers when working with mixtures, as this soft desorption ionization process produces only molecular ions. Also tandem MS reduces the chemical background resulting in improved sensitivity.

APPLICATION OF MS TO DYE ANALYSIS

The indentification of dyes by a single analytical procedure can pose special problems for the chemist. This is, in part, because synthetic dyes do not fall into one class or family of chemical compounds; rather, they are characterized by many chemical functionalities and range from very hydrophilic to very hydrophobic molecules. The attendant difficulties (solubility, volatility, ionization efficiency, etc.) inherent in analyzing such a wide variety of compounds are present here. Dye classification is based on the major chromophone of the dye, with azo (including benzidines), anthraquinone, polymethine, phthalocyanine, sulfur, arylmethane, stilbene, and coumarin being the main classes. This section contains examples of the application of the MS techniques presented in the previous section to the characterization of various types of dyes.

Electron Impact (EI) Mass Spectrometry

Several groups of dyes, including hydrophobic azo and anthraquinone, and ionic dyes have been analyzed by direct probe EI mass spectrometry (54–61). Anthraquinones having substituents such as hydroxyl, amino, and halogen groups are stable and produce intense molecular ions (55). The main mode of fragmentation is the loss of one or two carbonyl groups. When the substituents become more complex, such as anilino, toluidino, or alkylamino, more extensive fragmentation occurs.

Several studies have been conducted involving EI analysis of azobenzene (56) and substituted azobenzenes (57,58). The mass spectra produced contained molecular ions, the intensity of which depended on the molecular weight of the dye and the lability of the substituents. Cleavage was observed at the azo bond, with fragments resulting from cleavage at both C–N bonds adjacent to the azo linkage. Also observed were biphenylene or substituted biphenylene radical ions. A representative spectrum is shown in Figure 11.4, which is the EI mass spectrum of Disperse Orange 30 (58). Scheme 1 shows the main fragmentation pathways in EI ionization of Disperse Orange 30 (59).

Disperse Orange 30

High-resolution (HR) EI mass spectrometry was used to characterize 21 commercial synthetic dyes, including azo, disazo, and anthraquinone dyes (59). In addition to the typical fragment ions observed in the EI mass spectra of azo and disazo dyes, fragment ions were observed that suggested cleavage between the two azo nitrogens, followed by combination with two hydrogens to form the corresponding amines.

High-resolution MS was also used for the analysis of indigoid dyes (60,61). Samples were introduced through the solids probe and vaporized at temperatures up to 400°C. 6,6′-Dibromoindigotin, 6-bromoindigotin, and indigo itself were characterized by accurate mass measurement, at a resolving power of M/ΔM = 10,000.

Electron impact mass spectra of ionic dyes are difficult to obtain because these dyes are not readily vaporized. Organic salts produce volatile neutral compounds in the ion source, usually by thermal decomposition or rearrangements. Consequently, most of the ionic dyes do not produce molecular ions; however, their EI mass spectra prove valuable for structure confirmation. Electron impact mass spectra of a series of cationic dyes, including methylene blue, have been recorded (62).

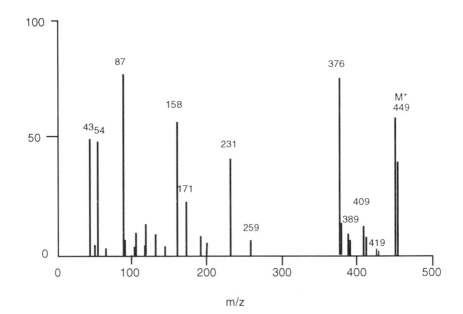

Methylene Blue

Methylene Blue and other cationic dyes produce intense molecular cation (C^+) and reduced (CH^+) ions. The base peak in Methylene Blue was due to ($C^+ - CH_3$).

FIGURE 11.4   EI mass spectrum of Disperse Orange 30.

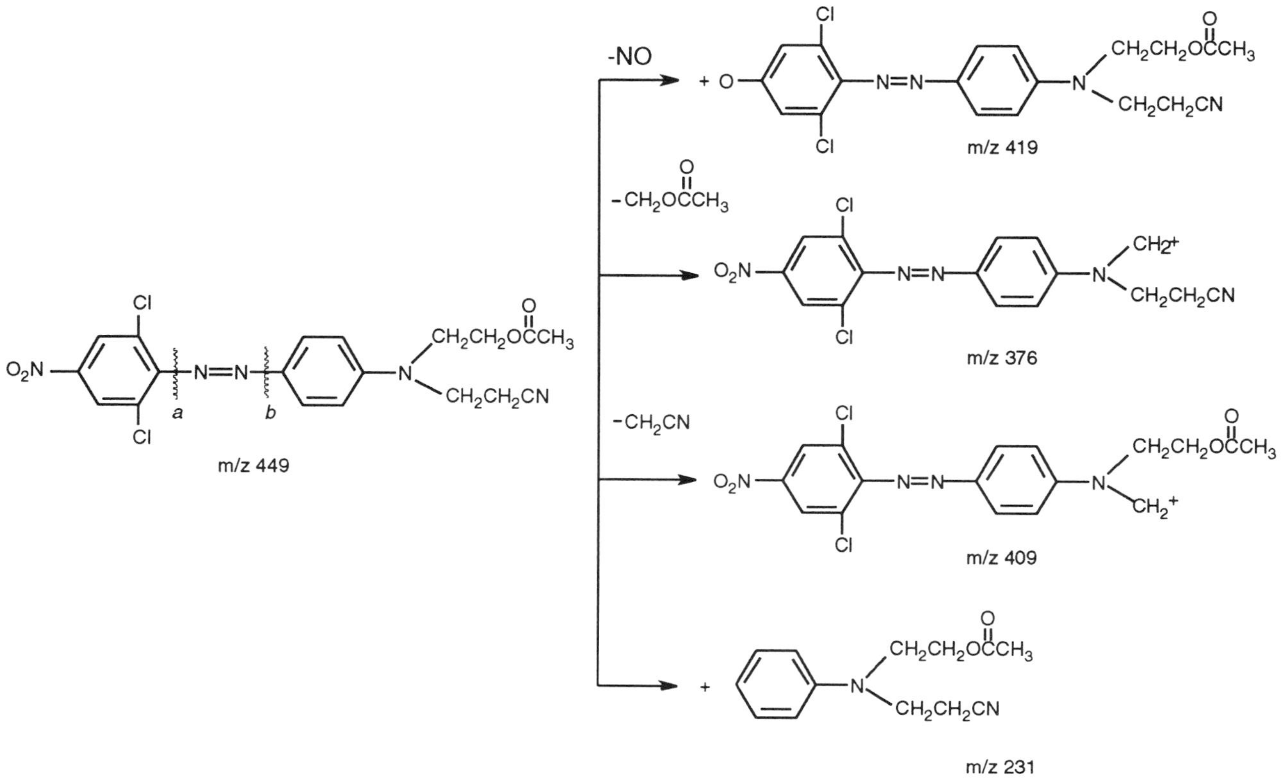

SCHEME 1

Similar results were obtained using a direct exposure probe (63). It was shown that the reduction process occurred also in the absence of a hydrogen-donating solvent.

Field Desorption (FD) Mass Spectrometry

Field desorption has been shown to produce mass spectra from organic and inorganic salts (62,64,65). Hence, this method has been applied to dyes and dye intermediates such as naphthalene and anthraquinone sulfonic acids and sulfonates (64). It was observed that ions of the type $[nM + 1]^+$, $[nM + Na]^+$, and $[nM + K]^+$ predominate at lower temperatures, $[M + 1]^+$, $[M + Na]^+$, or $[M + K]^+$ predominate at intermediate temperatures, and at higher temperatures, fragment ions formed by loss of SO_2 and SO_3 are dominant. For example, the ions observed in the FD mass spectrum of Acid Blue 62, which has a molecular weight of 422, are (together with their percent relative intensity and identity): 318 (100) − $[M − 1 − SO_3 Na]^+$, 319 (17) − $[M − SO_3Na]^+$, 320 (5) − $[M + 1 − SO_3 Na]^+$, 321 (2) − $[M + 2 − SO_3 Na]^+$, 341 (3) − $[M − 1 − SO_3]^+$, 422 (10) − $[M]^+$, 423 (3) − $[M + 1]^+$, 445 (4) − $[M + Na]^+$, 446 (2) − $[M + 1 + Na]^+$, 845 (1) − $[2M + 1]^+$.

Acid Blue 62

Secondary Ion Mass Spectrometry

Secondary ion mass spectrometry (SIMS), using 4.5 keV Ar^+ ions, was used for the analysis of three classes of cationic dyes (66). The dyes included: (i) fused heterocycles in which the cationic charge is delocalized over the molecule, (ii) dyes characterized by an azo bond, and (iii) triarylmethane salts. The spectra of most of the dyes in group (i) had a base peak corresponding to the parent cation (C^+) and a fragment ion due to $(C^+ - CH_4)$. The azo dyes showed extensive fragmentation, mainly resulting from the cleavage of a C–N bond and cleavage of the azo (N=N) bond. The SIMS spectra of triarylmethane dyes were characterized by a highly abundant parent cation peak (mostly the base peak). Major fragmentation pathways involved the loss of hydrocarbon residues from the substituent groups on the amino nitrogens, and cleavage of the triaryl structure with loss of benzyne or an aniline derivative.

Some cationic dyes, including Crystal Violet, Rhodamine B, and Methylene Blue, were found to be reduced by liquid SIMS in the presence of glycerol solutions (67). In addition to an intense C^+ peak, peaks at $[C + 1]^+$, $[C + 2]^+$, and $[C + 3]^+$

were also observed. These were explained as arising from oxidation–reduction reactions occurring in the liquid matrix.

A series of sulfonated azo dyes were analyzed by liquid SIMS in the negative-ion mode (68,69). The effects of matrix, concentration, primary beam energy, and mode of operation were investigated (68). Tandem mass spectrometry was used to characterize fragmentation pathways (69). 3-Nitrobenzyl alcohol matrix was found to produce SIMS spectra containing the most intense sample ions and the least interference from matrix ions. Minimum concentrations of 0.4 and 4 μg/μL (dye in matrix) were necessary to produce useful full-scan spectra of monosulfonated and disulfonated azo dyes, respectively. Ions observed included $[M - Na]^-$ ions as well as ions formed by cleavage of the C–N bond on the coupler side of the azo linkage and by cleavage of the N=N azo bond. In the disulfonated azo dyes, $[M - 2Na + H]^-$ ions were also observed. As an example, the fragmentation pathway of Acid Red 1 is shown in Scheme 2. The structurally important ions arise from the loss of SO_2, SO_3, ring substituents (such as $NHCOCH_3$), and product ions due to the cleavage of the azo bond. Liquid SIMS/tandem mass spectrometry was also used for structural characterization of a group of reactive dyes (70).

California-252 Plasma Desorption Mass Spectrometry

A group of 31 cationic, anionic, and nonionic dyes were analyzed by ^{252}Cf plasma desorption mass spectrometry (PDMS) in both positive- and negative-ion modes (71,72). Abundant positive ions were obtained for the respective cations in all the singly charged triarylmethane, phenazine, and bis (quinolinium) dyes, while negative-ion spectra produced only low-intensity peaks corresponding to anions of the parent dye. Anionic monosodium salts underwent cationization by sodium ions to form intense peaks in the positive molecular ion region, and formed sulfonate anions in the negative ion mode. No doubly charged ions were observed, even when the dyes themselves were dicationic or dianionic. The nonionic dyes provided abundant cations and anions corresponding to (M − H), M, (M + H), and were best observed by conversion to their salts, where they produced intense peaks in the positive ion mode for $(M + H)^+$. Figure 11.5 shows the positive- and negative-ion PD mass spectra of Acid Blue 25.

Acid Blue 25

A comparison of the spectra produced by liquid SIMS and PDMS analysis of a series of dyes (73) showed that both methods produce similar spectra. In liquid

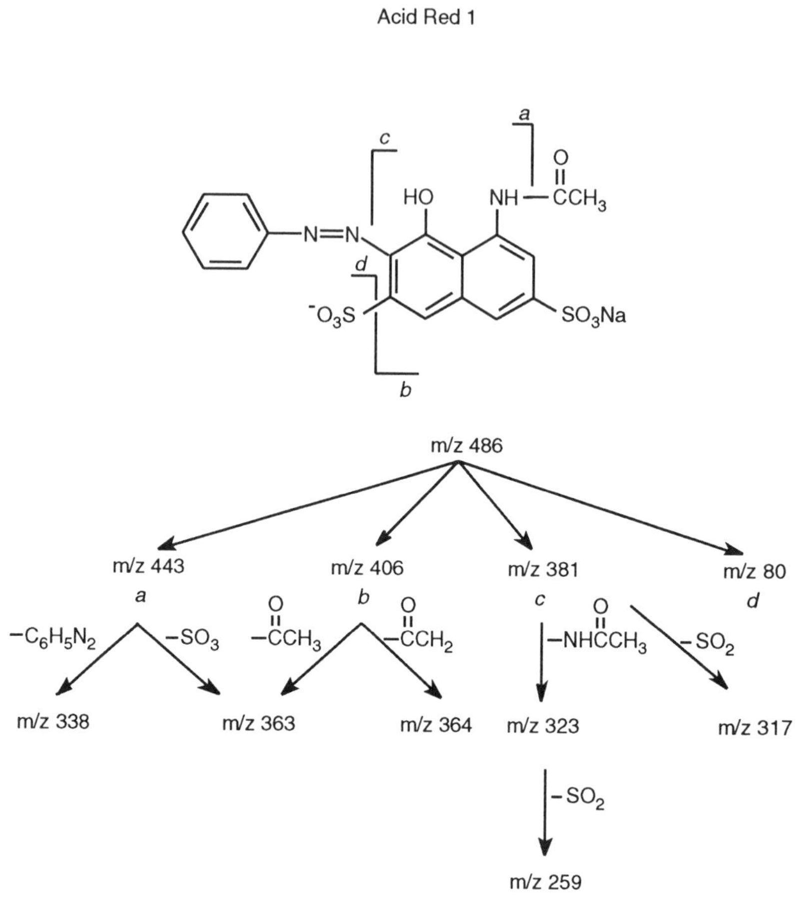

Acid Red 1

m/z 486

m/z 443   m/z 406   m/z 381   m/z 80

$-C_6H_5N_2$ $-SO_3$ $-\overset{O}{\overset{\|}{C}}CH_3$ $-\overset{O}{\overset{\|}{C}}CH_2$ $-NH\overset{O}{\overset{\|}{C}}CH_3$ $-SO_2$

m/z 338   m/z 363   m/z 364   m/z 323   m/z 317

$-SO_2$

m/z 259

SCHEME 2

SIMS, enhanced $[C + 1]^+$ and $[C + 2]^+$ ions were observed. These ions were attributed to reduced forms of the original cation, probably produced by radicals formed in the matrix by energetic particle bombardment.

Fast Atom Bombardment (FAB) Mass Spectrometry

Sulfonated azo dyes are more difficult to analyze by mass spectrometry than their unsulfonated counterparts. Indeed, some of the most difficult dyes to study by mass spectrometry are the acid dyes. Before the advent of modern MS techniques, the salts of the sulfonated dyes could not be volatilized without thermal degradation. A common approach to analyzing such dyes was to form derivatives of the sulfonic

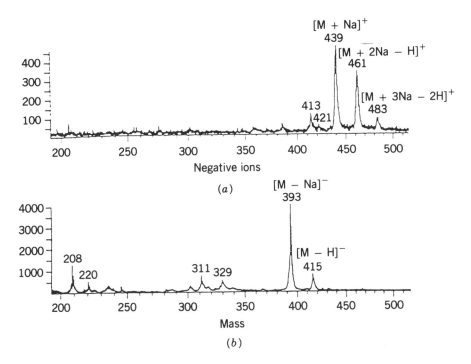

FIGURE 11.5 (a) Positive-ion and (b) negative-ion PD mass spectra of Acid Blue 25.

acid groups. However, this method was not always satisfactory since the volatility was still low, making it often difficult to generate a molecular ion. In the last 10 years, FAB, ion spray, and electrospray techniques have been successfully applied to sulfonated dyes (mainly azo dyes). Monaghan and co-workers employed FAB to analyze a group of sulfonated and phosphonated azo dyestuffs (74,75).

Fast atom bombardment mass spectra of a range of sulfonated azo dyes, generated from glycerol as a matrix, were recorded (74). The analyzed compounds produced abundant sample ions corresponding $[M + Na]^+$ and $[M - Na]^-$, thus helping to establish the molecular weights. Negative ions at m/z 80 − $[SO_3]^-$ and m/z 81 − $[HSO_3]^-$ were observed and were characteristic of the sulfonated species. In the positive-ion spectra, there were weak peaks corresponding to $[M + Na - SO_3]^+$ and $[M + H - SO_3]^+$, with an equivalent weak fragment ion $[M - Na - SO_3]^-$ in the negative-ion mass spectra. The negative-ion spectra contained more abundant fragment ions and fewer interfering ions at low mass; but, the positive-ion spectra provided useful supporting information. Ions due to the cleavage of the azo linkages were observed in both the positive- and negative-ion mode.

Negative-ion FAB mass spectra of monophosphonated azo dyes were also found to be superior to those arising from the positive-ion mode (75). This was attributed to the inherent stability of the resulting phosphonate anions and to a higher sensitivity of the sample ions compared to matrix ions. In the positive-ion mode there was greater interference by the matrix (glycerol) cluster ions, particularly at low mass. All free acids showed abundant $[M + H]^+$ and $[M - H]^-$ ions, while the sodium salts produced $[M + Na]^+$ and $[M - Na]^-$ ions. In the negative-ion mode, peaks at m/z 63 – $[PO_2]^-$ and m/z 79 – $[PO_3]^-$ were always present and were characteristic of phosphonated dyes. Less intense peaks were also observed at m/z 80 and m/z 81 and were assigned to $[HPO_3]^-$ and $[H_2PO_3]^-$, respectively.

Fast atom bombardment mass spectra of several cationic dyes, dissolved in glycerol, were recorded (76). Positive-ion spectra contained several molecular ions, as well as numerous singly charged fragment ions arising from charge separation, dealkylation, and elimination reactions.

Fragmentation pathways for a number of fluorescent xanthane dyes were determined by FAB mass spectrometry using linked scan techniques and accurate mass measurements (77). The dyes, prepared as glycerol solutions, included basic dyes (such as Rhodamine B) and acid dyes (such as Rhodamine WT). Positive-ion mass spectra of the basic dyes showed the cation C^+ as the base peak. The major fragment ions were due to loss of CH_2, O, C_2H_4, and CO_2 from the cation and loss of the substituted phenyl ring. In the mass spectra of the acid dyes, the ion corresponding to loss of the substituted phenyl ring from the cation constituted the base peak. Intense $[M + H]^+$, $[M + 2H]^+$, $[M + 2H + Na]^+$, and $[M + 2H - Na]^+$ were also observed.

A group of nitro, azo, disazo, trisazo, oxazine, and anthraquinone dyes, in glycerol solutions, was analyzed by FAB mass spectrometry (78). $[M + Na]^+$ ions, cleavage of azo groups, and loss of sulfonate from $[M + H]^+$ or $[M + Na]^+$ ions afforded the peaks most commonly observed. Figure 11.6 shows the FAB mass spectrum of Acid Blue 113 recorded in the positive-ion mode. In FAB spectra of sulfonated dyes, the ion intensity in the negative-ion mode was found to be higher than that observed in the positive-ion mode.

High-performance liquid chromatography fractionation of river water to which known amounts of selected dyes had been added were analyzed by FAB mass spectrometry (78). Figure 11.7 shows the FAB spectrum, in the negative-ion mode, of an HPLC fraction of river water spiked with Acid Blue 25, Acid Red 151, Acid Red 73, and Acid Blue 113. The complexity of the spectrum is due to the presence of other compounds, such as polyethoxypropoxy ether sulfates, in the water extract. The detection limits depended on the chemical structure of the dyes and on the presence of compounds in the samples that interfere with the analysis.

The FAB analysis of ionic disazo direct dyes, such as Direct Black 17, showed that only negative-ion FAB produced a molecular ion (M^-), an $[M - Na]^-$ ion, and structurally significant fragment ions resulting from cleavage of the azo linkages (79). In related work, thioglycerol was found to be the best matrix for analyzing mono-, di-, and trisulfonated dyes having a molecular weight in the range m/z 300–700 (80). Thioglycerol facilitated the formation of the molecular ion (M^-), $[M - H]^-$, and $[M - Na]^-$ and the formation of abundant fragment ions.

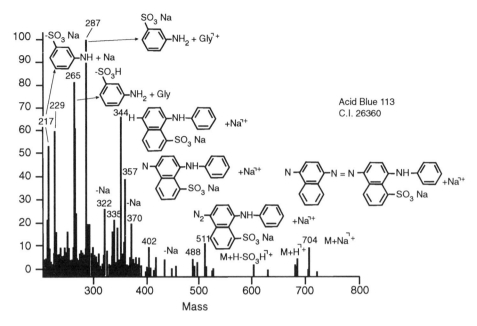

FIGURE 11.6 Positive-ion FAB mass spectrum of Acid Blue 113.

Thermospray and Particle Beam LC/MS

The characterization of commercial formulations of 14 solvent and azo disperse dyes using particle beam and thermospray LC/MS has been reported (81). The thermospray analysis of these dyes produced spectra containing primarily [M + H]$^+$ ions with very few fragment ions. In contrast, particle beam EI MS produced molecular ions and numerous fragment ions. Figure 11.8 compares the mass spectra obtained from those two techniques when Disperse Orange 13 was analyzed. The particle beam EI spectra of these hydrophobic dyes typically showed cleavage of the C–N bond on either side of the azo linkage and cleavage of the N=N bond with transfer of 1 or 2 protons to form an imine or amine.

Researchers have tried various techniques to overcome the problems associated with the analysis of sulfonated dyes using thermospray. Various groups have tried to maximize the ion signal generated by sulfonated dyes (82–84). Two of these studies (83,84) have reported the use of a needle tip or a wire repeller to enhance the signal generated from sulfonated azo dyes. Detection limits as low as 10 ng were reported for a simple monosulfonated monoazo compound, while good signals were produced by 500–1000 ng of disulfonated disazo compounds. Chromatographic separations were achieved with a minimum amount of ammonium acetate solution as an eluting buffer. It was found that too high a concentration of ammonium acetate suppresses ionization of sulfonated dyes.

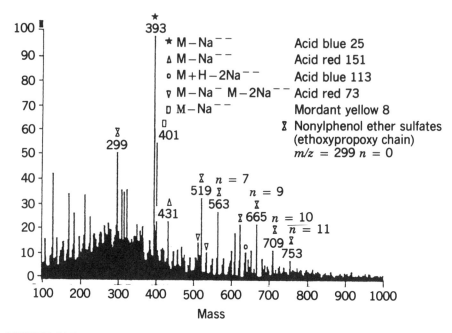

FIGURE 11.7 Negative-ion FAB mass spectrum of a mixture of acid dyes in wastewater effluents.

Atmospheric Pressure Ionization Using Electrospray or Ion Spray MS

Another technique that has been applied to the analysis of sulfonated dyes is atmospheric pressure ionization using three different LC/MS interfaces: a heated pneumatic nebulizer, an electrospray interface, and an ion spray interface (85). All three interfaces led to the detection of sulfonated azo dyes to some degree. The heated pneumatic nebulizer was used to detect the $[M - H]^-$ ion of the free acid of all of the monosulfonated dyes examined. Additionally, fragment ions of low relative abundances were observed. These fragment ions may be due to either thermal degradation or unimolecular fragmentation. This study was conducted using a triple quadrupole mass spectrometer, so product ion CID spectra were generated on these dyes. The disulfonated dyes were more difficult to analyze, and Acid Blue 113 did not show an $[M - H]^-$ ion peak. This dye did give fragment ions that may be due to thermal degradation in the heated nebulizer. Using electrospray, monosulfonated azo dyes gave $[M - H]^-$ ion peaks from the free acid, while the disulfonated azo dyes gave $[M - 2H]^{2-}$ as the base peak, together with a very weak $[M - H]^-$ ion peak. Very little or no fragmentation was observed in electrospray. The ion spray mass spectra of sulfonated azo dyes were identical with those obtained by electrospray. Additional structural information could be obtained on these acid dyes, using CID in the transport region of these atmospheric pressure interfaces. The electrospray CID spectrum of Acid Orange 13 (Figure 11.9) shows

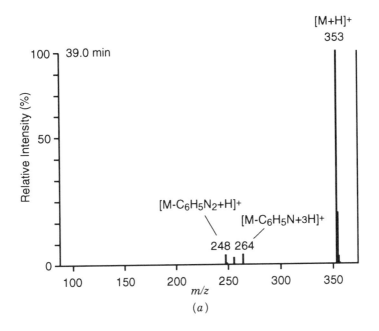

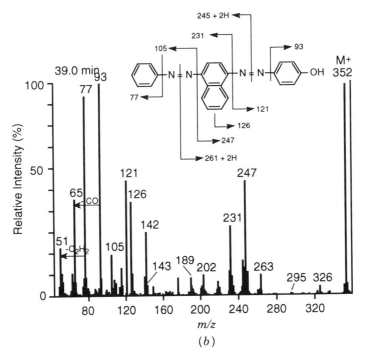

FIGURE 11.8 (a) Thermospray and (b) particle beam LC/MS mass spectra of Disperse Orange 13.

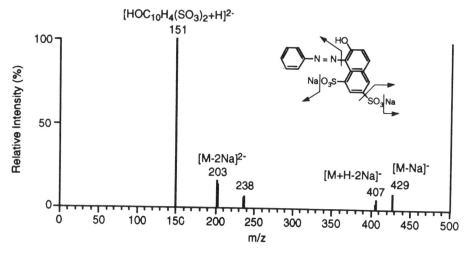

FIGURE 11.9   Electrospray mass spectrum of Acid Orange 10.

several fragment ions arising from losses of Na, SO_3, and diazobenzene. It is worth-while to note that in electrospray or ion spray, 10 ng is sufficient to record the full-scan mass spectrum of a monosulfonated azo dye (81).

Electrospray also proved very useful for obtaining both molecular weight and structural information on solvent dyes, disperse dyes, and organic pigments (86). The CID spectra obtained in the electrospray transport were very similar to the MS/MS spectra (selecting the $[M + H]^+$ ion for each dye, followed by CID of that ion to generate a product ion spectrum) obtained on an ion trap MS. The ions observed in either method were produced primarily from cleavage of the azo bond and the adjacent N–C bond, as well as from losses of –OH and alkyl substituent groups. Table 11.1 lists the major ions recorded by electrospray analysis of Pigment Red 3, Disperse Blue 3, Solvent Red 24, and Disperse Red 13.

LC/MS Analysis of Dye Residues in Mother Liquors and Wastes

Analysis of dye wastes by mass spectrometry is most effective when some knowledge of the dye classification is available. For example, the analysis of azo dyes could be straightforward if the dyes are disperse dyes. However, if the dyes to be analyzed are sulfonated, this makes the analysis limited to certain techniques. Furthermore, since dye molecules are often complex, a complete structure is seldom deduced from mass spectra alone. The complementary techniques of IR and NMR are often used in tandem with mass spectrometry. In fact, high-resolution mass spectrometry would be preferred, where possible, because it provides the elemental composition of the molecule being analyzed.

Similarly, thermospray LC/MS in tandem with MS/MS techniques, was an effective approach to determining the structure of Basic Red 14 (87). The analysis

TABLE 11.1 List of Product Ions Detected for Selected Dyes from CID in the Electrospray Transport Region

Dye	MW	CID in Electrospray m/z (Relative Intensity)	Tentative Identification
Pigment Red 3 [2425-85-6]	307	156 (100)	$C_{10}H_7NO^+$
		291 (8)	$[M + H - NO_2]^+$
		308 (61)	$[M + H - OH]^+$
			$[M + H]^+$
Disperse Blue 3 [2475-46-9]	296	235 (18)	$[M - NH_2CH_2CH_2OH]^+$
		252 (100)	$[M + H - C_2H_5O]^+$
		267 (20)	$[M + H - NHCH_3]^+$
		279 (16)	$[M + H - H_2O]^+$
		297 (34)	$[M + H]^+$

TABLE 11.1 Continued

Dye	MW	CID in Electrospray m/z (Relative Intensity)	Tentative Identification
Solvent Red 24 [85-83-6]	380	91 (100) 109 (25) 149 (38) 156 (20) 209 (10) 224 (34) 275 (11) 381 (94)	$C_7H_7^+$ $C_{10}H_8NO^+$ $[M - C_7H_7N]^+$ $[M + H]^+$
Disperse Red 13 [3180-81-2]	348	102 (35) 126 (20) 149 (27) 177 (11) 208 (12) 232 (48) 349 (100)	 $[M + H - CH_3O]^+$ $[M + H]^+$

was assisted by the presence of the precursor molecules that condense to form Basic Red 14. This is shown in the mass spectrum in Figure 11.10. Since thermospray is a soft ionization technique, the intense ions at m/z 174 and 189 would not be present if this were a pure compound. Individual product ion CID spectra were generated for m/z 344 (taken as the molecular cation of Basic Red 14), m/z 189 and 174. The peaks at m/z 189 and 174 arise from the protonated forms of the precursor molecules, 4-[N-2-cyanoethyl)methylamino]benzaldehyde and 1,3,3-trimethyl-2-methylene-indoline, respectively. Authentic samples of these two compounds were obtained, and their spectra were compared with the spectra of the unknowns, with a good match resulting. Based on these results, the structure of Basic Red 14 was proposed as that shown in Figure 11.11, along with its CID spectrum.

The preceding work (ref. 87) was an example of how impurities in the parent dye can sometimes aid in the analysis. Such is not normally the case, and dye purification remains an important step in the mass spectrometric analysis of synthetic dyes. Indeed, in the quantification of dyes in wastewaters or other matrices, it is imperative to calibrate using a pure dye standard. Thin-layer chromatography and column chromatography are two major procedures that have been utilized in the purification of dyes.

Since complete structural analysis of all components in a dye analysis is so time consuming, most analyses are, by default, target searches. A target search is an analysis in which several specific analytes are targeted for identification. An authentic sample of each target analyte is usually available, and the mass spectrum, solubility, and other characteristics are also known.

The analysis of wastewater for the presence of azo dye Disperse Red 1 is an example of a target study (88). However, since this work focused on the fate of this dye in a typical activated sludge wastewater treatment process, the other unknown analytes present were of interest as possible degradation products. The only analyte that was quantified was the parent dye; other compounds were characterized qualitatively. Since the activated sludge treatment of Disperse Red 1 was undertaken in a controlled environment and a control sample was available, it was a straightforward process to determine the major degradation products. The wastewater samples were analyzed by thermospray LC/MS and LC/MS/MS to measure the amount of Disperse Red 1 present following its exposure to various steps in the waste treatment process, and to identify the degradation products. The limits of detection were 0.6 ng in the single quadrupole scanning mode, 2.0 ng in the product ion CID mode, and 0.18 ng in the selected reaction monitoring mode (where one ion is allowed to pass through the first quadrupole, collide with an inert gas in the second quadrupole, and one of the fragment ions from this collision is monitored in the third quadrupole). The dye and other organic material were extracted by a methylene chloride shakeout procedure. Recoveries of Disperse Red 1 using this procedure ranged from 82 to 96% from spiked organic-free water and 80 to 102% from a control spiked wastewater. These numbers were based on both HPLC/UV-visible and MS analysis. The chromatography was accomplished with the aid of a 2.1-mm RP-8 column, using a flow rate of 0.3 mL/min of CH_3CN and

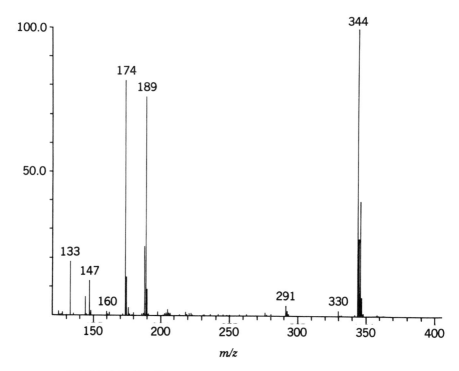

FIGURE 11.10   Thermospray LC/MS spectrum of Basic Red 14.

H_2O (gradient of 50/50 to 100/0 in 5 min) and a postcolumn flow of 1.0 mL/min of NH_4OAc solution. These conditions separated the parent dye (m/z 315) and one of the degradation products (m/z 285) as shown in Figure 11.12. The other major degradation product, which appeared at m/z 181 on a direct injection that bypassed the column, was absent. It was only when NH_4OAc was added to the eluting solvent that the peak at m/z 181 appeared. Both degradation products are believed to arise from reductions, m/z 181 being associated with the amine formed by azo reductive cleavage and m/z 285 being formed from the reduction of a nitro group to the corresponding amino group.

Another study involving azo dyes pertained to GC/MS and particle beam LC/MS analysis (89) of mother liquors and other samples from the dye industry. This study involved 24 target analytes, compounds that were either possible starting materials or degradation products from the production of various azo dyes. Since this target list encompassed a wide range of volatile and nonvolatile compounds of varying polarities, different techniques were required to make a complete assessment of these samples. Thus, GC/MS was utilized for the more volatile and less polar compounds, and particle beam and LC/MS was used for the more nonvolatile and polar compounds. Eight compounds (including acetoacetanilide, 3-

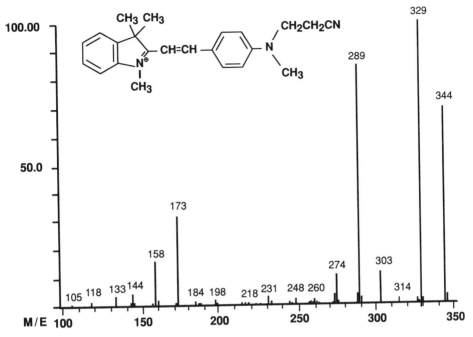

FIGURE 11.11 Proposed structure of Basic Red 14 and its CID spectrum (m/z 344 = molecular cation of Basic Red 14).

hydroxy-2-naphthoic acid) proved too difficult to chromatograph using the generalized HPLC conditions. Anion exchange columns and ion-pairing reagents or an acidic mobile phase were required to provide a sharp separation of some of these compounds. The detection limits for the other target compounds ranged from 0.05 to 0.2 μg/mL (liquid samples) by HPLC/UV, 0.1 to 5.0 μg/mL for GC/MS, and 0.25 to >20 μg/mL for particle beam LC/MS. Particle beam LC/MS has the advantage of generating an electron impact ionization spectrum with its resultant fragmentation for compounds that might not be detected by GC/MS. In this study, the authors found that dyes having a molecular weight greater than 350 daltons could not be detected by GC/MS but were detected by particle beam LC/MS. However, in another study nonionic azo dyes were examined using on-column capillary GC/MS (90). These authors used an on-column injector of their own design and were able to generate EI spectra on azo dyes having molecular weights up to 379 daltons. They then examined waste streams from an industrial dye facility. While most of the compounds identified were dye precursors such as substituted anilines and nitrogen heterocycles, several dyes were also identified.

In another study using thermospray HPLC/MS, azo and anthraquinone dyes in wastewater and gasoline were characterized (91). The wastewater samples were

HPLC/MS Chromatogram of Spiked Wastewater Sample UF05A-5

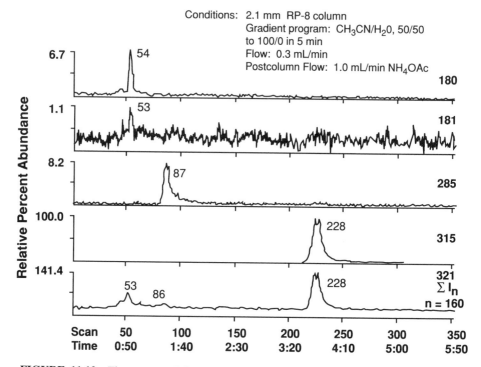

FIGURE 11.12   Thermospray LC/MS total ion current chromatograms derived from a wastewater sample of Disperse Red 1 and its degradation products.

spiked with five azo dyes in the 100-ppb to 10-ppt range. Gasoline was spiked with 1 ppm of Diazo Red. The dyes were extracted through solid-phase cartridges, and the analysis was conducted by thermospray introduction with the filament off or on. With the filament off, thermospray ionization (or buffer-assisted ionization), gave an $[M + H]^+$ peak for the major ion, with very few fragment ions. With the filament on, a chemical ionization process involved more energetic collisions with the analyte, and this resulted in more fragment ions. The fragment ions detected were indicative of simple cleavages resulting in loss of NO, OH, or an alkyl group from the protonated molecule. Analysis using the negative ion mode was also attempted. The sensitivity was poor in this mode, due to the lack of free thermal electrons in thermospray ionization. The HPLC/MS analysis protocol developed in this study proved to be an effective way to analyze for azo dyes in water. The mean percent recovery from the extraction and cleanup procedures was 85% relative to the direct analysis of an identical concentration of standards. A linear correlation coefficient of 0.976 from the 100 ppb to 10 ppt concentration range was determined from the analyses.

An innovative method for screening for azo dyes was developed by Voyksner and co-workers (92,93). This method was based on the chemical reduction of azo dyes to form aromatic amines, followed by the direct analysis of the amines formed. This method provides a way to estimate the total azo dye content of industrial effluents, in addition to identifying the amines formed from the reduction. With a judicious choice of a target list of 30–40 aromatic amines, most commercial azo, dyes can be identified after reduction, in a single analysis. Tin chloride, sodium hydrosulfite, and Pt/H$_2$ were evaluated as reagents for effecting reductive cleavage of the azo linkage in solvent, disperse, and acid dyestuffs. Figure 11.13 compares the effectiveness of tin chloride and sodium hydrosulfite in reducing Solvent Yellow 3. Both chemical reduction methods resulted in nearly complete reductive cleavage of most of the commercial azo dyes employed. However, the tin chloride was a more effective reducing agent, yielding a greater number of products as well as essentially complete cleavage of each dye. Both the parent dyes and the resulting amines were characterized by thermospray and particle beam LC/MS. The Pt/H$_2$ reduction products were determined by capillary GC/MS.

Yinon and Saar used thermospray LC/MS to analyze dyes extracted from textile fibers in a forensic science application (94). The authors extracted a group of disperse dyes, one basic dye, and one vat dye from single fibers of textiles. The disperse dyes were extracted using chlorobenzene, by heating in a sealed tube at 100°C for 15 min, and injected, after cooling, into the HPLC/MS system. The basic dye was extracted using formic acid at room temperature, and the vat dye was extracted using formic acid with heating at 80°C for 1 min. The parent dye was often a minor component of the mixture displayed in an HPLC/MS reconstructed chromatogram. Such a peak would be difficult to observe with an HPLC/UV-visible system because of lack of specificity. The amount of dye was very small, estimated at 2–200 ng.

Voyksner and co-workers (95) used a combination of thermospray LC/MS and GC/MS to characterize the products formed from the fading of Basic Yellow 2 on poly(acrylonitrile) (PAN) fabric. Liquid chromatography/mass spectrometry analysis of faded Basic Yellow 2 dyed fabric extracts implicated the reduction and hydrolysis of the C=NH$_2^+$ group to form primarily benzophenone derivatives, and N-demethylation as the chemistry associated with fading. Due to the limited fragmentation in the thermospray analysis (only [M + H]$^+$ ions were observed), the fabric extracts were analyzed by GC/MS to obtain EI spectra and confirm the identity of the aforementioned degradation products. Many of the degradation products were sufficiently volatile for analysis by GC/MS. Electron impact mass spectral of the photodegradation products of Basic Yellow 2 exhibited molecular ions and structurally important fragment ions complementary to thermospray data. The EI mass spectral data indicated that the most prevalent degradation product formed was [(CH$_3$)$_2$NC$_6$H$_4$]$_2$C=O) (Michler's ketone). Hydrolysis of the C=NH$_2^+$ group to C=O is the main color-destroying reaction in the photodegradation of Basic Yellow

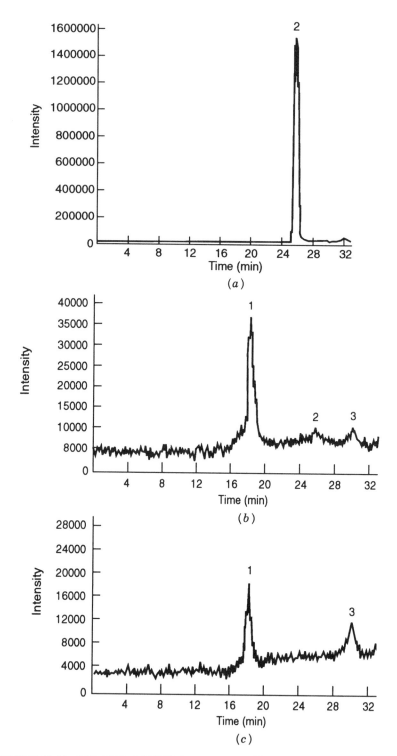

FIGURE 11.13 HPLC/PB–MS TIC chromatogram of identical quantities of Solvent Yellow 3 (A) before reduction, (B) after reduction with $Na_2S_2O_4$, and (C) after reduction with $SnCl_2$. Peak 1 = 2-aminotoluene ($M^{+\bullet}$ = 107); peak 2 = parent dye ($M^{+\bullet}$ = 225); peak 3 = 2-methyl-1,4-diaminobenzene ($M^{+\bullet}$ = 122).

TABLE 11.2 Estimation of the Success of Various MS Techniques for the Characterization of Selected Dye Classes

| | MS Techniques | | | | | | |
| | Gas-Phase Ionization | Desorption Ionization | | | | |
Dye Classes	EI/CI, GC/MS, Particle Beam–LC/MS	FAB/SIMS	LD[a]	FD[a]	Electrospray or Thermospray LC/MS
Acid	○	●	●	⊕	●
Basic	⊕	●	⊕	○	●
Direct	⊕	●	⊕	○	●
Disperse	●	⊕	○	⊕	⊕
Fiber reactive	○	●	⊕	⊕	●
Fluorescent Brighteners	⊕	●	⊕	⊕	●
Mordant	○	⊕	●	●	●
Solvent	●	⊕	○	⊕	⊕
Sulfur	○	⊕	⊕	○	⊕
VAT	○	⊕		○	⊕

[a]Limited information available to estimate success of the technique.

○ ⊕ ●
Increasing success for MS analysis.

287

2. Demethylated products, which give rise to shade changes, were also detected in the extracts of faded samples.

Basic Yellow 2

SUMMARY

In an attempt to help potential users match the MS techniques discussed with the class of dye to be analyzed, Table 11.2 has been compiled as a summary of prior experiences with various MS techniques in the generation of mass spectra on the dye classes listed. This table considers volatility, thermal stability, and polarity. Matrix complexity and quantity of dye to be measured must also be considered before a technique is chosen. When working with complex matrices or trace levels of analytes, chromatography followed by MS (i.e., GC/MS or LC/MS) would be the preferred route. Probe technique and off-line LC peak collection procedures usually limit the ultimate detection to 10–100 ng of sample. And GC/MS and LC/MS approaches can usually reduce the detection limit by 1–2 orders of magnitude.

Research pertaining to new methodologies in MS, notably electrospray LC/MS and new desorption ionization techniques, such as MALDI, has benefited dye chemists in their attempts to characterize increasingly complex, polar, and nonvolatile dyestuffs. Furthermore, the specificity and sensitivity achieved with GC/MS and LC/MS approaches enable monitoring for trace levels of dyestuffs in wastewater samples to ensure compliance with the Clean Water Act as it pertains to production processes in dye-related industries.

ACKNOWLEDGMENTS

This review has been funded, in part, by the U.S. Environmental Protection Agency, through its Office of Research and Development (ORD), and conducted in collaboration with the Research Triangle Institute and the Weizmann Institute of Science, under the management of the Environmental Monitoring Systems Laboratory in Las Vegas (EMSL-LV), to support the Hazardous Waste Issue. It has been subjected to ORD's peer and administrative review and has been approved as an EPA publication.

REFERENCES

1. S. V. Kulkarni, C. D. Blackwell, A. L. Blackard, C. W. Stackhouse, and M. W. Alexander, *Textile Dyeing Operations*, Noyes Publications, Park Ridge, NJ, 1986.

2. G. L. Baughman and T. A. Perenich, *Environ. Toxicol. and Chem.* **7**, 183 (1988).

3. O. Yoshida and M. Miyakawa, in W. Nakahara et al., eds., *Analytical and Experimental Epidemiology of Cancer*, University Park Press, Baltimore, 1973.

4. C. R. Nony and M. C. Bowman, *J. Chromat. Sci.* **18**, 64 (1980).

5. F. Joachim and G. M. Decad, *Mutation Res.* **136**, 147 (1984).

6. J. L. Radomski and L. S. Harrow, Ind. Med. Surg. **35**, 882 (1966).

7. O. N. Witt, *Ber. Deutsch. Chem. Ges.* **9**, 522 (1876).

8. D. K. Laing, R. Gill, C. Blacklaws, and H. M. Bickley, *J. Chromatogr.* **442**, 187 (1988).

9. L. D. Betowski, S. M. Pyle, J. M. Ballard, and G. M. Shaul, *Biomed. Environ. Mass Spectrom.* **14**, 343 (1987).

10. C. Pace and M. Roby, Characterization and Analysis of the Base/Neutral Fraction from Azo Dye Waste Samples, EPA 600/X-91/147, October, 1991.

11. J. F. Lawrence, F. E. Lancaster, and H. B. S. Conacher, *J. Chromatogr.* **210**, 168 (1981).

12. M. L. Puttemans, L. Dryon, and D. L. Massart, *J. Assoc. Off. Anal. Chem.* **64**, 1 (1981).

13. L. Gagliardi, G. Cavazzutti, A. Amato, A. Basili, and D. Tonelli, *J. Chromatogr.* **394**, 345 (1987).

14. J. Throck Watson, *Introduction to Mass Spectrometry*, Raven Press, New York, 1985.

15. G. W. Wood, *Mass Spectrm. Rev.* **1**, 63 (1982).

16. M. Barber, R. S. Bordoli, G. J. Elliott, R. D. Sedgwich, and A. N. Tyler, *Anal. Chem.* **54**, 645A (1982).

17. H. Grade, N. Winograd, and R. G. Cooks, *J. Am. Chem. Soc.* **99**, 7725 (1977).

18. M. Karas and F. Hillenkamp, *Anal. Chem.* **60**, 2299 (1988).

19. R. J. Colter, *Anal. Chem.* **60**, 781A (1988).

20. D. S. Cornett, M. A. Duncan, and I. J. Amster, *Anal. Chem.* **65**, 2608 (1993).

21. T. Covey, E. Lee, A. Bruins, and J. Henion, *Anal. Chem.* **58**, 1451A (1986).

22. E. C. Huang, T. Wachs, J. J. Conboy, and J. D. Henion, *Anal. Chem.* **62**, 713A (1990).

23. C. Blakely, M. McAdams, and M. Vestal, *J. Chromatogr.* **158**, 261 (1978).

24. C. Blakely and M. Vestal, *Anal. Chem.* **55**, 750 (1983).

25. E. Dodd, *J. Appl. Phys.* **24**, 73 (1953).

26. J. V. Iribarne and B. A. Thompson, *J. Chem. Phys.* **64**, 2287 (1976).

27. R. D. Voyksner and C. A. Haney, *Anal. Chem.* **57**, 991 (1985).

28. R. D. Voyksner, J. Bursey, and E. Pellizzari, *Anal. Chem.* **56**, 1507 (1984).

29. R. Willoughby and R. Browner, *Anal. Chem.* **56**, 2626 (1984).

30. P. Winkler, D. Perkins, W. Williams, and R. Browner, *Anal. Chem.* **489** (1988).

31. R. D. Voyksner, P. Knox, and C. Smith, *Biomed. Environ. Mass Spectrom.* **19**, 523 (1990).

32. T. Bellar, T. Behymer, and W. Budde, *J. Am. Soc. Mass Spectrom.* **1**, 92 (1990).

33. W. V. Ligon and S. B. Dorn, *Anal. Chem.* **62**, 2573 (1990).

34. R. M. Caprioli, T. Fan, and J. S. Coltrell, *Anal. Chem.* **58**, 2946 (1986).

35. Y. Ito, T. Takeuchi, D. Ishii, and M. Goti, *J. Chromatogr.* **647**, 13 (1993).

36. D. E. Games, S. Pleasame, E. D. Ramsey, and M. A. McDowall, *Biomed. Environ. Mass Spectrum.* **15**, 179 (1989).

37. A. E. Ashcroft, *Org. Mass Spectrum.* **22**, 754 (1987).

38. D. J. Bell, M. D. Brighwell, W. A. Neville, and A. West, *Rap. Comm. Mass Spectrum.* **4**, 88 (1990).

39. J. P. Gagne, A. Carrier, L. Varfalvy, and M. J. Bertrand, *J. Chromatogr.* **647**, 13 (1993).

40. J. E. Coutant, T. G. Chem, and B. L. Ackermann, *J. Chromatogr.* **529**, 265 (1990).

41. M. Dole, L. Marck, R. Hines, R. Mobley, and L. Ferguson, *J. Chem. Phys.* **49**, 2240 (1968).

42. C. Whitehouse, R. Dreyer, M. Yamashita, and J. B. Fenn, *Anal. Chem.* **57**, 675 (1985).

43. M. Yamashita and J. B. Fenn, *J. Chem. Phys.* **88**, 4451 (1984).

44. B. Thompson and J. Iribane, *J. Chem. Phys.* **71**, 4451 (1979).

45. J. B. Fenn, *J. Am. Soc. Mass Spectrom.* **4**, 524 (1993).

46. J. A. Loo, R. R. Ogorzalek-Loo, K. J. Light, C. G. Edmonds, and R. D. Smith, *Anal. Chem.* **64**, 81 (1992).

47. A. P. Bruins, T. R. Covey, and J. D. Henion, *Anal. Chem.* **59**, 2642 (1987).

48. C. M. Whitehouse, S. Shen, and J. B. Fenn, in *Proceedings of the 40th ASMS Conference on Mass Spectrom*, Allied Topics, Paper TOA-1:15, Washington, D.C., May 31–June 5, 1992.

49. S. Pleasance, J. Anacleto, M. Bailey, and D. North, *J. Am. Soc. Mass Spectrom.* **3**, 378 (1992).

50. K. Duffin, T. Wachs, and J. D. Henion, *Anal. Chem.* **64**, 62 (1992).

51. R. D. Voyksner, and T. Pack, Rapid Comm. Mass Spectrom., **5**, 263 (1991).

52. R. D. Smith, J. A. Loo, C. G. Edmonds, and C. J. Barinaga, *Anal. Chem.* **62**, 882 (1990).

53. K. L. Bush, G. L. Glish, and S. A. McLuckey, *Mass Spectrometry/Mass Spectrometry: Techniques and Applications of Tandem Mass Spectrometry*, VCH, New York, 1988.

54. T. E. Buekelman, in K. Venkataraman, ed., *The Analytical Chemistry of Synthetic Dyes*, Wiley, New York, 1977, pp. 255–267.

55. J. H. Beynon and A. E. Williams, *Appl. Spectrosc.* **14**, 156 (1960).

56. D. Srzic, M. Zinic, Z. Meic, G. Czira, and J. Tamas, *Org. Mass Spectrom.* **27**, 1305, (1992).

57. J. H. Bowie, G. E. Lewis, and R. G. Cooks, *J. Chem. Soc. (B)*, 621 (1967).

58. J. C. Gilland, Jr. and J. S. Lewis, *Org. Mass Spectrom.* **9**, 1148 (1974).

59. T. L. Youngless, J. T. Swansiger, D. A. Danner, and M. Greco, *Anal. Chem.* **57**, 1894, (1985).

60. P. E. McGovern, J. Lazar, and R. H. Michel, *J. Soc. Dyers Col.* **106**, 22 (1990).

61. P. E. McGovern, J. Lazar, and R. H. Michel, *J. Soc. Dyers Col.* **107**, 280 (1991).

62. C. N. McEwen, S. F. Layton, and S. K. Taylor, *Anal. Chem.* **49**, 922 (1977).

63. D. J. Burinsky, R. L. Dilliplane, G. C. DiDonato, and K. L. Busch, *Org. Mass Spectrom.* **23**, 231 (1988).

64. A. Mathias, A. E. Williams, D. E. Games, and A. H. Jackson, *Org. Mass Spectrom.* **11**, 266 (1976).

65. H.-R. Schulten and D. Kümmler, *Z. Anal. Chem.* **278**, 13 (1976).

66. S. M. Scheifers, S. Verma, and R. G. Cooks, *Anal. Chem.* **55**, 2260 (1983).

67. G. Pelzer, E. De Pauw, D. V. Dung, and J. Marien, *J. Phys. Chem.* **88**, 5065 (1984).

68. S. D. Richardson, A. D. Thruston, Jr., J. M. McGuire, and G. L. Baughman, *Org. Mass Spectrom.* **26**, 826 (1991).

69. S. D. Richardson, J. M. McGuire, A. D. Thruston, Jr., and G. L. Baughman, *Org. Mass Spectrom.* **27**, 289 (1992).

70. S. D. Richardon, A. D. Thruston, Jr., J. M. McGuire, and E. J. Weber, *Org. Mass Spectrom.* **28**, 619 (1993).

71. L. K. Pannell, E. A. Sokolowski, H. M. Fales, and R. L. Tate, *Anal. Chem.* **57**, 1060 (1985).

72. M. U. D. Beug-Deeb, J. A. Bennett, M. E. Inman, and E. A. Schweikert, *Anal. Chim. Acta* **218**, 85 (1989).

73. P. J. Gale, B. L. Bentz, B. T. Chait, F. H. Field, and R. J. Cotter, *Anal. Chem.* **58**, 1070 (1986).

74. J. J. Monaghan, M. Barber, R. S. Bordoli, R. D. Sedgwick, and A. N. Tyler, *Org. Mass Spectrom.* **17**, 569 (1982).

75. J. J. Monaghan, M. Barber, R. S. Bordoli, R. D. Sedgwick, and A. N. Tyler, *Org. Mass Spectrom.* **18**, 75 (1983).

76. D. F. Barofsky and U. Giessmann, *Int. J. Mass Spectr. Ion Phys.* **46**, 359 (1983).

77. R. M. Brown, C. S. Creaser, and H. J. Wright, *Org. Mass Spectrom.* **19**, 311 (1984).

78. F. Ventura, A. Figueras, J. Caixach, D. Fraisse, and J. Rivera, *Fres. Z. Anal. Chem.* **335**, 272 (1989).

79. H. S. Freeman, R. B. Van Breemen, J. F. Esancy, D. O. Ukponmwan, Z. Hao, and W.-N. Hsu, *Text. Chem. Color.* **22**, 23 (1990).

80. H. S. Freeman, Z. Hao, J. Sokolowska-Gajda, R. B. van Breemen, and J. C. Le, *Dyes Pigments* **16**, 317 (1991).

81. R. Straub, R. D. Voyksner, and J. T. Keever, *J. Chromatogr.* **627**, 173 (1992).

82. D. A. Flory, M. W. McLean, M. L. Vestal, and L. D. Betowski, *Rap. Comm. Mass Spectrom.* **1**, 48 (1987).

83. M. A. McLean, D. A. Flory, M. L. Vestal, and R. B. Freas, Proceedings of the 36th ASMS Conference on Mass Spectrometry and Allied Topics, San Francisco, CA, June 1988.

84. J. Yinon, T. L. Jones, and L. D. Betowski, *Biomed. Environ. Mass Spectrom.* **18**, 445 (1989).

85. A. P. Bruins, L. O. G. Weidolf, J. D. Henion, and W. L. Budde, *Anal. Chem.* **59**, 2647 (1987).

86. H. Y. Lin and R. D. Voyksner, *Anal. Chem.* **65**, 451 (1993).

87. L. D. Betowski and J. M. Ballard, *Anal. Chem.* **56**, 2604 (1984).

88. L. D. Betowski, S. M. Pyle, J. M. Ballard, and G. M. Shaul, *Biomed. Environ. Mass Spectrom.* **14**, 343 (1987).

89. C. Pace and M. Roby, Characterization and Analysis of the Base/Neutral Fraction from Azo Dye Waste Samples, EPA 600/X-91/147, October, 1991.

90. R. T. Rosen, L. H. Rovin, T. G. Hartman, L. B. Clark, J. Oxenford, C.-T. Ho, and J. D. Rosen, Proceedings 36th ASMS Conference on Mass Spectrometry, San Francisco, CA, June 1988, p. 34.

91. R. D. Voyksner, *Anal. Chem.* **57**, 2600 (1985).

92. R. D. Voyksner, R. Straub, J. T. Keever, H. S. Freeman, and W.-N. Hsu, *Environ. Sci. Technol.* **27**, 1665 (1993).

93. R. Straub, R. D. Voyksner, and J. T. Keever, *Anal. Chem.* **65**, 2131 (1993).

94. J. Yinon and J. Saar, *J. Chromatogr.* **586**, 73 (1991).

95. R. D. Voyksner, T. W. Pack, C. A. Haney, H. S. Freeman, and W.-N. Hsu, *Biomed. Environ. Mass Spectrom.* **18**, 1079 (1989).

PART 4

REGULATORY ISSUES

CHAPTER 12

U.S. SAFETY, HEALTH, AND ENVIRONMENTAL REGULATORY AFFAIRS FOR DYES AND PIGMENTS

HUGH M. SMITH
Sun Chemical Corporation, Cincinnati, Ohio

INTRODUCTION

A Warning

This chapter begins with a warning to our readers that it may be necessary to disregard a portion of the following text! The subject of safety, health, and environmental affairs is today so volatile and capable of change that it is anticipated that some of the following information will be obsolete in the months following publication of this survey. The chapter must, therefore, be viewed only as a "snapshot" of the state of affairs existing at the time of writing and not perceived as a static treatise on a most dynamic topic. Also, while every attempt has been made to maintain objectivity in the topics covered, it is natural that the background and focus of any author must, to some extent, flavor the treatment given to the subject matter, which is dealt with from the U.S. perspective.

Five Principles of Environmental Affairs within the Dye and Pigment Industries

In order to better understand the major issues facing today's dye and pigment industries, five general principles are presented for consideration:

Need for Workplace and Environmental Regulations . . . The Legacy of Our Fathers As one who grew up in an industrial society where the safety and health of the individual and the related impairment of the environment were occasionally submerged beneath the paramount goals of efficiency and profitability, the birth of today's regulatory maze was both logical and overdue. The "legacy

Environmental Chemistry of Dyes and Pigments, Edited by Abraham Reife and Harold S. Freeman.
ISBN 0-471-58927-6   © 1996 John Wiley & Sons, Inc.

of our fathers'' must thus be acknowledged as typical of at least a small portion of past chemical industry, and with it, the dye and pigment industries.

Two unfortunate, but well-documented examples that come to mind are 1) the multiple cases of bladder cancer that occurred some 30–40 years ago in an Ohio dyestuff intermediates facility, manufacturing the human carcinogen, benzidine, and 2) a strikingly similar incident in Georgia, during the same time period, involving the manufacture of another human carcinogen, β-naphthylamine, for dyestuff production. Ironically, at the very time during which employees were being inadvertently exposed to the two carcinogens, the danger of these agents to human health was well known to medical specialists but apparently was not known to the manufacturers concerned. National standards for worker protection were obviously overdue. Today we are thankful for a major paradigm shift that has occurred in recognition of the imperative of safety, health, and environmental protection, in most manufacturing industries within the developed world.

Multiplicity of Regulations and the Problems of Information Overload As a result of this paradigm shift, an ever-increasing number of worker, consumer, and environmental protection measures have been implemented in the United States by means of national, state, and occasionally, even by city ordinances. Sometimes, following a change in national administration, when political appointees have been known to gravitate from federal agencies to states sympathetic to their previous regulatory agenda, state regulations may be introduced well in advance of a federal counterpart. Again, from time to time it is not unusual to see individual state-to-state variations in regulatory language covering a common issue, to the point where detailed understanding of the differences frequently requires specialized insight beyond the in-house capabilities of most smaller dye and pigment manufacturing companies. And if we add to this often confused situation a consideration of the export activities of manufacturers, processors, and users of dyes and pigments, the many disparate requirements overseas provide a veritable maze through which today's manufacturer, exporter, importer, or consumer of dye and/or pigment has to pass. Currently, the sheer multiplicity of safety, health, and environmental regulations encountered represent an information overload of truly enormous proportions. If there will be any resolution to this dilemma, it surely must come from a second paradigm shift, involving regulators and regulated communities alike, as well as a concerned public, in finally accepting the necessity for regulatory harmonization. This would create a level ''playing field'' by which worker, consumer, and environment can equally and adequately be protected.

Globalization of Local Issues and the Problem of the Shrinking Universe At first sight, the probability of local workplace and environmental constraints in one part of the country impacting dye and pigment business thousands of miles away in another may seem remote. In recent years, however, the borders of regulatory enforcement have become somewhat blurred due to business rather than regulatory requirements. Two examples from the pigments industry will serve to illustrate the point. Model legislation, developed by the Source Reduction Council of the Coalition of North Eastern State Governors (CONEG), was enacted in the early 1990s by several states (though not yet by the federal government) setting limits on trace amounts of the four heavy metals mercury, cadmium, lead, and chromium (VI) in packaging material. As a result, pigment manufacturers in the United States were required to comply with this limitation in products sold into the packaging ink and ultimately packaging industries. But more recently, some European manufacturers selling pigments to European packaging ink houses who, in turn, sell products to European packaging concerns who may export their packages into the United States are requiring the same restriction on heavy-metal impurity levels that was originally conceived of as a purely local, U.S. state issue.

Again, commercialization of a new dye or pigment substance in the United States requires premanufacturing notification to the U.S. Environmental Protection Agency (EPA) before the new substance can be entered on the toxic substances "inventory" and be legally manufactured, marketed, or imported. But in some recent instances, it is not unusual for a dye or pigment maker to be required to furnish to the customer proof of presence on the inventories of Japan, Australia, Korea, the Philippines, or wherever the user plans to export downstream products.

Subjectivity In Risk Perception and the Problem of Balance   We have already discussed the paradigm shift in industrial values to one where protection of the environment and human health is now perceived as more important than the profitability and efficiency of a business. But we must also take into account a radical view espoused by community and national activist bodies, and beloved by the media, that industry must never be trusted to self-police its boundaries, that increased regulatory enforcement is the only way to hold industry accountable for its practices, and that use of the media is legitimate to popularize the activist viewpoint, even if somewhat exaggerated claims are made against industry. In such vein, legally permitted discharge of dye or pigment industrial effluent for treatment by a publicly operated treatment works (POTW) is invariably termed "dumping" of "pollution" and even the handling of "tainted toxins" by the media.

As might be expected, such checks and balances upon industrial practice are viewed as either an unbridled abuse of power (by industry) or a necessary evil (by environmentalists). Where does the truth lie? Rather than corresponding to the portrayal of the media, the truth is often hidden in scientific terminology that is difficult for the public to understand. It is hoped, however, that the fledgling science of risk assessment and risk communication will advance to the point where objective truth in safety, health, and environmental issues impacting dyes and pigments no longer need an interpreter.

Economic Issues, the Problems of Deep Pocket Perception and the Unlevel Playing Field   Dyes and pigments in the United States are regulated today from their birth, through their processing, to their use, and ultimate fate within the environment. Noncompliance with these regulations is no longer considered a trivial offense, the cost of which companies can easily afford to pay, as a normal cost of doing business. Several dye and pigment manufacturers have already found this out, much to their dismay! Several million dollars in fines are unfortunately now commonplace! It is clear that the U.S. chemical industry is perceived by the regulatory agencies, as having "deep pockets" containing limitless sums of money to pay for enforcement violations. Realistically, however, such is not the case with the majority of U.S. dye and pigment manufacturers, for fines of this magnitude could easily cripple their business. A second economic fact of life has to do with importation of dyes and pigments made offshore without the same restrictions levied on U.S. manufacturers. In other words, undue economic burden from U.S. regulatory requirements upon U.S. dye and pigment manufacturers could eventually make offshore products, provided their quality and impurity profiles are comparable, more and more attractive to U.S. customers.

PREMANUFACTURING CONSIDERATIONS

In 1976, the U.S. Congress enacted the Toxic Substances Control Act, popularly known as TSCA (PL94-469, 15 USC 2601 et seq.). Motivation for this action stemmed in part from awareness of widespread persistence of polychlorinated biphenyls (PCBs) and chlorofluorocarbons (CFCs) in the environment and the realization that the EPA must be empowered to control risks associated with hazardous and toxic chemicals.

Under Section 5 of TSCA, the EPA is empowered to assess the safety of all new substances, including dyes, pigments, raw materials or intermediates or additives, before their manufacture and/or importation into the United States. By definition, a new substance is one that does not appear on the *TSCA Inventory of Existing Chemical Substances*. In order to determine whether or not a premanufacturing notification (PMN) request must be filed, the nonconfidential portion of the TSCA Inventory should first be consulted, and if necessary, a bona fide letter of intent to manufacture the substance submitted, as a trigger to the EPA to search the confidential portion of the TSCA Inventory for the substance in question. After determining that the substance in question is truly "new" and does not qualify for exemption from the PMN review process (e.g., low volume, R&D only, impurity, by-product, site-limited intermediate, export only, etc.), a $2500 fee is submitted to the agency, together with a comprehensive profile on the substance, including projected manufacturing plans, and toxicological reports, if available.

Following a 90-day review period, and typically several phone discussions between the submitter and agency reviewers, the EPA will either decide to regulate manufacture of the substance, by means of a Section 5(e) order, or will raise no objection to manufacture. Within 30 days of first manufacture or importation, however, the submitting company must file an NOC (Notice of Commencement to Manufacture) with the agency, after which the EPA will add the new substance to the TSCA Inventory. Once the new substance appears on the TSCA Inventory, other companies are at liberty to manufacture or import it into the United States, provided that the agency does not impose additional restrictions through imposition of a SNUR (Significant New Use Rule).

Today, a number of new dye and pigment products have been placed on the TSCA Inventory after PMN review. In addition, a significant number of pigment "additives," used in association with pigments to enhance their working properties (e.g, the aluminum salt of quinacridone sulfonic acid, or phthalimidomethylated copper phthalocyanine, both used for rheology enhancement in paint and ink systems, respectively), have been synthesized and entered on the TSCA Inventory. Naturally, this area is one rife with confidential business information, but a careful reading of the recent patent literature should prove helpful in understanding the type of chemistry that may be involved.

PRODUCT SAFETY ISSUES

To manufacture and sell dyes and pigments today, several critical considerations must be reviewed by the manufacturer, importer, and purchaser: (1) the inherent toxicity of the dye or pigment, (2) any toxic impurities that may be present or could be produced upon breakdown of the product, and (3) the possibility of product misuse leading to undesirable effects produced on human health or the environment from over exposure. This area of concern is rightly called "product safety," and is a new discipline that has emerged in the United States following promulgation of TSCA. In general, the inherent safety of most classes of dyes and pigments is well attested (1,2), but several of the exceptions are noted in the following discussion.

Inadvertent Impurities

Included in this area is the inadvertent presence of trace impurities in the product, for example, PCBs in diarylide and phthalocyanine pigments, polychlorinated dibenzodioxins or dibenzofurans (PCDDs and PCDFs) in chloranil-derived dyes and pigments, and "CONEG heavy metals" in packaging-ink-grade dyes and pigments. Careful attention to the manufacturing process, including choice of raw materials, solvents, and materials of construction used in the manufacturing equipment, together with exhaustive analyses has reduced such issues to controllable levels, as will be illustrated.

Benzidine   In past years whenever pigments synthesized from benzidine were found to contain trace but detectable levels of this carcinogen, the products were discontinued from commerce and are no longer made.

Cyanide   In former years, salt milling of phthalocyanine "crudes," using salt containing sodium ferrocyanide as an anticaking agent, produced pigments containing analytically detectable amounts of cyanide. This practice has been discontinued in the United States.

PCBs and PCDDs   In Diarylide Yellow manufacture, the use of formate buffer to enhance pigment transparency for offset ink was discontinued in the 1970s when it was realized that PCBs were being produced as an undesirable by-product of the coupling reaction.

In Phthalocyanine Blue crude synthesis, the commonly used solvent, trichlorobenzene, was also discontinued in the United States to eliminate a potential source of PCBs. More recently, the use of chloranil, manufactured from chlorinated phenols, has been discontinued in the synthesis of dioxazine violet crude and sulfonated dioxazine acid dyes, so as to minimize by-product formation of PCDDs and PCDFs. A new grade of high-purity chloranil is now produced from hydroquinone for dye and pigment manufacture.

Product Breakdown

The possibility of product breakdown to regenerate undesirable starting materials can be illustrated in examples from both dyes and pigments.

Benzidine Dyes   The well-documented metabolic breakdown of benzidine dyes to the starting material and human carcinogen benzidine is today recognized by dyestuff companies in both Europe and the United States (1). Because of this, and concern over worker exposure to benzidine, a voluntary withdrawal of this type of dyestuff was enacted several years ago in the United States and more recently in Europe. Recently, EPA has proposed a Significant New Use Rule (SNUR), aimed at curtailing possible importation of 67 listed Benzidine Dyes, from Third World Countries.

Diarylide Pigments   A second example of product breakdown pertains to diarylide pigments, derived from the animal carcinogen 3,3′-dichlorobenzidine (DCB). Heating these colorants above 200°C in certain polymers and waxes partially solubilize the pigments and facilitates their breakdown to DCB (3). Because of the breakdown potential, this class of pigment is no longer recommended for the processing of plastics, such as polypropylene, polyamide, and polyesters, at temperatures in excess of 200°C.

Product Toxicity

Carcinogenicity   A few dyes and pigments are considered to be carcinogenic, by U.S. regulatory agencies. Among these are benzidine and benzidine-congener dyes, such as CI Direct Black 38, CI Acid Red 114, CI Direct Blue 15, and CI Direct Blue 218. Lead chromate pigments contain both lead and hexavalent chromium, and, as a consequence, are defined by the EPA as carcinogenic. Experimentally, however, lead chromate pigments have been found to be nonmutagenic and noncarcinogenic, due, no doubt, to their extremely low solubility (4).

Irritancy   Certain dyes and pigments are recognized as skin and eye irritants. For example, diarylide pigments are sometimes treated with primary aliphatic amines to enhance dispersibility in publication gravure ink systems; and since the amine treatment agents are themselves known skin and eye irritants, it is not surprising that some of today's commercial pigments may require careful handling as slight irritants. A related area, currently under consideration by the Food and Drug Administration (FDA), is that of skin sensitization from wood rosin, used in some cosmetic colorants.

Self-Heating of Dry Powders

A separate issue that is receiving much attention at the present time is the phenomenon known as self-heating. Certain dry pigments, including some heavily resinated diarylide yellows, metallized monoazos, and black iron oxides, may exhibit an internal heating phenomenon to temperatures in excess of 200°C, when maintained over a 24-hr period at a temperature of 140°C. Such products are designated as self-heating substances, not otherwise specified (n.o.s.), UN 3088 and 3190, Packing Groups II and III, according to Department of Transportation (DOT) requirements, which in turn are based on U.N. recommendations on the transportation of dangerous goods (5). Such designation mandates special labeling and reference on the material safety data sheet (MSDS). Presently, there is no universal method for predicting self-heating properties. This would appear to be an excellent field for further research.

Deflagration

A related, but different area of concern is that of deflagration, which is the ability of a dye or pigment to support its own combustion under fire conditions without the necessity of an external source of oxygen. An excellent example of a deflagrating pigment is Dinitraniline Orange (C.I. Pigment Orange 5), a colorant that has two nitro groups in its molecule. Fires involving such pigments must be handled carefully, even after dousing with water, as they have been sometimes perceived as apparently extinguished, only to progressively self-heat, through a dry stage, to a smoldering stage, to reburn conditions.

Reactivity Considerations

Most dyes and pigments are unreactive. One exception, however, is lead chromate, which can act as an oxidizer in the presence of certain monoazo pigments and, following intimate mixing, produce a fire (4).

Miscellaneous

Miscellaneous product safety concerns involving dyes and pigments that must be guarded against include accumulation of a static charge of electricity during transfer or processing operations on powdered dyes and pigments (2,4). Such a charge can give rise to spark conditions leading to a fire. The possibility of dust explosions must also be considered. For example, before a new dye or pigment product is spray dried, it is good practice to carry out a dust cloud/dust layer assessment, confirming suitability for this operation.

WORKPLACE CONSIDERATIONS

Four major workplace regulatory events have impacted the U.S. dye and pigment industries.

OSHA's Workplace Standards

First (in 1970), the federal Occupational Safety and Health Administration (OSHA) workplace standards (29 CFR 1910) were introduced, codifying and requiring a multitude of working practices by which dyes and pigments could be safely made.

OSHA's Carcinogen Standards

Second (in the mid-1970s), specific OSHA standards for the safe handling of 13 human and animal carcinogens were established. Included in the list is dichlorobenzidine, a new raw material for the manufacture of diarylide pigments, regulated under 29 CFR 1910.1007.

Hazard Communication Standard

Third, in the 1980s, as a result of the ''right to know'' movement for greater public awareness of the toxic and hazardous properties of materials to which workers might be exposed, OSHA's Hazard Communication Standard was promulgated as 29 CFR 1910.1200.

Process Safety Standard

Lastly, in the 1990s, OSHA's Process Safety Standard was introduced as 29 CFR 1910.119, requiring hazard analysis and process safety assessment of selected hazardous substances. For example, the use of methanol in dye and pigment syntheses now requires an extensive process safety analysis and audit.

As a result of such workplace regulatory developments, hazard assessment of all materials used in U.S. dye and pigment workplaces is now commonplace, understanding of safe handling practices, including proper engineering controls and personal protective procedures, is widespread, and reading of the freely available MSDSs is usually high. Compared with typical practices in the 1960s and before, it is obvious that workplace safety today in the United States pertaining to the manufacture and use of dyes and pigments has undergone significant improvement.

The reader is encouraged to read the recent Color Pigments Manufacturing Association's (CPMA) booklet, entitled *Safe Handling of Pigments* (4), for further treatment of this topic. A second edition is now under development, as well as multi-international versions, each customized with regard to prevailing national regulations.

THE ENVIRONMENT

Air Emissions

In 1990, Congress passed major amendments to the Clean Air Act (PL 101-549; 42 USC 7401), after expressing serious dissatisfaction with the EPA's prior regulation of airborne toxics and imposing upon the agency a requirement to regulate 189 listed hazardous air pollutants. Included in the list is dichlorobenzidine (DCB), a key raw material for Diarylide Yellow pigments. Today, the sole U.S. DCB manufacturer is required to control air emissions of DCB using maximum achievable control technology (MACT). At present, regulation of air emissions from dyes and pigments manufacturing plants is typically governed by state requirements, where raw materials as well as particulate matter from dyes or pigments production are regulated. Because of this, it is customary for state approval to be obtained before any new dye or pigment type can be introduced to a given manufacturing facility. It is also anticipated that in the near future MACT requirements will be placed on the control of volatile solvents used in the dye and pigment manufacturing industries.

Water Discharge Issues

Turning now to the issue of wastewater release from dye and pigment manufacturing and using facilities, four major issues should be considered. Before doing so, it should be understood that in the United States, wastewater may be discharged directly into navigable waters, following treatment to levels proscribed by an NPDES (National Pollution Discharge Elimination System) permit, set by the EPA or its state designee. Many dye and some pigment manufacturing facilities have NPDES permits. Alternatively, many U.S. pigment facilities discharge their industrial wastewater into POTWs, which regulate their influent to suitable "pretreatment" levels, thus enabling the POTW to meet their NPDES permit.

POTW Limits on Copper Phthalocyanine   One long-standing issue is the concern on the part of some POTWs, particularly on the West Coast, to limit the quantity of copper phthalocyanine in wash water from the cleanup of flexographic ink presses. Although the copper found in the wash water is in an insoluble, tightly molecularly bound form, the use of the EPA's documented analytical technique for copper determination uses a strong nitric acid solution, which destroys the pigment molecule, unbinding the metal in the process, and measures the unbound metal as if it were the soluble copper ion.

OCPSF Regulations   In addition to conventional effluent discharge limits for parameters such as biological oxygen demand (BOD), total suspended solids (TSS), acidity (pH), fecal coliform, or oil and grease, a second, regulatory restriction for organic pigment and dye manufacturing facilities is the Organic Chemicals, Plastics and Synthetic Fibers Categorical Standards Regulation (OCPSF, 40 CFR 414.80-85). This specifies discharge limits for a long list of organic and inorganic substances, regardless of direct discharge or discharge to a POTW. There have been instances in which compliance has required companies to address some unusual problems. One example that comes to mind, is a recent problem encountered by some U.S. azo dye and pigment makers involving trace but detectable levels of toluene in their wastewater above prescribed OCPSF limits. Since toluene was not known to be present in the facilities, an exhaustive search produced the reason for the overage. It was found that a major manufacturer of arylide couplers had been using toluene as the reaction solvent during their syntheses, and residual solvent was adhering to the coupler, thus finding its way into the wastewater.

POTW Sludge Fate   A third wastewater discharge issue is concerned with the ultimate fate of the sludge generated by a POTW. In cases where the POTW land applies the sludge for agricultural purposes, new restrictions on specific heavy metals now make it necessary for some POTWs to place further restrictions upon industrial users. Specifically, this issue is now restricting the manufacture of molybdated pigments and could possibly impact phthalocyanine dyes and pigments made with molybdenum oxide as catalyst.

EPA's Waste Stream Proposal   A fourth wastewater discharge issue has recently been proposed as a new regulation by the U.S. EPA (6). Responding to a legal challenge by the Environmental Defense Fund, the EPA has analyzed and evaluated the characteristics of wastewater discharge from a cross section of dye and pigment plants and has determined that wastewater and wastewater treatment sludge from the production of azo dyes and azo pigments will be regulated as hazardous wastes under Subtitle C of the Resource Conservation Recover Act (RCRA). Both the dyes and pigments industries are strenuously opposed to the proposed rule and have pointed out that while the incremental risk (in terms of cancer cases avoided) is near zero, two-thirds of U.S. pigments and dyes manufacturing facilities may incur significant costs, and one-quarter may face closure as a result (7).

Solid and Hazardous Waste Issues

Barium Salt Pigments   Under 40 CFR 261, SubPart C, a hazardous waste is one that exhibits ignitability, corrosivity, reactivity, or toxicity. The last mentioned characteristic is evaluated by the Toxicity Characteristic Leaching Procedure (TCLP). The procedure checks the presence of 39 organic and several inorganic substances, one of which is barium. This means that manufactured barium salt pigments may contain regulated levels of acid leachable barium, which may render any resulting waste hazardous.

Need for Education An important issue is the need for educating many U.S. waste haulers into recognizing that merely because a dye or pigment waste is highly colored does not render it hazardous. This truth, unfortunately, is still not well known enough, even today.

Environmental Fate A further issue impacting some disazo dyes and diarylide pigments is that of environmental fate. Is it conceivable that under worst-case conditions, waste dye or pigment might break down to the component starting amine. The EPA is currently considering this issue, and contrary evidence regarding the nonbiodegradability of diarylide pigments has recently been provided to the agency. Environmental fate has also been a consideration for recyclers of printed material, colored plastics, and other products containing dye and pigment.

Toxic Release Inventory

Each year, U.S. dye and pigment manufacturers are required to submit two itemized lists to local, state, and federal authorities: 1) specific information on the quantity and location of OSHA hazardous substances located at each facility and 2) a record of listed toxic chemical releases to air, water, and waste. The latter listing is known as the Toxic Release Inventory (TRI) and is deliberately given public access by the environmental agencies. As might be expected, debate on the significance of the TRI data sometimes follows "party" lines, with manufacturers perhaps claiming that the releases are modest for the type of processes involved, are all within legally permitted limits, and are often declining each year, due to pollution prevention activities within their facilities. On the other hand, many environmental groups perceive quite another story from the same numbers, and each year translate "legally permitted" as "dumped" or "polluted," believing that the only good level of release (pollution) is zero. Such a dialogue would seem healthy for U.S. industry were it not for the efforts of the media, who 1) often take an anti-industry bias, 2) may mislead the public, and 3) are ultimately responsible for a considerable financial outlay by industry, in an attempt to set this and similar records straight.

Within the U.S. dye industry, such a responsive role is played by the U.S. Operating Committee of the Ecological and Toxicological Association of Dyestuff Manufacturers (ETAD), while the North American Pigments Industry is ably served by the Color Pigments Manufacturers' Association (CPMA). One notable outcome in this area has been the recent "delisting" of copper phthalocyanine pigments from inclusion in the TRI Inventory. Among inorganic pigments that must be reported for the TRI are pigments containing lead, cadmium, chromium, cobalt, nickel, and zinc.

ANALYTICAL COMPLIANCE

Over the years, the U.S. dye and pigment industries have often perceived themselves as targeted by federal and state agencies. And sometimes, "technology forcing" has been applied by requiring industry to minimize environmental emissions or workplace levels of certain substances down to specified limits, without the regulating body having a clear idea as to the specific control technology necessary to either achieve such limits or, equally important, how to reliably analyze for the substances in question, in an industrial matrix, down to the levels required by regulation. As a consequence, recent analytical technology in both dye and pigment industries has been mainly focused on methods development in such areas as quantitating PCB impurities in diarylide and phthalocyanine pigments, CONEG heavy-metal impurities in packaging grade dyes and pigments, and OCPSF-listed substances in dye and pigment wastewater.

CONCLUSION

As will now be apparent to our readers, a careful review of health, safety, and environmental affairs pertaining to the U.S. dyes and pigments industries reveals major change and growing complexity over the last 25 years. In fact, regulatory concern can possibly be considered as the major force that has shaped today's industrial workplace and environment. With continued focus on dyes and pigments assured over the foreseeable future by state and federal agencies, it is certain that this area of endeavor will continue to merit close attention.

REFERENCES

1. BNA, *Chemical Regulation Reporter*, April, 1, 8–9, 1994.
2. National Fire Protection Association (USA), *Prevention of Fire and Dust Explosions in the Chemical, Dye, Pharmaceutical and Plastics Industries*, NFPA, Quincy, MA, 1988.
3. R. Az, B. Dewald, and D. Schnaitmann, *Dyes Pigments*, **15**, 1 (1991).
4. Color Pigments Manufacturing Association, *Safe Handling of Pigments*, CPMA, Alexandria, VA, 1993.
5. United Nations, *Recommendations on the Transport of Dangerous Goods*, United Nations, New York, 1990.
6. Hazardous Waste Management System, *Fed. Reg.*, **59** 66072 (1994).
7. CPMA News Release, Color Pigments Association Opposes EPA Proposed Pigments Waste Stream Rule as Unnecessary, Unfair, and Contrary to EPA Policy, December 9, 1994.
8. H. M. Smith, *Am. Paint Coatings J.* October, 1993.

CHAPTER 13

REGULATORY AFFAIRS (INTERNATIONAL PERSPECTIVE)

E. A. CLARKE ETAD
Basel, Switzerland

INTRODUCTION

Given the complexity of the health and environmental regulations to be complied with in most developed countries, the task of providing an international perspective is formidable. Chapter 12 by H. Smith presented a much more detailed review of the regulations of importance to the pigments industry in the United States. This section aims at providing a perspective of the regulatory situation for a supplier of organic colorants marketing these products internationally. Particular emphasis is given to the regulations in the European Union (EU), the United States, and in Japan, as these are the main manufacturing locations of the major traditional manufacturers of organic colorants. Although it is important not to overlook pioneering regulations developed by such small industrialized countries as, for example, Switzerland (Poison Law, 1969), the comparison of the regulatory requirements in United States, the EU, and Japan provides a wealth of opportunity to illustrate the diversity of regulations to be complied with by a company marketing organic colorants internationally. Such a comparison is given in Table 13.1.

This perspective of the vast array of health and environmental regulations is greatly influenced by the following important characteristics of the colorants industry:

- *Multiproduct Industry* An estimated 2500–3000 organic colorants are marketed.
- *Small Volume* Almost 80% of these colorants are marketed worldwide in amounts of less than 50 tons/year.
- *Wide Range of Applications* Textiles, paper, leather, and plastics are the predominant user industries, but there is a multitude of specialized outlets (e.g., photochemicals, food, drugs, and cosmetics).

Environmental Chemistry of Dyes and Pigments, Edited by Abraham Reife and Harold S. Freeman.
ISBN 0-471-58927-6 © 1996 John Wiley & Sons, Inc.

- *Innovative Industry*   Organic colorants constitute a significant proportion of the new substances notified in the United States and the EU. The proportion ranges from almost 10% of premanufacturing notifications in the United States to 20–25% of premarketing notifications in the EU. The higher EU figure is expected due to the differences in notification requirements.
- *Highly Visible*   Even trace amounts of colorants in the environment can be readily seen when discharged at point sources, and prompt public concern.

Europe still plays a dominant role in the introduction of new colorants. This brief profile of the industry highlights several facets that explain the particular vulnerability of the colorants industry to chemical control regulations. Like other sectors of the specialty chemicals industry, the colorants industry has had to absorb significant notification costs on small-volume new substances, many of which, by virtue of the innovative nature of the industry, will have a limited commercial life time (Table 13.2).

The dramatic acceleration of health and environmental regulation of the chemical industry occurred in the late 1960s to early 1970s. It is beyond the scope of this chapter to examine in detail the reasons for this explosion of regulations. Clearly the growing movement of environmentalism was a significant influence in awakening public awareness and creating a political climate for such regulation. Pubic opinion was shaped not only by such passionate and widely read or at least quoted books as Rachel Carson's *Silent Spring* but also by much publicized accidents involving chemicals and the chemical industry. For example:

Japan: Poisoning incidents involving cadmium (at Toyama), mercury (Minimata and Niigata), and PCBs (at Yusho)

United States: Asbestos, Kepone, PCBs, and vinyl chloride

European Union: Dioxin release in Seveso, Italy; major explosion at Flixborough, England

Such incidents, combined with public concern about cancer and the widespread but erroneous perception by the pubic that most carcinogenic exposures are to synthetic rather than natural carcinogens, assured the passage of the comprehensive legislation required for chemical control. In the United States, the Toxic Substances Control Act (TSCA) was the most far-reaching chemicals control legislation ever enacted. It certainly greatly exceeded in scope legislation in place at that time in other countries and also provided a benchmark for subsequent legislation. Requirements subsequently introduced in other countries certainly reflected many of the elements covered by TSCA; but it was apparently politically unacceptable to accept any harmonization that could have been interpreted as a derogation of national authority. As a consequence, the regulated industry, which operates internationally, is faced with substantial additional costs of meeting diverse requirements in an increasing number of countries, without any payback in terms of increased health and environmental safety. A prime example of this significant waste of resources

TABLE 13.1 Some Key Environmental Legislation in the United States, Japan, and the European Union

Aspect[a]	United States	Japan	European Union
Chemical control	Toxic Substances Control Act (1976)	Chemical Substances Control Law (1974)	Directive 79/831/EEC (6th Amendment) Directive 92/32/EEC (7th Amendment)
Occupational safety	Occupational Safety and Health Act (1970) Hazard Communication Standard (1987)	Labor Safety and Sanitation Law (1979)	Directive 89/391/EEC Directive 80/1107/EEC Directive 88/642/EEC
Air quality	Clean Air Act (1970 and amendments)	Air Pollution Control Law (1970)	Directive 92/72/EEC Directive 84/360/EEC
Water quality	Federal Water Pollution Control Act (1972) Safe Drinking Water Act (1974) Pollution Prevention Act (1990)	Water Pollution Control Law (1971) Sewage Law (1958) Marine Pollution Prevention Law (1976)	Directive 76/464/EEC Directive 91/271/EEC
Hazardous waste	Resource Conservation and Recovery Act (1976) Comprehensive Environmental Response, Compensation and Liability Act (1980) Emergency Planning and Community Right-to-Know Act (1986)	Waste Management Law (1970) Agricultural Land Soil Pollution Prevention Law (1970)	Directive 91/689/EEC

[a]Some of the laws listed cover several aspects but for simplicity are listed only once.

has been the creation of inventories of existing substances in numerous countries (e.g., United States, the EU, Australia, Canada, Korea, and the Philippines). International cooperation within the Organisation for Economic Cooperation and Development (OECD) shows some signs of securing harmonization through:

- Mutual acceptance of health and environmental safety data
- Standardization of testing methods
- Coordination of testing of existing chemicals.

Closer cooperation between at least the major regulatory agencies could lead to a more harmonized regulatory framework in the future, although it must be anticipated that any benefits from such harmonization will be more than offset by the other countries introducing their own unique regulatory requirements.

Comprehensive coverage of the detailed regulatory requirements in the various major industrialized countries is far beyond the scope of this chapter. The aim of the following paragraphs is to address certain aspects that particularly impact the colorants industry and to discuss the different ways in which these issues have been addressed in various countries, notably in the United States and the EU (Table 13.3).

The aim of the data in Table 13.3 is to identify the EU regulatory instruments corresponding to key sections of TSCA. Equivalence is not implied, and specific differences in the U.S. and EU requirements are abundant.

CONFIDENTIALITY

The industry's ability to prevent disclosure of its commercial secrets to its competitors has been eroded by current chemical control regulations to an extent that has greatly exceeded all expectations. This problem has been more acute in the EU than in the United States. Under TSCA (Section 14) the legitimate need for business to retain confidentiality of certain data is acknowledged, although the law broadly mandates public disclosure of data received. In practice, the U.S. EPA has permitted

TABLE 13.2   Life Expectancy of Commercial Dyes[a]

Dye Class	Life Expectancy (years)
Acid	15.0
Basic	14.5
Disperse	13.4
Direct	19.5
Reactive	15.4

[a]Based on a review of dyes introduced from 1965 to 1984 (Horning, unpublished report, 1985).

TABLE 13.3 Cross Reference of Key TSCA and EU Requirements

Aspect	United States	European Union
Chemical testing	TSCA Sect. 4	Directive 92/32/EEC (new substances)
		Regulation 793/93/EEC (existing substances)
Notification	TSCA Sect. 5	92/32/EEC, Art. 7-18
Banning, restrictions	TSCA Sect. 5(e) and 6	76/769/EEC
Imminent hazards	TSCA Sect. 7	92/32/EEC, Art. 31
Reporting		
Production, use	TSCA Sect. 8(a)	Regulation 793/93/EEC
Health and safety data	TSCA Sect. 8(d)	
Significant risk	TSCA Sect. 8(e)	
Export notification	TSCA Sect. 12	Regulation 2455/92/EEC
Confidentiality	TSCA Sect. 14	92/32/EEC, Art. 19
Preemption	TSCA Sect. 18	92/32/EEC, Art. 30
Classification		92/32/EEC, Art. 4
Labeling	TSCA Sect. 6	92/32/EEC, Art. 23-25
Packaging		92/32/EEC, Art. 22
Definitions	TSCA Sect. 3	92/32/EEC, Art. 2
Safety data sheet		92/32/EEC, Art. 27

the more important confidential information to remain protected but has discouraged excessive confidentiality claims by making the task of asserting confidentiality claims more burdensome. In contrast, within the EU all substances classified as dangerous, and this covers the full range of hazard categories (physical, toxic, and environmental), including such mildly dangerous substances as eye irritants, are published with full disclosure of chemical structure. Limited protection is provided for in the case of preparations (Council Directive 88/379/EEC) and the main source of confidentiality protection is the tardiness of the classification process in the case of existing substances, that is, substances included in EINECS (European Inventory of Existing Commercial Chemical Substances). In the case of new substances, the disclosure of specific chemical name within 2 years of marketing the substance for the first time makes the ELINCS (European List of Notified Chemical Substances) a gift for the competition without any persuasive evidence that such disclosure is needed to protect human health or the environment.

TESTING COSTS

The testing costs involved in notification of a new substance in the United States or the EU can be substantial. In the United States there is no minimum set of data that must be submitted. This advantage is at least partially offset by the fact that once the substance is added to the TSCA inventory a second manufacturer (or importer) is not subject to notification. Furthermore, as TSCA requires premanufacturing notification rather than premarketing notification (as in the EU), new isolated intermediates in the manufacturing chain are also subject to notification and may also lead to requests for data during the review process. In contrast, in the EU a base set of data must be provided at an appreciable cost (approx. $150,000), prior to placing a new substance on the market. This data set must also be provided by any subsequent notifier.

STRUCTURE–ACTIVITY RELATIONSHIP

The application of structure–activity relationship considerations is acknowledged by the U.S. Environmental Protection Agency (EPA) as an important aspect of its review of premanufacture notifications. Confidence in the value, but not necessarily infallibility, of this technique has in the past bolstered resistance by the EPA to accept the need for a standard minimum data set. Within the EU notification system, there has been less enthusiasm for the use of structure–activity relationships for purposes of hazard assessment, although it is expected that nonconforming data points would be subject to close scrutiny. Under German regulations, it is concluded that azo dyes, which would yield carcinogenic aromatic amines on reductive cleavage of the azo group(s), should be treated as carcinogens. This has led to a regulation in Germany banning the use of such dyes in certain consumer goods, including colored textiles, the use of which involves more than brief contact with the skin.

RISK VS. HAZARD

Comparison of TSCA with the EU 6th Amendment (Council Directive 79/831/ EEC) reveals a very fundamental difference in approach. Whereas TSCA is aimed at controlling unreasonable *risks* without unduly impeding innovation, the EU approach is *hazard* oriented, with the emphasis on harmonization of classification, packaging, and labeling. Only in the more recent 7th Amendment (Directive 92/ 32/EEC, Art 3.2) does the concept of assessment of potential risk to human health and the environment appear, although regulatory action under the Restrictions Directive (Council Directive 76/769/EEC) is risk based. The principles to be followed by the member states in conducting these risk assessments are laid down in Council Directive 93/67/EEC. This shift in emphasis from hazard to assessing potential risk is strongly supported by ETAD, which for many years has advocated reduction of occupational exposure and environmental releases as the most effective means of reducing any human health or environmental risks that may be posed by colorants. The European approach to risk assessment tends to be much more qualitative than in the United States, reflecting a reluctance to disguise the inherent uncertainties in quantification of risk.

HAZARD COMMUNICATION

Two primary means of hazard communication, both health and environmental, are product labels and safety data sheets. In the United States these hazard communication requirements are governed mainly by the OSHA Hazard Communication Standard; but, the EPA may also require labeling under TSCA Section 6. In contrast to the EU, U.S. labeling is "performance based," that is, certain information must be communicated; but, how this is achieved is a matter for the chemical supplier. Product liability considerations tend to encourage overlabeling. In the EU the labeling system, based on symbols and a series of fixed standard risk (R-) and safety (S-) phrases, aims not only to communicate the hazard information but also to eliminate differences in labeling that could lead to a distortion of the harmonized market.

   In Japan, there are no regulations on safety data sheets, although many companies make them available on request or with the first delivery of a product, and when revised. Similarly, hazard labeling is not generally required in Japan. Labeling is required for "specified chemical substances" and "designated chemical substances" under the Chemicals Control Law administered by the Ministry of International Trade and Industry (MITI). An interesting development, in terms of harmonization, has been the initiative by ETAD members to use the EU classification and labeling system for dyes in Japan. Warning and safety phrases are given in

Japanese. A similar initiative has been taken in Brazil, since mid-1993, also following consultation with the competent authorities. Such initiatives may succeed in discouraging the mandating of different but by no means better systems of labeling.

Although harmonized labeling can certainly help avoid market distortions, in practice the EU system has the following weaknesses:

- Only a small portion of substances fulfilling the criteria for "dangerous substances" have been entered into Annex I of Council Directive 67/548/EEC.
- The labeling of a dangerous substance, as mandated by the Annex I entry, may not be modified by the supplier to reflect new information, that is, the objective of harmonization is given a greater priority than hazard communication. The new information should be provided by the authorities, so that the Annex I entry can be amended.
- Customers, understandably, favor colorants not labeled as dangerous. This results in product discrimination against new substances (which are automatically added to Annex I if dangerous) in favor of existing substances for which adequate data are not yet available. Responsible suppliers who test and label according to the hazard criteria are placed at a disadvantage to suppliers who do not test or label. Whereas the 6th Amendment required that the supplier classify on the basis of available data, the 7th Amendment (Article 6) requires an investigation to determine relevant and accessible data relating to substances not yet in Annex I.

PENALTIES AND COMPLIANCE

The heavy penalties imposed by the EPA for failure to comply with TSCA requirements are unparalleled elsewhere, especially in terms of the large civil penalties sought in the absence of any health or environmental harm or evidence of willful violation. This aggressive approach has provided a large incentive to all companies to establish comprehensive TSCA compliance programs to avoid:

- Massive fines ($25,000/violation/day)
- Lost business
- Litigation initiated by customers
- Adverse publicity

The EU 7th Amendment does not set penalties for noncompliance. These are incorporated in the implementing laws, regulations, and administrative provisions enacted by the member states.

ANALYTICAL TECHNOLOGY

Remarkable advances in analytical instrumentation have been made in recent years enabling ever smaller traces of toxic environmental contaminants to be measured. This provides indispensable tools for the regulator and the environmental scientist. However, the ability to detect minute traces raises serious problems also for the regulator and legislator who is increasingly faced with setting a specific limit rather than establishing the requirement of "not detectable." A further downward turn is provided to limits for inadvertent impurities in colorants (e.g. heavy metals, aromatic amines etc.), and resources have to be dedicated to meet ever lower limits without due consideration of cost effectiveness or actual risk.

BENZIDINE-BASED DYES

The situation with benzidine-based dyes is worth some comment as it exemplifies the need to reinforce voluntary industry initiatives with regulations, the need for enforcement measures, and the need for international cooperation.

The major dye manufacturers ceased the manufacture of benzidine-based dyes in the early 1970s, and they were classified as carcinogenic (Group 2A) by the International Agency for Research in Cancer in 1982. Although there has been a significant overall reduction in usage levels, these dyes are still widely used for the dyeing of leather, offering excellent technical properties at low cost. The dyes are manufactured by small- and medium-sized companies in South America, Europe, and Asia.

Only a few of the current suppliers indicate a possible cancer risk on labels or safety data sheets. Also, there are indications that some importers conspire with manufacturers to provide no hazard warning, so that workers are unaware that they are handling carcinogens. India, a major source of these dyes, banned their manufacture and use from January 1993, following a 3-year phase-out period. In spite of this ban, they are still widely manufactured in India and indeed still included in the sales catalogs of Indian suppliers. In Europe the manufacture of these dyes, but not their use, was banned in the United Kingdom in 1967, and their manufacture and use are effectively banned under current German regulations.

CONCLUDING REMARKS

Comparison of the requirements in various countries is only one aspect of the situation, as the relationship between regulator, the regulated industry, and public-interest groups plays an important role. For example, in the United States the major environmental groups are politically strong, well organized, and show little reluctance to pursue their objectives in the courts. In contrast, in Japan environmental groups tend to be active at local level, to focus on narrower issues, and only rarely to resort to the legal system. Europe because of its more heterogeneous nature is more difficult to characterize.

In the developed regions, the regulatory structure established to protect human health and the environment is now comprehensive, and compliance costs are a substantial part of the cost of doing business. There is a deep-rooted commitment to the concepts of sustainable development and to responsible care programs in the major companies and many of their smaller counterparts. The customer industries have shifted eastward, and it is inevitable that they will be followed by the colorant manufacturers, to a greater or lesser extent. Lower labor costs are the major incentive, but environmental costs are also a factor that can be decisive for the viability of a manufacturing location or a company. Unless the regulatory framework, including enforcement, provides a reasonably level playing field, the end result will be merely the relocation of sources of pollution rather than a reduction of global pollution. It is hoped that this broader perspective, and the recognition that consultation between the regulators and the regulated industries can enable environmental objectives to be met in a more cost-effective way, will encourage closer cooperation so that limited resources are not wasted.

The contribution of the OECD efforts on harmonization have already been mentioned. The further expansion of the EU, the formation of larger free-trade areas such as North American Free Trade Agreement (NAFTA), and the various examples of regional cooperation such as the Nordic Council will increasingly help tackle environmental issues at least at regional level. This should enable pressure to be brought to bear more effectively on countries that do not enforce adequate standards of environmental protection, so that a global solution to global pollution problems can be found.

BIBLIOGRAPHY

M. Richardson, ed., *Chemical Safety; International Reference Manual*, VCH Press, Weinheim, Germany 1994.

EU Environment Guide, the EC Committee of the American Chamber of Commerce, Brussels, 1994.

TSCA Handbook, 2nd ed., McKenna & Cueno, Government Institutes Inc., Washington, D.C., 1989.

V. T. Covello, K. Kawamura, M. Boroush, S. Ikeda, P. F. Lynes, and M. S. Minor, *Risk Analysis*, **8**(2), 247, 1988.

L. Kramer, *Focus on European Environmental Law*, Sweet and Maxwell, London, 1992.

INDEX

317